Real, Mechanical, Experimental

REAL, MECHANICAL, EXPERIMENTAL

Francesco G. Sacco

More information about this series at http://www.springer.com/series/5640

Francesco G. Sacco

Real, Mechanical, Experimental

Robert Hooke's Natural Philosophy

 Springer

Francesco G. Sacco
Humanities Department
CATS College
Canterbury, UK

ISSN 0066-6610   ISSN 2215-0307   (electronic)
International Archives of the History of Ideas Archives internationales d'histoire des idées
ISBN 978-3-030-44453-2   ISBN 978-3-030-44451-8   (eBook)
https://doi.org/10.1007/978-3-030-44451-8

This Springer imprint is published by the registered company Springer Nature Switzerland AG.
The registered company address is: Gewerbestrasse 11, 6330 Cham, Switzerland

Acknowledgments

Over more than a decade and half of work on a subject dealing in many directions, I have accumulated many obligations. Massimo Bucciantini served as examiner for my doctoral dissertation, and his comments and suggestions have helped me turn it into something more. Thanks to a postdoctoral grant, I had the opportunity to spend two academic years at the Warburg Institute, where Guido Giglioni encouraged me to carry forward my work on Hooke's unpublished papers and supported the project of this book.

To many libraries, I am obliged in many ways. I especially wish to thank the director, Keith Moore, and the courteous staff of the library and archives of the Royal Society. At the Rare Books and Manuscript rooms of the British Library, I had the opportunity to spend long periods of pleasant and productive work. Tim Warrender and Wendy Hawke facilitated the access to Hooke's manuscripts at the London Metropolitan Archives and Guildhall Library while Melissa Grafe to Towneley's manuscripts at the Cushing/Whitney Medical Library of Yale University.

At the Royal Society, in many instances, Felicity Henderson generously provided her great expertise on manuscript culture and Hooke's manuscripts and shared a draft of an unpublished seminar she held at the University of Exeter in 2013. Maurizio Sciuto allowed me to read his dissertation on Hooke's methodological program written under the supervision of the late Paolo Rossi at the University of Florence.

Guido Giglioni and Andrea Asioli kindly helped with Hooke's Latin manuscripts. Michael Hunter's criticism of my early ventures in philology forced me to rethink my approach to the transcription of early modern manuscripts. I hope that the mistakes I have not been able to avoid do not prejudice the reading and understanding of the manuscripts here published for the first time. Jim Bennett kindly read a draft of Chap. 6 and gave valuable comments. Timothy Miller, Michael Carlos, and Bruce Comens patiently read early drafts of the whole manuscript; their suggestions have significantly improved its style.

This book exists because of the belief and support of Roberto Bondì. In July 2002, Professor Bondì suggested me to spend the summer studying Hooke's ideas on the history of the Earth. He gave me an electronic copy of *Micrographia* and a long list of readings. I soon realized that the assignment was far beyond the reach of an undergraduate student, but I was confident that a short decent paper could be put together before dropping the topic after graduation. I was mistaken. In the following decade, Roberto guided me through masters and doctoral dissertations on Hooke's natural philosophy. From a supervisor, he became a mentor and a friend; I owe a special debt to him.

I conducted the last stage of the project of this book, involving the preparation and revision of the final manuscript, back and forth through the Atlantic between Great Britain and the United States. Along this bumpy ride, only the enduring support of Aparajita allowed me to complete, despite the delay, the work. This book is dedicated to Delia, who was born during this difficult time but has made our lives much happier.

Introduction

After his death in 1703, Hooke's numerous books and papers were scattered among different archives and libraries. Some of Hooke's papers were collected and published by William Waller in 1705 and William Derham in 1726, but many remained unpublished and almost neglected by historians.[1] The fate of Hooke's books and papers mirrors that of his natural philosophy. The emergence of Newtonianism and empiricism in the eighteenth century has influenced the historiography of early modern philosophy and science for at least two centuries. In this historiographical paradigm, the place for Hooke has been marginal, focused as it has been only on some innovative scientific contributions often described as anticipations of later discoveries. The rest of Hooke's work was neglected.

This book aims to reconstruct the larger scheme of natural philosophy that Hooke drew up since the early 1660s.[2] Such a reconstruction rests on the integration of manuscripts and published works composed by Hooke in more than three decades of investigations. Since Hooke's unpublished papers have proved decisive to define the emergence and evolution of Hooke's natural philosophy, some of these have been transcribed and published here for the first time. Far from providing a definitive edition, the transcriptions have been included to offer the reader the possibility to access some of the sources often referred to in the book. For this reason, throughout the book, extensive quotes have been employed. Parts of the manuscripts that neither Waller nor Derham included in their publication have been edited in journal articles and dissertations since the 1950s.[3] Many others are still unpublished, and a complete edition of Hooke's works remains a desideratum.

Hooke's larger scheme originated from "the real, the mechanical, the experimental philosophy" introduced in the preface to *Micrographia* and evolved over more

---

[1] Henderson (2009), 92, 95, 97, 106; Rostenberg (1989), 124–30; cf. Keynes (1960), 81–84

[2] Patterson (1950), 44; cf. Cooper and Hunter (2006), XIX

[3] Hall (1951); Oldroyd (1980); Kassler and Oldroyd (1983); Hunter and Wood (1986); Oldroyd (1987); Henderson (2007)

than three decades of intense and variegated activity.[4] If the description of experimental philosophy as real was not uncommon in the Restoration,[5] Hooke's bold association of the latter with mechanical philosophy contributed to locate his work at the crossroads of two cultural traditions. A clear demarcation between a speculative mechanical philosophy and a real experimental philosophy has proved a procrustean bed in which Hooke's project did not fit.[6] In a seminal essay, Jim Bennet has shown how the acknowledgment of the influence of mechanical arts and mixed mathematics on both mechanical and experimental philosophies has been frustrated by a broadly internalist framework of research among historians of early modern science.[7] Neither Descartes's metaphysical physics, in other words, nor Newton's experimental philosophy should be considered the exclusive models for the experimentally practiced corpuscular philosophy of nature that gained significant consensus in the second half of the seventeenth century in England and on the Continent.[8] Before both mechanical and experimental philosophies crystallized in a framework unquestioned by historians, the early modern philosophical landscape was much more variegated. In this context, Hooke's real, mechanical, experimental philosophy was not the puzzling anomaly that it seems today.

In a short undated note kept in the archives of the Royal Society, Hooke sketched the main components of his larger scheme of natural philosophy. A "Theory of Motion, of Light, of Gravity, of Magneticks, of Gunpowder, of the Heavens" was in Hooke's view the most significant theoretical result that he reached, whereas the improvements of "shipping, watches, opticks, engines for trade, engines for carriage" were the main applied outcomes. The note significantly also mentions an "inquiry into the figures of bodys [and] qualitys of bodys."[9] Hooke did not refer to a consistent and definitive result in this field, but he clearly reached a dynamic theory of matter in the late 1670s. The structure of this book broadly follows Hooke's note.

Hooke's revised Baconian set of ministrations to human faculties was a decisive component of his larger scheme of natural philosophy. It was not, however, its foundation. Even in Hooke's work, a methodological programme played different roles. Historians are now aware that the way scientific research was carried out rarely corresponded to the methodological statements of modern scientists. Methodological claims were often used as a posteriori justification of new hypotheses conflicting with old theories.[10] But methodological programmes sometimes worked even as general guidelines for research, especially when they were part of a wider image of

---

[4] Hooke (1665), sig. a2r

[5] Harrison (2011), 426–27

[6] Anstey (2005), 215, 218, 223, 226; Anstey and Vanzo (2012), 500–1, 513–5, 518; Janiak (2015), 55–7

[7] Bennett (1986), 1

[8] Cf. Garber (1992), 75–7; Shapiro (2004), 205, 206

[9] Royal Society Classified Papers, vol. XX, f. 67r

[10] Floris Cohen (1994), 153; Bellone (1980a), 57; Schuster and Yeo (1986), X–XI; Lynch (2005), 172–6; Zemplén (2011), 125–28, 131

science that set the field of problems and possible solutions for natural philosophers.[11] The mathematical-deductive and the experimental-inductive approaches echoed opposite views of nature. On the one hand, a quantitative approach, based on hypotheses and abstractions from ordinary experience, often rested on a belief in the existence of a mathematically structured cosmos. On the other hand, the image of nature as a labyrinth or a forest led to emphasis on the role of patient observations and the repetition of experiments.[12]

Hooke's methodological programme and his theory of matter mutually evolved. For this reason, Chap. 1 reconstructs Hooke's views on the origins, sources, and nature of human knowledge through a comparison between published and unpublished papers. Building on this theory, Hooke drew up a project of a new philosophical history of nature wherein hypotheses and matters of fact were crucially linked. By focusing on this project, Chap. 2 shows that a clear demarcation between matters of facts and hypotheses was far from an unquestioned principle of the experimental form of life in Restoration England. Chapter 3 is dedicated to the enigma of Hooke's philosophical algebra. Rather than a definitive solution to the enigma, the chapter provides insights into Hooke's use of symbolic algebra as a tool to develop a form of deductive, demonstrative, and quantitative knowledge independent of the assumption of a mathematical structure of nature.

Hooke's dynamic matter theory is discussed in Chap. 4, while Chap. 5 focuses on the role of congruity in Hooke's hypotheses on the nature of ether and subtle fluids. Chapter 6 reconstructs the genesis and impact of Hooke's system of the world. Since the 1660s, Hooke introduced a dynamic approach to the analysis of the compound nature of curvilinear motion. This proved decisive to unraveling some fundamental problems of planetary motion. Thus, in 1674, Hooke described the three main principles of a new system of the world based on the attractive action of the centre of the solar system. Chapter 7 focuses on Hooke's ideas about fossils and the introduction of the time dimension within the traditionally static approach to the history of nature. Hooke's geological ideas proved to be part of an equally dynamic cosmology. Chapter 8, the last chapter, offers a historiographical alternative to current views on the relationship between Hooke and Newton. It aims to show the irrelevance of the question of priority in the assessment of the exchange of ideas on planetary motion. Focusing on Newton's early notes on Hooke's *Micrographia*, the chapter reconstructs the influence of Hooke's congruity on the idea of associability employed by Newton from 1675 till the last edition of *Opticks*. Although it was nor the only neither the main source of influence over Newton's natural philosophy, Hooke's larger scheme played a role in the development of Newton's dynamic mechanical philosophy. The conclusion discusses the impact of the previous outline of Hooke's work on current views of experimental philosophy and wider trends in the history of early modern science. It suggests a revision of the current prevalent view of Restoration science as a form of legitimation of a new political order based

---

[11] Bellone (1980b), 6

[12] Rossi (1984), 23–4, 124–5

on an aristocratic moral economy. Surprising as it might seem, this view has revived a positivistic view of science that considered Baconianism and experimental philosophy as strictly inductive forms of inquiry, albeit for political rather than strictly methodological reasons. By readdressing the question of Hooke's social status in the light of the previous reconstruction of his work, the conclusion intends to support an alternative view of British experimental philosophy.

Bibliography

Anstey, Peter. 2005. Experimental versus speculative natural philosophy. In *The science of nature in the seventeenth century: Patterns of change in early modern natural philosophy*, ed. Peter Anstey and John Schuster, 215–241. Dordrecht: Springer.

Anstey, Peter, and Vanzo, Alberto. 2012. The origins of experimental philosophy. *Intellectual History Review* 22: 499–518.

Bellone, Enrico. 1980a. Elogio di Galilei. In *Scienza e storia*, a cura di Antonio Di Meo, 50–64. Rome: Editori Riuniti.

———. 1980b. *Il sogno di Galileo: oggetti e immagini della ragione*. Bononia: Il Mulino.

Bennett, Jim. 1986. The mechanics' philosophy and the mechanical philosophy. *History of Science* 24: 1–28.

Cohen, Floris H. 1994. *The scientific revolution: A historiographical inquiry*. Chicago and London: University of Chicago Press.

Cooper, Michael, and Hunter, Michael (eds.). 2006. *Robert Hooke: Tercentennial studies*. Aldershot: Ashgate.

Garber, Daniel. 1992. *Descartes' metaphysical physics*. Chicago and London: University of Chicago Press.

Hall, Alfred Rupert. 1951. Two unpublished lectures of Robert Hooke. *Isis* 42: 219–230.

Harrison, Peter. 2011. Experimental religion and experimental science in early modern England. *Intellectual History Review* 21: 413–433.

Henderson, Felicity. 2007. Unpublished material from the Memorandum Book of Robert Hooke, Guildhall Library Ms 1758. *Notes and Records of the Royal Society of London* 61: 129–175.

———. 2009. Robert Hooke's archive. *Script and Print* 33: 92–108.

Hooke, Robert. 1665. *Micrographia*. London.

Hunter, Michael, and Wood, Paul. 1986. Towards Solomon's house: Rival strategies for reforming the early Royal Society. *The British Journal for the History of Science* 24: 49–108.

Janiak, Andrew. 2015. *Newton*. Chichester: Willey Blackwell.

Kassler, Jamie, and Oldroyd, David. 1983. Robert Hooke's Trinity College 'Musick Scripts', his music theory and the role of music in his cosmology. *Annals of Science* 40: 559–595.

Keynes, Geoffrey. 1960. *A bibliography of Dr Robert Hooke*. Oxford: Clarendon Press.

Lynch, William T. 2005. A society of Baconians? The collective development of Bacon's method in the Royal Society of London. In *Francis Bacon and the refiguring of early modern thought: Essays to commemorate the advancement of learning (1605–2005)*, ed. Julie Robin Salomon and Chaterine Gimelli Martin, 173–202. Ashgate: Aldershot.

Oldroyd, David. 1980. Some 'Philosophical Scribbles' attributed to Robert Hooke. *Notes and Records of the Royal Society of London* 35: 17–32.

———. 1987. Some writings of Robert Hooke on procedures for the prosecution of scientific inquiry, including his 'Lectures of Things Requisite to a Ntral History'. *Notes and Records of the Royal Society of London* 41: 145–167.

Patterson, Louise Diehl. 1950. Hooke's gravitation theory and its influence on Newton II: The insufficiency of the traditional estimate. *Isis* 41: 32–45.

Rostenberg, Leona. 1986. *The library of Robert Hooke: The scientific book trade of Restoration England*. Santa Monica (CA): Madoc Press.

Schuster, John, and Yeo, Richard (eds.). 1986. *The politics and rhetoric of scientific method: Historical studies*. Dordrecht: Springer.

Shapiro, Alan. 2004. Newton's "experimental philosophy". *Early Science and Medicine* 9: 185–217.

Zemplén, Gábor. 2011. The argumentative use of methodology: Lesson from a controversy following Newton's first optical paper. In *Controversies within the scientific revolution*, ed. Marcelo Dascal and Victor Boantza, 123–147. Amsterdam and Philadelphia: John Benjamins Publishing.

Contents

Chapter 1
Human Understanding

1.1 Corpuscular Ideas

Hooke's ideas on the nature and origin of human knowledge are mainly to be found in the text of a lecture delivered at the Royal Society on June 14, 1682 and in a short manuscript kept in the library of Trinity College, Cambridge.[1] In Hooke's later writings there is no place for innate Cartesian ideas, as all human knowledge originates from the senses.[2] The human mind is described as a "piece of soft wax,"[3] both impressions and ideas as corporeal alterations of the brain. "These ideas," Hooke writes, "I will suppose to be material and bulky, that is to be certain bodies of determinate bigness, and impregnated with distinctive motions, and to be in themselves distinct."[4] Like Kenelm Digby, Hooke thinks of ideas as the stamps left on the brain by material impressions. These stamps take the shape of "little images."[5] Although every sense conveys to the brain different kinds of impressions, Hooke does not distinguish them because all "are generally comprised under one name." The only difference among ideas is due to their different degrees of complexity and composition. Some of them are "more immediate," others "more mediate and complex."[6] Immediate ideas "are the more first and more simple, such as are the results of the impressions of the senses." Complex ideas have different degrees of composition, being the result of the comparison of a variable number of simple ideas. Therefore, "the ideas that are made from fewer and more simple ideas, are less compounded ideas"; on the contrary "those which are made from a greater number, and those

[1] Birch (1756–57), vol. IV, 153; cf. Sutton (2013), 299.

[2] Oldroyd (1980), 19; cf. Clarke (1989), 43–4, 48–9, 50, 71, 100.

[3] Oldroyd (1980), 17.

[4] Hooke (1705), 142.

[5] Ibid, 144; Oldroyd (1980), 17; cf. Digby (1644), 282–4; see also Henry (1982), 223–4.

[6] Oldroyd (1980), 17–8.

© Springer Nature Switzerland AG 2020
F. G. Sacco, *Real, Mechanical, Experimental*, International Archives
of the History of Ideas Archives internationales d'histoire des idées 231,
https://doi.org/10.1007/978-3-030-44451-8_1

more compounded ideas, are yet more and more compounded, and more and more accomplish'd and perfect."[7]

Gassendi, Harvey, and Bacon seem the main sources of Hooke's theory of ideas. The view of perception as the source of all human knowledge was part of Harvey's legacy, and shared by the seventeenth-century community of Oxford physiologists to which Hooke and young Locke belonged. In the preface to *De generatione animalium,* Harvey undertook an empiricist interpretation of Aristotelian psychology.[8] "Sensible things," he pointed out, "are of themselves and antecedent; things of intellect, however, are consequent, and arise from the former, and, indeed we can in no way attain to them without the help of the others."[9] By denying the existence of innate ideas, Hooke connected Harvey's empiricist interpretation of Aristotelianism with Bacon's decisive aversion for the abstractions of the Schools. Bacon rejected the Platonic belief in the existence of innate ideas. In his view, even Aristotelians contributed to this "alienation of the mind." They did acknowledge the primary role of senses, but "they have based themselves in everything on the agitation of their own wit, content to circle round and round for ever amid the darkest idols of the mind under the high-sounding name of contemplation."[10]

Harvey, on the contrary, held a different opinion about Aristotelian principles. From the original texts of Aristotle on the source of knowledge, according to Harvey, it plainly appears that "there is no perfect knowledge of which can be entitled ours, that is innate; none but what has been obtained from experience, or derived in some way from our senses."[11]

From different points of view, therefore, Harvey and Bacon offered to Hooke two examples of the rejection of innate ideas and the defence of direct and active observation in the study of nature. Influenced by both, Hooke acknowledged the leading role of experience in acquiring natural knowledge and did not neglect how this contrasted with the practice of many Aristotelians.[12] By avoiding the direct study of nature, these latter denied Aristotle's principles. "Though some of them affirm *Nihil esse in intellectu quod prius non fuit in sensu,*" he notes, "yet upon the whole, we may find that in their manner of proceeding they did otherwise."[13]

Gassendi's works are perhaps another relevant source for Hooke's theory of ideas. The French savant described the human brain as a "satin table" which receives stamps and impressions from the sensory organs by means of animal spirits. Knowledge is based on universal conclusions reached by means of the comparisons

---

[7] Hooke (1705), 146.

[8] Wear (1983), 223, 231–2; Schmitt (1983), 7, 10, 32; Id. (1985), 16–7, 75, 93; Kessler (2001), 83; Bianchi (2003), 136–45; Sgarbi (2012), 103–9.

[9] Harvey (1651), B3v; Id. (1847), 157.

[10] Bacon (1857–74), vol. III, 600–1; Farrington (1964), 82.

[11] Harvey (1651), B4v; Id. (1847), 171.

[12] Cf. Sgarbi (2013), 156, 166.

[13] Hooke (1705), 4.

and compositions of singular ideas in the mind.[14] Following Gassendi, Hooke distinguishes two faculties of the mind. Since simple ideas are passively received by the senses and complex ideas are formed by the human mind, two different faculties are involved in these processes. The "passiue faculty" is responsible for the acquisition of impressions by the senses. Both the comparison of different singular impressions and the formation of complex ideas need an "active faculty, which collates and compares these impressions, and is thereby enabled to compound, and compose new ones." Furthermore, by means of comparisons and collations the active faculty is able to correct perception's mistakes. Simple ideas can be incorrect when they are directly formed from singular and isolated sensory data. A false simple idea is a notion of an object based only on a single and isolated experience of the object itself. A singular experience of the sun, for instance, cannot give a correct idea of its distance from the earth. Comparing different singular impressions of the same object is the only way to obtain the true simple idea of it.[15]

These observations on the accuracy of ideas highlight the extent to which Hooke's late theories depend on his early thoughts on experimental work. Since the 1660s Hooke described scientific knowledge in terms of a continuous comparison between experimental data and increasingly general hypotheses. Experiments and observation are the bedrock of natural philosophy, but there is more in the study of nature. Queries, conjectures, and hypotheses continually originate from and are revised against that experimental basis. In very similar terms, in the 1680s Hooke maintained that the formation and correction of general ideas about objects and natural phenomena depend on a process of comparison of singular sense impressions. Such a process is performed by an "active" faculty of the mind. Mistakes in sense perception, for instance, can be corrected by comparing different sensory impressions of the same object. The validity of complex ideas and general concepts that this faculty produces ultimately depends on what the senses provide it with:

> For we are only ascertaind of the truth or certainty of any axiome merely by the information receiu'd from the senses. For as every axiom is a proportion and every proposition an arbitrary mark impos'd upon a sensible object, by which we would signify our sensation of it soe or soe to an other, Soe we cannot be sure that a proposition is true unless we are sure that all those sensations upon which it is grounded and from which it is abstracted be true.

Even mathematical knowledge originates from sense experience. That $2 + 2 = 4$ is for Hooke true by experience. The only criterion of truth we dispose of consists in "distinguishing between the words, the signa, and the thing, the signata," and then in testing both the truth of the ideas of the things expressed by signs and the relationships between concepts of the mind and objects of experience.[16]

The active faculty is also responsible for the position of ideas in the brain, thence for the functioning of human memory. Hooke discusses the nature of memory in a lecture delivered at the Royal Society on June 14, 1682. As in the Cambridge

---

[14] Gassendi (1658), vol. I, 92–3.

[15] Oldroyd (1980), 17, 19.

[16] Ibid, 19–20.

manuscript, here too, human understanding is described as consisting of a passive and an active faculty. The formation of ideas and their conservation in the memory are corporeal processes that take place in the brain and depend on the action of the active faculty or soul. Describing these processes, Hooke largely uses analogies and metaphors.[17] The retentive action of memory is compared with the retentive nature of the Bononian Stone and Baldwin's phosphorus. These substances release the light of the sun in the dark as memory releases the ideas of objects of past experience.[18] Hooke describes the structure of the human brain in cosmological terms. Like planets in the Copernican system, ideas surround a central source of motion and activity, and receive a constant radiation that reflects their place and distance from the centre. Since their position defines the proximity to the centre of the activities of thinking, old ideas stored farther away are sometimes more difficult to remember.

It had been claimed that Hooke's analogies support an interpretation of his theory in terms of a hermetic microcosm.[19] But, as Hooke notes, those analogies are "examples" used "for explication only." Despite the use of metaphors, Hooke's description of memory and brain maintains an indubitable mechanical character. Ideas are "actual locomotions" and memory is "really organical." If they were to represent his cosmological ideas, the Copernican metaphor could rather refer to the new mechanistic image of the universe that Hooke helped to develop. Moreover, not all the analogies employed by Hooke in this respect are cosmological. Memory is also compared to the "unison-toned strings, bells or glasses, which receive impressions from sounds without, and retain that impression for some time answering the tone by the same tone of their own." This might be read, perhaps, as a reference to his theory of matter.[20] The "sympathetic agreement" among ideas described by Hooke could entail the concepts of congruity and incongruity, along with the different geometrical configurations of matter at microscopic level.[21]

Although incorporeal, the human soul has "its principal and chief seat" in a "certain place or point somewhere in the brain."[22] Since it is described as a centre of orbicular radiations, like the sun, the soul can only have place in the middle of the space occupied by ideas, i.e. the brain. Hooke is well aware of the dualistic consequences of his theory. "I cannot conceive – he affirms – how the Soul, which is incorporeal, should move and act upon the ideas that are corporeal, or how those on the other side should by their proprieties, qualifications and motions, re-act upon and influence the soul."[23] Even for Joseph Glanvill, it was still a "mystery in nature"

---

[17] Cf. Draaisma (2000), 64; Id. (2006), 111, Singer (1976), 119.

[18] Hooke (1705), 141.

[19] Draaisma (2000), 41, 58–59; cf. Id. (2006), 116.

[20] Hooke (1705), 140–1.

[21] Ibid, 145.

[22] Ibid, 141.

[23] Ibid, 147.

that something "that it self hath neither bulk nor motion" could move the mate-rial ideas.[24]

Nonetheless, this dualistic aspect does not confer on Hooke's theory a plain Cartesian character.[25] The co-existence of an active immaterial soul and a passive material faculty was not a Cartesian prerogative. Gassendi, who opposed Cartesian metaphysics, defended the immateriality of the soul.[26] As the "holy faith" and "just reason" show, for Gassendi the human soul is infused by God and wholly "incorporeal."[27] Like Hooke's theory of memory, Gassendi's physiology of human understanding was not limited to an immaterial principle. Images, concepts and ideas are formed by the material soul. This is an inferior physiological principle common to men and animals.[28] Apart from the formation of vital heat and its distri-bution in the entire body, this corporeal "very thin substance" is responsible for sensation, imagination and memory.[29] Animals, therefore, can perform activities often considered exclusively human. "Reason," Gassendi writes, "is not only a power of man, but even of other animals, which we call brutes, just as they share fantasy, they share reason too." According to Gassendi, there are "two different kind of reason, the one sensitive and the other intellectual." Human being and animals share the former, which is "improperly and only by analogy called reason," because it is sensitive and "the same as fantasy." Only mankind has a superior non-corporeal faculty, "that is the same as intellect and mind, and is properly and particularly called reason."[30]

Gassendian theories influenced the anatomical work of Thomas Willis,[31] who employed the young Hooke as assistant from 1655 to 1657.[32] Hooke maintained Willis' idea that the soul is located in the brain, along with the rejection of the Cartesian pineal gland.[33] Willis's seminal work on the anatomy of the brain, *De cerebri anatome*, was published in 1664, the same year as the publication of Descartes *Traité de l'homme*, two years before Malpighi's *De cerebro*, and five years before Steno's *Discours sur l'anatomie du cerveau*. By employing anatomical observations of the brain, a new generation of medical scholars questioned the cen-tre of Cartesian rational neurophysiology, the pineal gland. The fibrous nature of the brain emerged as one of the few anatomical facts accepted by anatomists and phi-losophers.[34] According to Willis, memory consists in the preservations of animal

---

[24] Glanvill (1665), 16.

[25] For a different opinion see Oldroyd (1980), 21; MacIntosh (1983), 349–50.

[26] Michael (2000), 168; Osler (1994), 200.

[27] Gassendi (1658), vol. II, 255.

[28] Ibid, vol. I, 92.

[29] Ibid, vol. II, 250.

[30] Ibid, 411.

[31] Willis (1680), vol. II, 2–10, 51–4, 62.

[32] Chapman (2005), 9, 20; Wright (1991), 245.

[33] Cf. Lewis (2009), 157–8.

[34] Meschini (1998), 122, 125; Adelman (1966), vol. I, 300–2.

spirits in the "circles and cramps" of the brain, "principal home of the rational soul in Man, and sensitive soul in brutes."[35] Memory and imagination are carried out by the inferior material soul. Reason, judgement, and will, on the other hand, are the prerogative of the immaterial human soul, although this latter operates exclusively on the impression conveyed by the senses and stored in the brain.[36] These innovative elements in Willis' work influenced the young Robert Hooke. While maintaining the dualistic enigma of the interaction between immaterial and material soul, he stressed the material and mechanical aspect of human understanding. Willis's distinction between the lower corporeal functions of the *anima brutorum* and the higher rational functions of the immaterial soul was rejected by his former assistant. Hooke describes all intellectual operations, from memory to thinking and reasoning, as mechanical processes. The mechanical operation of human understanding depends on sensory impressions and, more importantly, cannot be reduced to a one-sided or direct action of an immaterial principle on a passive substratum. Sensation, cognition, remembering, and ratiocination "are plainly the results of the conjunct influences of the soul, and the ideas or bodies placed within the repository or sphere of its activity."[37]

This latter is a radiation from the centre which produces modifications of shape, dimensions, and movement of the bulky ideas. "Attention" and "Reasoning" are two similar actions by means of which the soul produces new ideas in the brain. These actions consist in the comparison of different simple ideas. For this reason, the soul or active principle has a sort of apprehension of the nature of simple ideas by means of a process described as the reaction or reflection of central radiation from bulky ideas. This is what Hooke defines as "thinking."[38]

Hooke's emphasis on the corporeal operations of human understanding was not unnoticed. After listening to Hooke's lecture in 1682, some of the virtuosi of the Royal Society objected that his "discourse seemed to tend to prove the soul mechanical." Even though Hooke did not question the immaterial nature of the human soul, his papers draw a mechanical image of mind. Memory and all other thinking processes are mechanical, since they take place by means of material interactions of corporeal bodies, i.e. ideas. Hooke's answer to the objections of some of the fellows aimed to clear any possible allegation of religious heterodoxy while maintaining his previous position:

> Mr. Hooke answered, that no such thing was hinted, or in the least intended in it; it being only designed to shew, that the soul forms for its own use certain corporeal ideas, which it stored up in the repository or organ of memory, and that by its power of being immediately sensible of those ideas, whenever it exerts its power for that end, it thereby becomes sensible of those ideas formerly made, as if they were made at that instant, but with this difference, that the farther they were removed from the center or seat of its more immediate

---

[35] Willis (1680), vol. I, 48–9.

[36] Ibid, vol. II, 48.

[37] Hooke (1705), 147.

[38] Ibid, 144–6.

momentary residence, the more saint are the reflections or reactions from them, and that this occasions the notion of distance of time.[39]

The reaffirmation of the immaterial and active nature of the soul does not affect the corporeal character of the soul's actions on ideas. This led Hooke's critics to emphasize the similarities between his theory and Thomas Hobbes's materialist description of human nature.[40] But in some of his notes on the nature of time, probably written in the last decade of the seventeenth century, Hooke seemingly renounced a materialistic conception of human soul. As an incorporeal being and an active faculty, the human soul has an internal source of activity. Hooke defines it as *"Primum movens."* This "self-moving principle," he notes, "has in itself a power of radiating every way *in orbem* from its centre of being every where as it were actually present, in every point of the sphere of its radiation." However, "it may be supposed to be more immediately and powerfully present in the centre of its being." The similitude with the sun is presented as the only possible way to understand the nature of this complex entity. The human soul cannot "be truly understood or described, but only by similitude; and the best similitude for that purpose is the sun in the great world."[41] Hooke's ideas on the sun and gravitational attraction, developed in the same decade, support this analogy. The radiation of the sun, on which universal attraction depends, is described by Hooke as an ethereal phenomenon. As in the case of the active soul of the internal microcosm, the origin of solar radiation is posited in the body of the sun, and consists in the vibratory movements of its constituent particles. Hooke's matter theory, in other words, seems the key to understanding the active nature of the human soul and its interaction with bulky corporeal ideas in the brain.

1.2 Brain, Mind, and Soul

An unpublished manuscript of 1665 sheds new light on Hooke's ideas on the nature of the soul and human understanding. This is the Latin version of a lecture on symbolic algebra delivered at Gresham College on June 10, 1665.[42] The view of knowledge and its origins expressed in this lecture is similar to Hooke's later theory of the 1680s. Hooke denies the innate nature of "axioms," or general principles of knowledge. Far from being "innate or infused," these are rather "acquired habits, deductions drawn by a continual series of ratiocination by comparing, compounding and separating, and general ways of examining and applying the most sensible

---

[39] Birch (1756–57), vol. IV, 154.

[40] Cf. Kassler (1995), 135.

[41] Hooke (1705), 146.

[42] Respectively Guildhall Library, London MS 1757.12 and Royal Society Classified Papers, vol. XX, ff. 65–66.

proprieties of bodies." The formation of general principles is an example of the ordinary operations of human understanding:

> Here we see evidently before our eyes, how invention is prosecuted and carried on in the braine, how from such plaine and obvious slight truths, as that 2 and 2 make 4 we proceed to find out the most abstruse mysteries and; how, when a matter is propounded to be found out, the brain or reason of man works and contrives and turns itself as it were to find it out.

This passage raises some questions. Does it suggest that the axioms are not innate or infused in an immaterial soul, but acquired by experience and ratiocination in the brain? Is there any relationship between the rejection of "innate or infused" axioms and the existence of a similarly innate or infused immaterial soul? A comparison between the two versions of the lecture shows that Hooke initially used a clear set of terms, which he lately emended. Some of the words used in the Latin text, in fact, are missing in the English one. Why did it need this kind of amendment? A possible answer can come from the use of the terms, "brain" and "reason." In the Latin text Hooke refers to the human intellective faculty as "ipsum cerebrum, ipsa ratio, immo quam ipsa anima humana." This corresponds to the English expression "the brain, reason or the very soule of man." But in the Latin text there is a correction. The words "mens vel" are inserted before "anima humana." The emended version thus reads: "ipsum cerebrum, ipsa ratio, immo quam ipsa mens vel anima humana." The insertion aims to emphasize the immaterial nature of "anima humana." "Mens" is here employed as synonymous of the human immaterial soul. However, the same word is employed also to replace two different expressions occurring in the remaining part of the Latin text. These are "cerebro" and "cerebrum vel ratio," translated respectively as "braine" and "brain and reason" in the English version. In both cases Hooke replaced them by "mens." It seems, then, that the emendation aims to reduce the extensive use of the term "brain" as the subject of human understanding, and its association with "reason," which make the two words synonyms.[43] In the same period, in fact, Hooke uses the word reason as the higher faculty of man. In *Micrographia* "the three faculties of the soul" are sense, memory and reason.[44] The expression "cerebrum vel ratio" suggests an identification between the intellective faculty and its substratum. In contrast, "mens" clearly underlines the existence of an immaterial component of human understanding. The replacement of "cerebrum" by "mens" seems due to this reason.

Hooke's early writings on human understanding are close to Willis's *De cerebri anatome,* and to his collaboration in Willis's experimental activities of 1655–57. In the same years as his algebraic lectures, Hooke was engaged in the composition of his masterpiece *Micrographia*, published in 1665, and in composing a *General Scheme* on the improvement of experimental procedures. In these works, building on Willis's anatomical observations, he develops a mechanical version of the chain

---

[43] Guildhall Library, London MS 1757.12, f. 113r-v; Royal Society Classified Papers, vol. XX, f. 65r-v.

[44] Hooke (1705), 12.

of being.[45] In *Micrographia* four kingdoms of nature are listed: elemental, mineral, vegetable and animal. The elemental kingdom includes air, water and earth. As Hooke notes, "there is no curiosity" in these bodies because air and water have "not form at all, unless a potentiality to be form'd into globules, and the clods and the parcels of earth are all irregular." In the mineral kingdom nature "begins to geometrize, and practise, as 'twere, the first principles of mechanicks, shaping them of plain regular figures, as triangles, squares, etc., and tetrahedrons, cubes, etc." Geometrical shapes can also be found in the vegetable kingdom, wherein nature adds "multitudes of curious mechanick contrivances." In the last kingdom, the animal one, "we shall find, not onely most curiously compounded shapes, but most stupendious mechanism and contrivances."[46] The four orders form a scale of ascending degree of mechanical complexity.[47] In her "transitions" by "degrees and steps" nature "passes from one thing to another in the formation of species."[48] Hooke's chain of being begins "from fluidity, or body without any form" and ascends to "the highest form of a brute animal's soul." From the first to the last kingdom of nature, the "steps" distinguished in this ascent are "fluidity, orbicularitation, fixation, angularization, or crystallisation, germination or ebullition, vegetation, plantamination, animation, sensation, imagination."[49] The dominant characters of Hooke's mechanical scale are graduation and continuity. "As has been excellently well observed by the learned physician Dr. Willis," monkeys and baboons have a different bodily constitution, they "seem to want the use of reason and speech." But, according to Hooke, "there is no doubt but that a diligent observer may by accurately anatomizing each, and comparing them together, find divers other considerable variations which are of a kind of middle constitution, between those of man and those of other the most brute creatures."[50] In natural productions, "harmony, consent and uniformity" can be easily observed. In Hooke's eyes, fossil remains fill the lack of conjunction in the "gradual transition" between species.[51] Anatomical and physical variations are the effects of changing environmental conditions in the course of time. Entire species have been annihilated, whereas "new varieties of the same species" may have been generated as effects of alterations in "climate, soil and nourishment."[52] This is the common course in all natural kingdoms. Man, as part of the animal kingdom, is not excluded. From the observation of "such variations as happen to bodies from their being produc'd at differing times, or in differing places, or the same medium, or in differing mediums, or with a differing quantity," Hooke expects to enlighten his reader on the "difference between the stature, age, strength,

---

[45] Ibid, 53; cf. Lovejoy (1936) 186–9.
[46] Hooke (1665), 154.
[47] Ibid, 127.
[48] Hooke (1705), 52.
[49] Id. (1665), 127.
[50] Id. (1705), 53.
[51] Ibid, 341.
[52] Hooke (1665), 327.

shape etc. of men at the beginning of the world and now."[53] The "history of man,"
part of the new experimental history of nature, includes "the anatomical history of
internals parts of man, compar'd with those of other animals," and comparative
descriptions "of sensation, motion in the mind, memory, reason, folly, madness,
sleeping, and dreams, etc."[54]

Meanwhile, reproductive processes of living species are part of the more general
"formation and configuration of bodies" in the chain of being. Here nature "seems
to act yet more secretly and farther remov'd from the detection of our senses." If, by
means of new mechanical instruments as microscopes, it was possible to inquire
into reproductive processes, these also would be "detected to be mechanical."
"These kind of actions" are considered "more spiritual" because inaccessible to our
senses. The use of immaterial agents to draw the boundaries of human investigation
in these matters is considered by Hooke "an opprobrium to philosophical inquiry."
Despite the lack of powerful microscopes, there are other means to overcome these
boundaries. "Yet," Hooke affirms, "if we more seriously consider the progress of
nature from the more simple and plain operations to the more complicated and
abstruse, we may from them deduce a great argument of incouragement."[55] By
means of a microscope, the shape of frozen urine and that of a fern look similar.[56]
Does it mean that the shape of frozen urine is the product of a similar "seminal
principle"? Or, rather, that both are nothing else than the effect of mechanical configurations of matter? "Not that I am more inclined," Hooke writes, "to this hypothesis then the seminal, which upon good reason I ghess to be mechanical also, as I
may elsewhere more fully shew."[57] New animal or vegetable bodies can originate
from putrefactive substances. Spontaneous generation, one of the most discussed
topics in seventeenth-century biology, takes place by means of dissolution of putrefying matter and its composition into new "homogeneous" bodies.[58] The passage
from inanimate to living matter, from fluidity to crystallization, from minerals to
vegetables and then to animals, does not need spiritual or immaterial principles,
because it can be reduced to the reconfiguration of particles of matter in more complex structures. "I must conclude," Hooke affirms, "that as far as I have been able to
look into the nature of this primary kind of life and vegetation, I cannot find the least
probable argument to perswade me there is any other concurrent cause then such as
is purely mechanical."[59] "Sensation" is the first step of the animal kingdom. It can
be found in the reaction of a simple hair to atmospheric variations observed in a
hygroscope and in the "zoophyts" or "sensitive plants."[60] What else is sensation if

---

[53] Id. (1705), 56.

[54] Ibid, 23.

[55] Ibid, 47.

[56] Hooke (1665), 90.

[57] Ibid, 134.

[58] Ibid, 123, 134; cf. Gunther (1968), vol. VII, 592–3.

[59] Hooke (1665), 130.

[60] Ibid, 124.

not "the most plain, simple, and obvious contrivance that nature has made use of to produce a motion" in living bodies?[61]

Drawing on Willis's distinction between *anima brutorum* and man's immaterial soul, Hooke describes the human place in nature as the upper part of a mechanical scale of bodies. The distinction, defended by Gassendi and Willis, between two kinds of intellectual activities, the lower attributed to a material animal soul and the higher to the immaterial Man's soul, is not supported by Hooke. All intellectual functions are mechanical, because these are nothing other than alterations of matter and motion. The soul's actions are also mechanical because they take place on corporeal ideas by means of physical radiations. Furthermore, they are conceived as reciprocal because the actions of the human soul are influenced by the results of its corporeal radiations on ideas.

The early identification of brain and reason in the first algebraic lecture emphasises a mechanical conception of the brain's physiology. In light of Hooke's insistence, in the 1680s writings, on the immaterial nature of the soul, the late intervention on the first draft of the 1665 lecture signals a significant shift from his early ideas. The presence of an immaterial soul was, nonetheless, still consistent with the image of Man emerging from those early writings, wherein the existence of such a principle was never questioned. On the other hand, Hooke's formal assertion of a superior and immaterial principle might have also been caused by both the pressing genuine anxiety concerning natural philosophy and the changing nature of orthodoxy in the second half of seventeenth century.[62] To orthodox criticism, reflected in the reactions of the fellows of the Society to the 1682 lecture, Hooke answered by reaffirming the existence of a divine immaterial principle in Man, absent in the remaining animal kingdom.[63] But such an inexplicable principle does not represent an alteration of the mechanical description of the brain's physiology and human intellectual activities previously outlined. As in the remaining part of the Hookian chain of being, there is space only for a divine principle compatible with the mechanical images of the world as a big clockwork and of God as the big clockmaker.[64] God is part of Hooke's natural philosophy from the beginning, because nature is conceived as a divine work showing the signs of the creator. In Hooke's eyes, the existence of a mechanical order is the primary argument against the epicurean domain of chance.[65] There is no contradiction between divine providence and the mechanical order of nature:

> Nature does not onely work mechanically, but by such excellent and most compendious, as well as stupendious contrivance, that it were impossible for all the reason in the world to find out any contrivance to do the same thing that should have more convenient properties. And can any be so sottish, as to think all those things to be the productions of chance?

---

[61] Ibid, 151.

[62] Cf. Hunter (1990), 444, 448; Id. (1995), 178–9, 229–30, 241.

[63] Cf. Petty (2012), 96, 103, 122–4.

[64] Hooke (1665), 124.

[65] Ibid, 177.

Certainly, either their ratiocination must be extremely depraved, or they did never atten-
tively consider and contemplate the works of the Al-mighty.[66]

In the scale of nature there are different degrees of "contrivance and perfection
of organisation and mechanism" because all bodies and beings have been "furnish'd
with those faculties which are requisite to perform those functions which are necessary to their preservation." Everything in each of them is "contrived on purpose, and
with design, respecting the end of their being and well being, and continuation
either in the individual or the species propagated." Even human senses and reason
had been "formed and act with design and respect to an end, and not fortuitously
and by chance." Because of this, human reason can be defined as a "spark of the
divine influence."[67] Thanks to the strict association between divine providence and
the mechanical order of nature, the material description of the brain's physiology
and the intellectual activities of Man are not in contrast with their divine origin.

1.3 Prejudices

By now it is apparent that Hooke accorded a preeminent role to sensory experience
in human knowledge. In this respect Harvey and Gassendi represent an important
frame of reference. "Diligent observation is therefore – Harvey pointed out – requisite in every science, and the senses are frequently to be appealed to." Personal
direct experience, namely sensory experience, is the foundation of science.[68] In contrast to Cartesian arguments, Gassendi described the senses as "quite passive"
because they "report only appearances, which must appear in the way they do owing
to their causes." Falsity is not a propriety of sensory perception, but of the mind's
judgment "which is not circumspect enough" in making propositions.[69] According
to Gassendi, a proposition is true only when it reflects the real relationships between
the objects to which it refers. Consequently, the certainty of judgement depends on
sensory evidence.[70] Hooke's approach to truth and falsity is consistent with Gassendi
and Harvey, but his theory contains also significant differences. Knowledge, "in the
highest idea of it," is for Hooke "a certainty of information of the mind and understanding founded upon true and undeniable evidence." The latter "is afforded either
immediately by sense without fallacy, or mediately by a true ratiocination from such
sense." Reason alone is not able to confer certainty on ideas or propositions.
"Evidence" is intended by Hooke as sensible evidence. Notwithstanding, truth and
falsity are considered attributes even of sensory data, and not only of the mind's
judgement. A false conclusion concerning the nature of things can originate either

[66] Ibid, 171–2.

[67] Hooke (1705), 120.

[68] Harvey (1651), B2v; Id. (1847), 157.

[69] Descartes 1964–74, vol. VII, 332; Id. (1984–91), vol. II, 230.

[70] Gassendi (1658), vol. I, 100, 103.

from an incorrect "ratiocination from such sense" or from the sense itself, or rather from the fallacies of senses.[71] According to Hooke, "there is no method of information so certain and infallible as that of sense," but only "if rightly and judiciously made use of."[72]

Because of Hooke's rejection of innate ideas and the mechanical description of brain physiology, senses remain the only foundation of human knowledge and the primary criterion of truth and certainty. Despite their passive nature, they are a source of errors and falsity as much as reason is. The judge of our conclusions concerning the nature of things is not the sense itself, but only the "sense without fallacy." Sensory evidence has to be "examined and found to be free and clear of all such fallacies" by reason, which compares several different affections of the same object.[73] In this way sensible experience can represent the source of certainty and evidence. It is also, in Hooke's eyes, a source of evidence of mechanical philosophy. Hooke's mechanical description of human perception and brain physiology is consistent with the mechanical order of nature. The rules of mechanics "are easily to be understood and imagined, and are most obvious and clear to sense, and do not perplex our minds with unintelligible ideas of things."[74] Hooke therefore dismisses the Cartesian metaphysical foundation of mechanism. Descartes's search for "the principles or first causes of everything" ended with "certain seeds of truth which are naturally in our souls." Senses and imagination can confer no evidence to mechanical principles.[75] On the contrary, building on the refusal of innate ideas, Hooke thinks that matter and motion fall under the reach of our senses.[76] Their evidence, therefore, is empirical, not metaphysical.

Hooke's emphasis on the fallacies of the senses is linked to the assumption of the existence in Man of obstacles to the correct understanding of nature. The limits and defects of human knowledge were debated topics among English Baconians. Under the influence of Bacon's idols, Joseph Glanvill for instance listed five "affections" hindering Man's understanding: "natural disposition," "custome and education," "interest," "love of our own productions," and the "homage which is paid to antiquity and authority."[77] Like Glanvill, Hooke does not follow strictly the Baconian distinction of four idols or "false notions." He describes them as "prejudices" darkening and clogging human faculties, thereby giving rise to imperfections and mistakes in human knowledge.

The first prejudice listed by Hooke reflects Bacon's idol of the tribe. It is, perhaps, the most relevant, as it affects the source of human knowledge, the senses. Man lacks "an intuitive faculty to see further into the nature of things at first." He is

---

[71] Hooke (1705), 330–1.

[72] Id. (1726), 172.

[73] Id. (1705), 330–1.

[74] Id. (1677), 32.

[75] Descartes (1964–74), vol. VII, 63–4; Id. (1984–91), vol. I, 143–4.

[76] Hooke (1677), 33.

[77] Glanvill (1665), 89; cf. Id. (1675), 23.

just "indued with organs as are capable of taking information of the operations of nature only by some peculiar ways of sensation." Ideas and sensory impressions do not reflect the nature of things themselves because they are appropriated by human organs and therefore are irremediably shaped in a way that distinguishes them from the things themselves. "Had we others kinds of organs," Hooke observes, "we should have other kinds of conceptions":

> So that our apprehensions of things seem to be appropriated to our species; and that if there were another species of intelligent creatures in the world, they might have quite another kind of apprehension of the same thing, and neither perhaps such as they ought to be, and each of them adapted to the peculiar structure of that animal body in which the sensation is made.

Perceptions of qualities are relative to human anatomical constitution, and change according to the organ conveying them to the brain. Hooke stresses the importance of this "prejudice," because it represents the main obstacle to the achievement of a sure and certain knowledge of things "as they are part of, and actors or patients in the universe, and not only as they have this or that peculiar relation or influence on our senses."

This first kind of prejudice equally affects all men, as part of the human species; the second, on the contrary, is peculiar to individual constitutions, which incline some men more or less to "contemplation and speculation" or to "operation, examination and making experiments."[78] Like Bacon, Hooke sees the harmful effect of this prejudice on past philosophies, ancient and modern:

> Thus Aristotle's physics is very much influenced by his logick; Des Cartes' philosophy favours much of his Opticks; Van Helmont, and the rest of the chymists of their chimical operations; Gilberts of the loadstone; Pythagoras's and Jordanus Brunus's, Kepler's etc. of arithmetic and the harmony of numbers.[79]

"Confident dogmatizing" and the imposition of personal opinions are effects of an individual "temper and disposition."[80]

Bacon distinguished between the idols of the marketplace and the idols of the theatre, whereas Hooke joined obstacles of education and of language in a single kind of prejudice.[81] He shifted the emphasis from natural languages to the language of science received as a constituent part of learning. To Hooke "the philosophical words of all language yet known in the world seem to be for the most part very improper marks set on confused and complicated notions." Traditional learning and education consist mainly in the "instilling" of "confus'd notions" into "young and tender" minds, incapable "to distinguish between assertions and demonstrations, and between opinions and realities." Man's ratiocination, therefore, is "puzzled and disturbed" by these notions imposed dogmatically.[82]

---

[78] Hooke (1705), 8–9; cf. Bacon (1857–74), vol. VIII, 76–7.

[79] Hooke (1705), 9; cf. Bacon (1857–74), vol. IX, 369–370.

[80] Hooke (1665), sig. a1v.

[81] Cf. Bacon (1857–74), vol. VIII, 78.

[82] Hooke (1705), 10–1.

Following Bacon, Hooke considers the new method of acquiring knowledge the chief remedy to all prejudices. He lists also specific remedies to each of them. It is relevant here to note the remedy suggested by Hooke to sensory fallacies:

> The best remedy therefore that seems to be against this prejudice is, to compare the several information we receive of the same thing, from the several impressions it makes on the several organs of sense and (by a rejection of what is not consonant) by degrees to find out its true nature and thereby to inform the intellect with a notion of the thing which is not according to this or that idea, rais'd from the impression of this or that sense, but by a comparative act of the understanding from all the various informations 'tis capable of receiving, more immediately by any of the senses, or more mediately by various other observations and experiences.[83]

It can be argued, therefore, that in Hooke's theory of ideas and human understanding the real nature of things is not beyond sensory perceptions but at their intersection. Since sensible experience is the only source of knowledge, its mistakes and fallacies can be emended only by perfecting human perception by means of a new method.

References

Adelman, Howard. 1966. *Marcello Malpighi and the evolution of embryology*, 5 vols. Ithaca/London: Cornell University Press.

Bacon, Francis. 1857–74. *Works*, 7 vols. ed. Robert L. Ellis, James Spedding, Douglas D. Heath. London: Longman.

Bianchi, Luca. 2003. *Studi sull'aristotelismo del Cinquecento*. Padua: Il Poligrafo.

Birch, Thomas. 1756–57. *The history of the Royal Society of London*, 4 vols. London.

Chapman, Alan. 2005. *England's Leonardo: Robert Hooke and the seventeenth-century scientific revolution*. Bristol/Philadelphia: Institute of Physics Publishing.

Clarke, Desmond. 1989. *Occult powers and hypotheses: Cartesian natural philosophy under Louis XIV*. Oxford: Clarendon.

Descartes, René. 1964–74. *Oeuvres*, 12 vols., ed. Charles Adam and Paul Tannery. Paris: Vrin.

———. 1984–91. *The philosophical writings*, 3 vols., Trans. John Cottingham, Robert Stoothoff and Douglas Murdoch. Cambridge: Cambridge University Press.

Digby, Kenelm. 1644. *Two treatises*. London.

Draaisma, Douwe. 2000. *Metaphors of memory: A history of ideas about mind*. Cambridge: Cambridge University Press.

———. 2006. Hooke on memory and the memory of Hooke. In *Robert Hooke: Tercentennial studies*, ed. Michael Cooper and Michael Hunter, 111–121. Aldershot: Ashgate.

Farrington, Benjamin. 1964. *The philosophy of Francis Bacon*. Liverpool: Liverpool University Press.

Gassendi, Pierre. 1658. *Opera omnia*, 6 vols. Leiden.

Glanvill, Joseph. 1665. *Scepsis scientifica*. London.

———. 1675. *Essays on several important subjects in philosophy and religion*. London.

Gunther, Robert T. 1968. *Early science in Oxford*, 15 vols. London: Dowsons of Pall Mall.

Harvey, William. 1651. *Exercitationes de generatione animalium*. Amsterdam.

———. 1847. *The works of William Harvey*. Trans. Robert Willis. London: Sydenham Society.

---

[83] Ibid, 9.

Henry, John. 1982. Atomism and eschatology: Catholicism and natural philosophy in the inter-regnum. *British Journal for the History of Science* 15: 211–239.

Hooke, Robert. 1665. *Micrographia*. London: J. Martyn and J. Allestry.

———. 1677. *Lampas*. London: J. Martyn.

———. 1705. *Posthumous works*, ed. Richard Waller. London.

———. 1726. *Philosophical experiments and observations*, ed. William Derham. London.

Hunter, Michael. 1990. First steps in institutionalization: The role of the Royal Society of London. In *Solomon's house revisited: The organization and institutionalization of science*, ed. Tore Frängsmyr, 13–29. Canton: Science History Publications.

———. 1995. *Science and the shape of orthodoxy: Intellectual change in late seventeenth-century Britain*. Woodbridge: Boydell Press.

Kassler, Jamie. 1995. *Inner music: Hobbes, Hooke and North on internal character*. London: Athlone.

Kessler, Eckhard. 2001. Metaphysics or empirical science? The two faces of Aristotelian natural philosophy in the sixteenth century. In *Renaissance readings of the corpus aristotelicum*, ed. Marianne Pade, 79–101. Copenhagen: Museum Tusculanum Press.

Lewis, Rhodri. 2009. A kind of sagacity: Francis Bacon, the ars memoriae and the pursuit of natural knowledge. *Intellectual History Review* 19: 155–175.

Lovejoy, Arthur. 1936. *The great chain of being*. Cambridge, MA: Harvard University Press.

MacIntosh, Jack. 1983. Perception and imagination in Descartes, Boyle and Hooke. *Canadian Journal of Philosophy* 13: 327–352.

Meschini, Franco. 1998. *Neurofisiologia cartesiana*. Firenze: Olschki.

Michael, Emily. 2000. Renaissance theories of body, soul, mind. In *Psyche and soma: Physicians and metaphysicians on the mind-body problem from antiquity to enlightenment*, ed. John P. Wright and Paul Potter, 147–172. Oxford: Clarendon.

Oldroyd, David. 1980. Some 'Philosophical Scribbles' attributed to Robert Hooke. *Notes and Records of the Royal Society of London* 35: 17–32.

Osler, Margaret. 1994. *Divine will and the mechanical philosophy: Gassendi and Descartes on contingency and necessity in the created world*. Cambridge: Cambridge University Press.

Petty, William. 2012. Of the scales of creatures. In *William Petty on the order of nature: An unpublished manuscript treatise*, ed. Rhodri Lewis, 93–199. Tempe: Arizona Center for Medieval and Renaissance Studies.

Schmitt, Charles. 1983. *Aristotle and the renaissance*. Cambridge, MA: Harvard University Press.

Sgarbi, Marco. 2012. Towards a reassessment of British Aristotelianism. *Vivarium* 50: 85–109.

Singer, B.R. 1976. Robert Hooke on memory, association and time perception. *Notes and Records of the Royal Society of London* 31: 115–131.

Sutton, John. 2013. Body and soul. In *The Oxford handbook of British philosophy in the seventeenth century*, ed. Peter Anstey, 285–307. Oxford: Oxford University Press.

Wear, Andrew. 1983. William Harvey and the 'way of anatomists'. *History of Science* 21: 223–249.

Willis, Thomas. 1680. *Opera omnia*, 2 vols. Cologne

Wright, John. 1991. Locke, Willis and the seventeenth-century Epicurean soul. In *Atoms, pneuma and tranquillity: Epicurean and Stoic themes in European thought*, ed. Margaret Osler, 239–258. Cambridge: Cambridge University Press.

Chapter 2
Ministrations

2.1 A Matter of Method

Hooke's project to reform natural knowledge aimed at the restoration of the original human condition lost in the Fall. Since then, humans have no more direct access to the hidden nature of things through their bare natural faculties. The only way to recover human "command over things" and "some degrees of those former perfections" consists in "rectifying the operations of the sense, the memory and reason."[1] As noted earlier, these are the "three faculties of the soul," which are "to be examined how far their ability and power, when in the greatest perfection extends, and wherein each of them are deficient, and by what means they may be assisted and perfected."[2] Like Bacon, Hooke thought of the means for reforming natural philosophy as a set of artificial *ministrationes* directed to the three human faculties.[3] Devoid of any assistance, these faculties have produced a "knowledge very confus'd and imperfect, and very insignificant as to the enabling a man to practise or operate by it."

Past philosophies prove this failure. For Hooke, as for Bacon, the unproductive nature of knowledge was a sign of its fallacy. Hooke's criticism was directed not only towards the ancients and their medieval followers, but also to Renaissance restorers of ancient learning. They all ignored the limits of human faculties, and their minds dominated over the bare and imperfect senses. "From a very few and uncertain histories they usually rais'd the most general deductions, and from them though never so imperfect," they prescribed "laws to the universe and nature itself."[4] In Hooke's eyes, natural "philosophy, though of almost as great an age as the world,

---

[1] Hooke (1665), sig. a1r.
[2] Id. (1705), 12.
[3] Bacon (1996), 215.
[4] Hooke (1705), 3–4.

© Springer Nature Switzerland AG 2020
F. G. Sacco, *Real, Mechanical, Experimental*, International Archives
of the History of Ideas Archives internationales d'histoire des idées 231,
https://doi.org/10.1007/978-3-030-44451-8_2

is yet as much in its infancy as ever." It looks like Cupid, "forever childish."[5] "Modern writers" erred as much as their predecessors. They "have a little varied from the receiv'd opinions," envisaging "new hypotheses or opinions of their own instead of the old." But their efforts have proven irrelevant, because they did not alter the fundamental approach to nature. "It may be questioned," Hooke observed, "whether piecing or mending will serve the turn." There are too many impediments to remove and too many artificial helps to supply as to expect any increase in knowledge.[6] Despite the existence in the past of as many philosophies as philosophers, the knowledge inherited from them has to be entirely rejected. The reform of natural knowledge does not just proceed from the adoption of new hypotheses and theories about specific phaenomena. On the contrary, it depends on the adoption of a new method of inquiry. The limits of ancient philosophers and their modern followers are due to the limits of their method. Human reason "if rightly imployd" can reach "a much higher pitch of knowledg[e] concerning naturall agents and causes." "Our predecessors have not arrived at it" because they did not dispose of the right method of using human faculties.[7]

Hooke described this method as a mechanical instrument, because it is a "guide and assistant" to regulate senses, memory and reason. By means of this "engine" naturalists can "perform much more than 'tis possible to do without that assistance."[8] To illustrate his ideas, Hooke referred to the case of a child "which has a strange naturall faculty in the speed resolving and answering any question propounded about arithmeticall multiplication so as to answer it sooner by note then any man is able to cast it with his pen." The child, Hooke noted, "is to be admired [more] then the most skilful arithmetician in the world." But "the skilfull artist" is to be "higher valued," because "whereas the child could only resolve questions propounded in one arithmeticall operation, namely multiplication, the artist is able to answer all."[9] Only a mechanical proceeding could help in disclosing all the secrets of nature, avoiding the uncertainty of the intuitive knowledge of both empiricists and dogmatists.[10]

---

[5] Oldroyd (1987), 151.

[6] Hooke (1705), 4, 6.

[7] Guildhall Library, London MS 1757.11, f.107r.

[8] Hooke (1705), 7.

[9] Guildhall Library, London MS 1757.11, f.107r.

[10] To avoid any confusion with the labels empiricism and rationalism, frequently used to describe a misleading distinction between continental Cartesianism and a supposedly Baconian British tradition, I do not follow Rees and Farrington in translating *rationales* as 'rationalists', see Farrington (1964), 97 and Bacon (2004), 153. I'm confident that 'dogmatists' and 'empirics' better conveys what Bacon described respectively as *rationales* or *dogmaticos*, and *empirici*.

2.2 Old and New Ministrations

Although Hooke adopted Bacon's tripartition of method, he did not maintain Bacon's order of ministrations. In his hands the Baconian project was significantly renewed. Bacon's ministrations consisted of a new "natural and experimental history," its "tables and structured sets of instances," and the "true and legitimate induction." Hooke gathered all of them in a single one, the ministration to memory. Scientific instruments and a new "philosophical algebra" were the new ministrations to sense and reason respectively.[11] Hooke's alterations of Baconian methodological programme were influenced by his ideas about the nature of matter and the structure of nature. In Hooke's eyes, "the works of nature are a great labyrinth" defended by "impenetrable walls." This labyrinth "is already built and perfected" and "there are not new passages to be made, other than what are already fixt." In the past, philosophers "immediately" have flown over these walls by means of their imagination. Unable to get out of this natural labyrinth, they "have thereupon feigned a way," and turned their minds to confused labyrinths. The new experimental philosopher, on the contrary, "must first find some open and visible entry, and there enter with his clew and instruments." In all his trials he has to take notes of turnings and passages. Only "by comparing of all which together, he will at last be able to give you the true ground plat of the whole labyrinth."[12] Truth, for Hooke, is not an inhabitant of the human mind; it cannot be attained through self-scrutiny. This method produced the "philosophy of discourse and disputation," a form of knowledge that "chiefly aims at the subtlety of its deductions and conclusions, without regard to the first ground-work, which ought to be well laid on the sense and memory." Hooke's new ministrations were meant to lead to a "right ordering" of human faculties, for the first time "serviceable to each other."[13]

In spite of the similarities to Bacon's image of science, Hooke's innovative ministrations clearly show some relevant differences. Hooke, for instance, elaborated a clearly corpuscular view of matter, whereas Bacon always balanced atomistic principles with his vitalist conception of nature.[14] Hooke's corpuscular commitment influenced his methodological programme.[15] Hooke described the new philosophy, at the same time, as real, mechanical and experimental. By means of experiments and observations, it aims at finding out "the true nature and properties of bodies," their "inward texture and constitution," and their "internal motions, powers and energies."[16]

Hooke's new ministrations consist of material and logical instruments. Mind as much as the senses should be guided beyond the external appearances of bodies,

---

[11] Bacon (2004), 214–7.

[12] Hooke (1705), 84.

[13] Hooke (1665), sig. a2r; cf. Harrison (2011), 426–7.

[14] Giglioni (2013), 50.

[15] Cf. Gaukroger (2006), 356; Laudan (1966a), 96.

[16] Hooke (1705), 3.

where the bare senses and the unaided intellect are stopped by the subtlety of nature. Ministrations to memory and reason are logical instruments of inference from the visible to the invisible, from the object of experience to the hidden proprieties of bodies and the unreachable causes of phenomena. Hooke's programme was essentially directed to the construction of the best inference from the domain of experience to the aspects of nature that exceed it. As will be apparent in the next chapters, for Hooke new knowledge could be achieved only if the limits of human perception were exceeded. These natural boundaries are both spatial and temporal, in Hooke's opinion.

Bacon and Hooke refused the abstract character of traditional logics.[17] In their view, even the ministration to reason aimed at the discovery of the simple natures of the bodies investigated.[18] The inability of past logic to discover these hidden proprieties was the necessary consequence of its abstract and rhetorical nature. Thus, natural philosophy became sterile and useless. Without knowledge of the hidden nature of bodies no operation on them is possible. "Talking and contention of arguments" should be turned into "labours." When "the fine dreams of opinions and universal natures, which the luxury of subtil brains has devis'd" will vanish and give place to "solid histories, experiments and works," man's place in nature will be restored. According to Hooke, mankind could be "restor'd" to its original state by the same way it fell, i.e. "by tasting too those fruits of natural knowledge, that were never yet forbidden."[19]

Practise and theory are both part of this new method of inquiry. Hooke's programme aimed to fill the gap between theoretical knowledge and practical arts. Strictly following Bacon, Hooke described the knowledge of the hidden features of bodies as the only useful instrument to operate on them. A comparison between the first aphorism of the *Novum Organum* and the first paragraph of the *General Scheme* leaves no doubt in this respect. "The work and aim of human power," Bacon wrote, "is to generate and superinduce a new nature or new natures on a given body." The discovery of "the form, or true difference, or *natura naturans*, or source from which a given nature arises," is the basis for the "transformation, within the bounds of the possible, of concrete bodies from one into another."[20] Echoing Bacon, Hooke affirms that the knowledge of the nature and proprieties of bodies "is not barely acquir'd for it self," but in order to allow man to join "fit agents to patients according to the orders, laws, times and methods of nature." By means of this knowledge, "he may be able to produce and bring to pass such effects, as may very much conduce to his well being in this world, both for satisfying his desires, and the revealing of his necessities." This new method is instrumental in "advancing his state above the

---

[17] Cf. Clucas (1994), 56–8.

[18] Hesse (1968), 132; Gaukroger (2001), 133–4; Giglioni (2013), 53–4.

[19] Hooke (1665), sig. b1v–b2r.

[20] Bacon (2004), 200–1; cf. Rossi (1996), 31.

common condition of men, and make him able to excel them as much, almost, as they do brutes or ideots."[21]

One of the tasks of the new mechanical and experimental philosophy was to "inquire and attempt by what means bodies may be changed or transmuted from one thing to another by a real change of all their former proprieties."[22] The old alchemical dream of transmutation is revived here in clearly corpuscular terms.[23] The link between operation and knowledge, between practice and theory, was founded on the mechanical nature of bodies. According to Hooke, "we may perhaps be inabled to discern all the secret working of nature almost in the same manner as we do those that are productions of art." The latter "are manag'd by wheels, and engines, and springs, that were devised by human wit," whereas the earlier consist of microscopic structures.[24] Unlike alchemists, Hooke maintained a clear corpuscular continuity between visible and invisible properties of bodies. The small machines operating under the visible surface of bodies follow the universal laws of mechanics. The only difference between visible and invisible level pertains to their dimensions.

2.3 Mechanical Instruments

Hooke's new ministration to senses was designed to rectify the "infirmities" in human perception. In ordinary experience, bare senses do not reach "an infinite number of things," either too small or too far. Furthermore, "many things, which come within their reach, are not received in a right manner." In *Micrographia* there are some echoes of the Baconian ministration to the senses.[25] But in the *General Scheme* Hooke shifted the emphasis to the "artificial helps" intended to correct the senses' defects. As discussed in the previous chapter, Hooke greatly valued direct and first-hand experience in the process of knowledge acquisition. Proper and correct ideas of things can be attained through a strict and continuous comparison of diverse sensible perceptions of them. Artificial instruments can improve the accuracy and power of the senses. By means of "instruments or standards" we may be able to receive "all degrees" of which "proprieties, powers and affections of bodies" are capable. Mechanical instruments might also introduce quantitative criteria in sensible perception. "By making the standard receive the same degree of the property with that in the body to be measur'd," Hooke noted, "the division of the standard may give the determinate quantity or degree. Hooke was aware of the difficulties involved in this programme." Quantitative standards would not easily apply to every

---

[21] Hooke (1705), 3.

[22] Ibid., 44.

[23] This was not unusual; see for instance Principe (1998), 48.

[24] Hooke (1665), sig. a2v.

[25] Ibid., sig. a1v–a2r.

perception produced by each of the five senses. Some seemed particularly resistant to quantification, "such as the smells and tastes of bodies, which never have been brought to any kind of theory." This programme was evidently based on a mechanical conception of bodily qualities. For this reason, Hooke maintained that as much as colours all the sense perceptions could be reduced "to a theory and standard." "The variety of colours," he claimed, "is not less than the varieties of tastes and smells, and yet 'tis not difficult to derive them all from two heads, and degrees of them."

The ministration to senses aimed to discover "those sensible proprieties in bodies, which our senses are not able to reach, and define them also."[26] Traditional natural histories were limited to the "superficiall and outward description" of bodies. No information of "the inward fabric, operations, vertues and uses of their parts" could be found in these histories. Bare eyes and hands are inadequate to attain any true knowledge. "Even where anatomy has been applyed for that purpose," Hooke notes, "we are gone noe further then to the forme and marks of the greater constituent parts such as are big enough to be visible to the eye and tractable to the hand." The smallest parts remained in a "primitive obscurity."[27] Only "artificial organs" can reach the inner level of the material structures which produce external appearances and proprieties of bodies.[28]

Like Bacon, Hooke did not see any contradiction in the application of artificial devices to the study of natural bodies. "People," Bacon wrote, "should instead have become inured to the idea that artificial things differ from natural things not in form or essence, but only in the efficient" cause.[29] Grounding himself on Bacon's refusal of the traditional ontological distinction between artificial and natural objects, Hooke described man's role in the creation in active terms:

> It is the great prerogative of mankind above other creatures that we are not only able to behold the works of nature, or barely to sustein our lives by them, but we have also the power of considering, comparing, altering, assisting and improving them to various uses.

Art can supply what humans naturally lack.[30] In Hooke's view, the human ability to know nature is not disjoined from the ability to intervene in it. These are two aspects of the same process. As Bacon claimed, only art "strips the mask and veil from natural things which generally lie concealed and hidden beneath a variety of shapes and outward appearances."[31] Nature does not show the inner causes of her productions. The only way to reach this level is to alter her ordinary course by means of artificial instruments and techniques. "Such experiments therefore,"

---

[26] Hooke (1705), 36.

[27] Royal Society Classified Papers, vol. XX, f. 183v.

[28] Hooke (1665), sig. a2r.

[29] Bacon (1996), 102–3; cf. Giglioni (2013), 46; Grafton (2007), 188; Da Costa Kaufman (1993), 193–4; Gaukroger (2010), 22–3.

[30] Hooke (1665), sig. a1r; Id. (1705), 532.

[31] Bacon (2004) 462–3; cf. Farrington (1964), 93, 99; Bacon (1857–74), vol. III, 612, 617–8; Id. (1996), 100–1; Rossi 1996, 27.

Hooke notes, "wherein nature is as 'twere put to shifts and forc'd to confess, either directly or indirectly the truth of what we inquire, are the best if they could be met with."[32]

For this reason, mechanical arts have a special place in Hooke's plan for a philosophical history. Information about "trades, arts, manufactures, and operations," especially containing "some physical operation, or some extraordinary mechanical contrivance" is a "philosophical treasury."[33] Mechanical arts provide an "extraordinary help" to discover the "true nature of the efficient and material cause of things." The knowledge of what art can change, divert or promote in natural bodies, is decisive to find "what they are that are thus wrought upon, since it seems very probable that they must be somewhat of the nature of those in art which promote them, and somewhat of a contrary nature to those that do alter and impede them."[34]

In sixteenth- and seventeenth-century Europe a new concept of experiment emerged, and the traditional notion of observation was transformed.[35] Hooke's interventionist approach contributed to this transformation of scientific experience. He considered experiments and instruments as mechanical alterations of the ordinary course of nature. The introduction of instruments, for instance, was a watershed in astronomy. Ordinary experience, largely consonant with common sense, was thus distinguished from observations through instruments. Hooke was part of a new generation of telescopic astronomers who aimed to bring the observational data to unprecedented standards of accuracy. This project contrasted with one of Tycho's most respected followers, the Danzig astronomer Johannes Hevelius. While Hevelius rejected telescopic sights, Hooke considered them as decisive for the growth of observational accuracy.[36] Rejecting telescopic sights equalled establishing a "bound or non ultra to human industry" and knowledge.[37] This belief echoed the Baconian idea of progress influenced by the craftsmen's world. According to Bacon, philosophy had been engaged in sterile disputations on words, whereas mechanical arts "grow and get better by the day."[38] Thinking that knowledge had already reached its highest perfection, scholars turned to books, whereby they thought to obtain "more easily and fully" any truth. On the contrary, by the constant application to practice mechanical arts produced great inventions, such as printing, gunpowder, the magnetic needle, microscopes, telescopes, and pendulum clocks.

---

[32] Hooke (1705), 34.

[33] Ibid., 24.

[34] Ibid., 58.

[35] Dear (2006), 106, 109, 121; Garber (2010), 2–16.

[36] Van Helden (1983), 54–5, 65–6, 69.

[37] Hooke (1677), 180; cf. Bacon (1857–74), vol. III; 584; Bennett (1989), 24–5; Applebaum (1996), 462–4; Vertesi (2010), 230–6.

[38] Bacon (2004), 12–3.

And much more could be expected.[39] "How many things," Hooke asked, "are there in the possibility of meckanics that the world is knowingly desirous of performing"?[40]

Hooke shared a critical view of the role of experience in the Aristotelian tradition. This view can be traced in Bacon's work. Despite referring to observations, Bacon noted, Aristotle "made up his mind beforehand, and did not take experience into due account when he framed his decrees and axioms." Instead, "having made up his mind to suit himself, he bends experience to his opinions." Consequently, Aristotle's followers, "the family of scholastic philosophers," abandoned experience altogether.[41] Hooke maintained that the "few observations they had read or collected" had been accommodated to their theories. "Instead of an indeavour to rectify and regulate those so receiv'd theories by those intimations, which careful and accurate observations would afford," they made them consonant with their preconceived opinions.[42] As Boyle noted, they used experience as a means of illustration but not of discovery.[43] The prevailing theoretical nature of experience in the Aristotelian tradition contrasts with the interventionist character of modern experiments. The former produced a library empiricism, the latter a laboratory science.[44] The traditional separation between practice and experience on one hand and between theory and speculation on the other, was seen by Bacon and many Baconians as the cause of both the decadence of philosophy and the limitation of mechanical arts.[45] Excluded from the field of theoretical knowledge, the world of craftsmen was limited to experience and practice. Because of this, Hooke observed, technicians were "very unwilling to be put out of their common road of working." "Science" and "art," theory and practice, "both ought to be call'd in for assistants in the prosecution of experimental philosophy."[46] The practical skills of mechanics and the theoretical subtlety of learned men should join in the new figure of the experimental philosopher. In the study of nature "'tis not enough to know how to manage an instrument, or to have a good eye, or a dextrous and steady hand, but with these there must be joined a skilfulness in the theoretical and speculative part." "Great diligence and circumspection" are needed to make experiments and take note of observations, since "it is not less considerable as to the prosecution of this designe to be able to judge when a new experiment or observation is requisite then to be able to make it such."[47]

---

[39] Hooke (1705), 4, 535.

[40] Royal Society Classified Papers, vol. XX, f. 174r.

[41] Bacon (2004), 100–1.

[42] Hooke (1705), 4–5.

[43] Boyle (1999), vol. II, 221. Focusing on these very same aspects of medieval science, some historians have described it as "empiricism without observations" and "a natural philosophy without nature," see Grant (2002), 141–6, 166; Murdoch (1982), 198–9.

[44] Crosland (2005), 234, 238; Van Helden (1983), 50; Daston (2011), 81, 84–6; Dear (1995), 21, 28, 161–2; for a different view see Anstey and Vanzo (2012), 509–15.

[45] Bacon (2004), 130–1; cf. Bacon (1857–74), vol. III, 600–1. Hooke (1705), 6.

[46] Hooke (1705), 135, 532; cf. Bacon (1857–74), vol. III, 591, 601–6, 608–9, 612–7.

[47] Guildhall Library, London MS 1757.11, f. 104r.

   The work of mechanical arts could bring philosophy "from words to actions."[48] Experimental philosophers "should be conversant with all kind of tradesmen to learne their operations and so practise their manner of working." They should not "attest barely upon the words or reports of some cosening workmen," but examine and perform mechanical operations "with circumspection and diligence."[49] In his capacity as curator of experiments and secretary of the Royal Society, as well as in his proposals of reform of the Society, Hooke supported the active involvement of mechanics and tradesmen. For Hooke, the Society's task was to accomplish the Baconian transformation of workshops and factories into laboratories.[50]

   According to Hooke, Galileo "has given diverse specimens of the benefit that may be made of very common and obvious experiments where industry and art is made use of, for the improving and perfecting of them to scientific knowledge."[51] The admiration for the work of "very conversant and subtle" craftsmen influenced Galileo's revolutionary decision to use a mechanical instrument for the inquiry of heavenly bodies.[52] His mathematical science, as has been insightfully claimed, was also concerned with "purposely contrived and controllable artefacts," such as scientific instruments.[53] In this respect, Galileo can be considered a relevant reference for Hooke's development of Bacon's instauration, in particular because of the importance attributed to optical instruments. In contrast with Galileo, Bacon evidently expressed a preference for experiments compared with instruments. Since "all interpretation of nature starts from sense," Bacon aimed to fill the distance separating bodies' properties and human perception. He included the new optical instruments among the supports to the senses listed in the "instances of the lamp or first information." Microscopes and telescopes enlarge the power of senses "by greatly increasing the size of the species" and "reduce the imperceptible to the perceptible."[54] The "glasses which Galileo's unforgettable enterprise gave us" enlightened the real nature of the moon and the Milky Way, and showed new celestial bodies around jupiter. Nonetheless, Bacon received the Galileian instrument and discoveries with expressions of doubt.[55] "For myself," he wrote in the *Novum Organum*, "I am chary

---

[48] Hooke (1665), sig. g1v; The aspiration to a different relationship between philosophers and craftsmen was shared by many Baconians. According to William Petty, for instance, mechanical arts provide philosophers with a "better matter to exercise their wits upon, whereas now they pulse and tire themselves about mere words and chimericall notions. Petty (1647), 22; see also Glanvill (1665), sig. b4v.

[49] Oldroyd (1987), 156–7.

[50] Birch (1756–57), vol. I, 316, 490; Hunter and Wood (1986), 90; cf. Espinasse (1974), 365–6; Thomas (2011), 1.

[51] Royal Society Classified Papers, vol. XX, f. 171v.

[52] Galilei (1890–1909), vol. III.1, 99; vol. VII, 49.

[53] Palmieri (2008), 10–1; cf. Lefévre (2001), 11–3; Valleriani (2010), 157–62; Strano (2012), 9–10, 18; Gal and Chen Morris (2010), 122, 128–9.

[54] Bacon (2004), 340–3.

[55] Such as *si fides constet* and *si iis credimus quae Galilaeus tradidit*. Bacon (1996), 132–3, 154–5, 156–7, 164–5.

of them mainly because the experiment has stalled on these few things, and many others just as worthy of investigation have not been found out by the same procedure." He expressed similar doubts about microscopes, which were expected to "let us see the latent and invisible minutiae of bodies, and their inner schematisms and motions." Their limitations "except for very small things (and for these too if they are situated in a large body) undermine their usefulness."[56] According to Bacon, optical instruments "do not count for much," because they let eyes directly judge the nature of things. Senses are "weak and wandering." Only "experiments and appropriated instances" can avoid perception's mistakes. Senses can judge only the experiments, whereas these judge "the nature and the thing itself."[57]

The Baconian discussion of optical instruments was linked to the doctrine of idols and the limitations of senses. Bacon's acknowledgement of human sensible limitations did not entail sceptical conclusions. Baconian and sceptical arguments, as Bacon himself notes, "are rather alike in their beginnings but end up in being very far apart and opposite to each other." Whereas sceptics "destroy the authority of the sense and intellect," Bacon aims to "think up and furnish helps for them."[58] Although failing and sometimes deceiving us, senses may still conduce to real knowledge, but only by means of experiments rather instruments. The disappointment of Galileian discoveries, therefore, is one with the prominence attributed to experiments over instruments. Both depend on the Baconian labyrinthine image of nature. The subtlety of its pneumatic components seemed to Bacon unreachable by the instruments then known.[59]

Galileo's pioneering introduction of optical instruments and his emphasis on their philosophical value was probably the major model for Hooke. He also benefitted from significant improvements in the magnifying power of the instruments. This evidently appears in the observation of the Pleiades, the last of *Micrographia*, where Hooke compares his results with those of the Pisan mathematician. "The famous Galileo" was able to see no more than 36 stars, while Hooke distinguished 78 of them.[60] The belief that there are no limits to the improvement of optical instruments, along with a corpuscular view of nature, is the main reason that led Hooke to adopt a new ministration for the senses. Hooke was confident that through optical instruments "the parts of the picture at the bottom of the eye" will be seen "as nice and particular as the body has distinct parts." Along Baconian lines, Hooke distinguished two ways to assist the senses, "either first by enlarging their power, or sphere of activity, and extending it much farther than that assign'd them by nature, or else secondly, by reducing other things to such a constitution, as to bring them within" their power. The improvement of optical instruments "is the most likely way to

---

[56] Id. (2004), 342–5.

[57] Id. (1996), 86–7; cf. Manzo (2001), 63–6; Id. (2006), 22, 175; Wilson (1995), 50.

[58] Bacon (2004), 78–9. For the image of Bacon as "sceptical empiricist, biased in respect to senses" and defender of a "sensory relativism" see Hamou (2001), 35–7.

[59] Cf. Bacon (2004), 346–9; Rossi (2003), 142; Giglioni (2013), 46.

[60] Hooke (1665), 241.

afford us the greatest help for the detection of the nature of bodies."[61] For Hooke, in this inquiry the chemical way outlined by Bacon did not offer the advantages of the microscope. Both were directed to the constituent parts, but "by fire or chymistry" these are "wholy vitiated and destroyed, and torne all to peices and scarce soe much lost intire as may deserve the name of the reuines of the body." Besides, fire mixes the "attomes" of the body with "other heterogeneous substances insinuated and mingled with them, that are properly parts of the fire or menstruum that dissolved them."[62]

Hooke maintained the Baconian comparison between the expansion of the geographical knowledge and the expected enlargement of the intellectual one. "Comparing what is now known and describ'd," Hooke wrote, "with what was known to the antients, we shall find more than a new world has been discover'd." The "new visible world discovered to the understanding" by telescopes and microscopes was "indeed the very world itself."[63] By "producing new worlds and terra-incognita to our view," the improvements of optical instruments will "make us, with the great conqueror, to be affected that we have not yet overcome one world when there are others to be discovered."[64] These "new parts" yet unknown of nature were "dreamt of as America was before the suggestion of Columbus." Optical instruments offered "reasons enough to remove despair." It seemed "much more easy to improve a thing already invented then twas to invent the instrument now made use of as the discovery of the rest of the parts of the earth have been more easy then that first of Columbus in the west Indies."[65]

The rejection of the Newtonian theory of light in 1672, along with the limitations it entailed to refraction-based optical instruments, confirmed Hooke in the belief of a possible correction of chromatic aberration by means of new shapes of lenses. As Hooke showed in *Micrographia*, aberration was not the only obstacle to the improvement of microscopes. Long before the composition of achromatic lenses, the first phase of microscopic researches was restrained by technical limitations as much as by the lack of adequate observational standards and systematic programmes of research.[66] The works of Leeuwenhoek and Hooke illustrate these limits. Their results can mainly be attributed to the refinement of observational techniques, in particular of the illumination systems.[67] In Hooke's mechanico-experimental programme, the study of living beings was not the main objective of research, as it was for Leeuwenhoek. Microscopes were mainly employed by Hooke in the quest for "the figures of the compounding particles of matter, and the particular schematisms

---

[61] Id. (1705), 12–3, 38–40.

[62] Royal Society Classified Papers, vol. XX, f. 183r.

[63] Hooke (1705), 478; Id. (1665), sig. a2v; cf. Gascoigne (2013), 219–20.

[64] Hooke (1665), sig. d2v.

[65] Royal Society Classified Papers, vol. XX, f. 172r.

[66] Ruestow (1996), 39; Bradbury (1976), 151–2.

[67] Fournier (2007), 212; Cf. Bennett (1997), 65.

and textures of bodies."[68] It is significant that *Micrographia* begins with the observation of a physical point, proceeds to greater bodies, and concludes with the moon and the stars.[69] Optical instruments in general and microscopes in particular, were the epistemological foundation of Hooke's mechanical theory of matter.

2.4   Logical Instruments

The new optical instruments did not provide any decisive proof of the new ideas on matter and cosmos, viz. Copernican astronomy and corpuscular philosophy. At the same time, they did offer decisive arguments against Aristotelian philosophy. As the telescope made untenable the ontological distinction between terrestrial and celestial bodies, the microscope showed the insufficiency of hylomorphism.[70] For Hooke, the "two general and (unless further explain'd) useless words of matter and form" should give way to new principles, such as "the subtility of the composition of bodies, the structure of their parts, the various texture of their matter, the instruments and manner of their inward motions."[71] Microscopes offered arguments "to suspect that those effects of bodies, which have been commonly attributed to qualities, and those confess'd to be occult, are performed by the small machines of nature, which are not to be discern'd without these helps." Although mechanical and hylemorphic causes are both invisible, significant differences persist. It can be claimed, perhaps, that the invisible status of natural causes played a different role in the mechanical worldview than it did in hylomorphism. In the mechanical philosophy there is no place for any ontological distinction between the realms of the visible and the invisible. A continuous scale joins the big machines perceptible by senses and the invisible machines not reached by them, both "seeming the meer product of motion, figure and magnitude."[72] Following Galileo, Hooke believed that by means of the mechanical intervention in nature the "real" and "true philosophy" could be distinguished from the "philosophy of discourse and disputation," thanks to the certainty of direct observations.[73] In the Galileian astronomical work, as Massimo Bucciantini has pointed out, truth was not just understood or discussed, rather for the first time it could be seen by means of telescopes.[74] The truth Galileo was interested in was that of the Copernican astronomy, while Hooke looked first for the mechanical

---

[68] Hooke (1665), sig. b2v.

[69] For a different view, see Böhme (2005), 385–9.

[70] Meinel (1998), 68, 72–6, 101; Lüthy (1996), 14; Van Helden (1983), 51. Like geocentricism, hylomorphism survived the advent of the new optical instrument for almost a century, cf. Buchwald and Feingold 2013:25–6.

[71] Hooke (1665), sig. a2v.

[72] Ibid., sig. g1r; cf. Wilson (1995), 41–3, 56; For a different view see Hutchinson (1982), 233, 242–3, 246

[73] Hooke (1665), sig. a2r;

[74] Bucciantini (2003), 174.

structures of matter. Nevertheless, until the microscope will allow us to "discover the true schematism and texture of all kind of bodies," Hooke wrote, "we must grope, as it were, in the dark, and onely ghess at the true reasons of things by similitude and comparisons."[75] Like Bacon and Descartes, Gassendi and Newton, Hooke not only largely used analogies in his work, but also explicitly theorised their role in the new science[76]:

> As ther[e] may be a propriety of a body which I would willingly know, but I cannot think of any way of making tryall on that body how to finde it, but knowing an other body which has divers other things common with it and seems to have the same with that on which I can make tryall by this meanes I say 'tis possible many quaerys may be fully answerd.[77]

Hooke outlined the analogical argument as an inference from the well-known causes of some phenomena to the unknown causes of similar phenomena. This inference needs two kind of relations: similarity and causality. From the similarity between the observable effects A and B, the causal relation between A and its known cause C is extended to B and its unknown cause D. Any such analogical inference is built on the principle of uniformity of nature. In Hooke's labyrinthine image of nature, this principle was limited by the existence of "monadic instances."[78] Many things "may be imagined and guessed at by analogy and the uniformity of the proceedings and productions of nature," but sometimes analogical inferences cannot apply. Indeed, "there are certain non-pareils of nature, of which kind possibly, nothing like them have been produced in all those particulars which are more common and obvious." Saturn, among the planets, is an example of a monadic instance or a nonpareil of nature.[79]

Bacon limited the role of analogy also because of its links with the magical and symbolic idea of the signatures that he rejected.[80] Hooke was mainly influenced by what he considered the primary role of the direct sensible apprehensions in human knowledge. Without a "full sensation" of an object "we must be very lame and imperfect in our conception about it, and in all the propositions which we built upon it." In fact, Hooke notes, "we often take the shadow of things for the substance, small appearances for good similitudes, similitudes for definitions." For this reason, "many of those, which we think to be the most solid definitions, are rather expressions of our misguided apprehensions then of the true nature of things themselves."[81] Without direct visual observations, true and adequate ideas of natural phaenomena cannot be achieved. Following Harvey, Hooke defined the direct apprehension of

---

[75] Hooke (1665), 114.
[76] Rossi (1986), 126, 138–8, 141–3, 146; Mandelbaum (1964), 61–3; McGuire (1970), 4; Shapiro (2004), 214; Gabbey (1985), 12–4; Id. (2001), 454–7; Buchwald (2008), 6.
[77] Guildhall Library, London MS 1757.11, f. 104r.
[78] Bacon (2004), 296–7; cf. Id. (1996), 100–1.
[79] Hooke (1726), 262–3; cf. Pyle (1995), 533.
[80] Vickers (1984), 95; Rossi (1986), 129–31, 145; Wilson (1988), 90–1; cf. Bianchi (1987), 32–44.
[81] Hooke (1665), sig. a1v.

things as *autopsia*.[82] Harvey adopted a Latin literal transliteration of an original Greek medical term to stress his empirical interpretation of Aristotelian philosophy.[83] In Harvey's opinion "the method of investigating truth commonly pursued at this time" was erroneous, because "many inquire what others have said, and omit to ask whether the things themselves be actually so or not." The lack of empirical evidence was largely filled by recourse to analogies, producing "mere verisimilitudes" instead of "positives truths."[84] Hooke similarly noted that "autopsia is not only necessary for directing the mind and intellect, in its progress to be made, for what is to be gone thro' with; but 'tis necessary also for reducing it to its right way." In the study of nature, inquirers can be "misguided by the false and erroneous suggestions" of famous authors "or some other person reputed eminently skilful in this, or that part of knowledge." Among these are the defenders of "the four Aristotelian elements, or the four chymical principles, or the three Cartesian materia's, or his mundane vortices."[85]

Hooke was aware that the new philosophy would lack adequate means of inquiry until optical instruments were fully developed. For this reason, he introduced logical instruments that could replace the mechanical ones not yet available. He followed Bacon in envisaging analogy as "surely useful but less certain" than the direct observation of the phenomena.[86] "Similitude, harmony and uniformity in the operation of nature" provide only a probable knowledge of the non-perceptible features of phenomena.[87] The "data" upon which the analogical inference is built are "uncertain and only conjectural," for things not reached by senses "remain in their primitive obscurity and are only the objects of conjecture and imagination." The analogical inference "can be no other than probable." Nevertheless, it plays an important role in Hooke's methodological programme. Deductions drawn by analogy, like hypotheses, can "become more and more probable as the consequences deduc'd from them appear upon examinations by trials and design'd observations to be confirmed." Their truth-content does not depend only on theoretical assumptions, such as the uniformity of nature. They can be connected to experience by the indirect verification of hypotheses, "so that the effect is that which consummates the demonstration of the invention itself; and the theory is only an assistant to direct such an inquisition."[88]

Hypotheses, analogies, and inductions were part of the new ministration to memory, the new experimental history of nature. Human memory, like other human faculties, seemed to Hooke "shallow and infirm." Relevant "circumstances" are easily

---

[82] Id. (1726), 263; cf. Hankins and Silvermann (1995), 41–2.

[83] Pomata (2011), 65–6.

[84] Harvey (1651), sig. B4r; Id. (1847), 162.

[85] Hooke (1726), 263–4. Interestingly, Hooke defines these schools as "sects" because of their "division from the true philosophy. Cf. Voltaire (2004), 374: "Every sect of whatever kind is the rallying point for doubt and error.

[86] Bacon (2004), 362–3.

[87] Hooke (1705), 172.

[88] Ibid., 573; Royal Society Classified Papers, vol. XX, f. 183v.

forgotten, while "frivolous or false" ones are treasured up. "Good and substantial" information is "overwhelmed and buried under more frothy notions." Memory seems to "prefer some things first in order before others, and some things with more vehemence and greater concern." Human reason is thereby "more apt to be sway'd to this or that hand, according as it is more affected or prest by this or that instance, and is very liable to oversee considerable passages, or to neglect them."[89] Following Bacon's statement that "history answers to memory," Hooke thought that the human "retentive faculty" could be assisted by "the committing to writing things observable in natural operations."[90] Ancient and Renaissance natural histories seemed pandering memory's natural defects, rather than correcting them. These histories focused just on the "outward shape and figure, or beauty" of natural things, lacked descriptive accuracy, and deal only with wonderful and extraordinary phenomena. Many seemed to Hooke "design'd more for ostentation then publique use." Devoid of "the more subtile examinations of natural bodies, by dissections, experiments or mechanical tryals" these works proved useless for natural philosophy.[91] In other words, they were composed of inaccurate descriptions of the ordinary course of nature. By contrast, Hooke described the new natural and experimental history as a book providing "the orthography, etymologia, syntaxis, and prosodia of nature's grammar." Without this book, the new philosophy could not learn the "language and sense of nature."[92]

In Hooke's plan the new history was something more than a detailed and accurate description of observations and experiments. It was a "philosophical history." Natural history could be "a description of the things themselves, whereby inquisitive persons that are ignorant of them, may come to a more perfect knowledge of them." But it could also be composed and arranged "in order to the use of philosophical inquiry, for the invention of causes, and for the finding out the ways and means nature uses, and the laws by which she is restrain'd in producing divers effects."[93] Hooke's experimental history was not a simple collection of matters of facts or experimental protocols, as has been claimed, but "the repository of materials out of which a new and sound body of philosophy may be raised."[94] The description of observations and experiments cannot and should not be separated from the inquiry of causes. For Hooke the new "naturall and experimentall history" was rather intended as a "magazine of naturall philosophy." Experiments and observations are the "foundation stones on which the whole structure should be raised." They "ought to be proportioned according to the rest of the materials. Superfluous and insignificant instances are not to be stored, repetitions are to be avoided."[95]

---

[89] Hooke (1665), sig. a1v; Id. (1705), 6.

[90] Hooke (1665), sig. d1r; Bacon (1996), 96–7.

[91] Hooke (1705), 3–4; Id. (1665), sig. b2v.

[92] Id. (1705), 338; cf. Bacon (2004) 468–69.

[93] Hooke (1705), 26; Bacon (2004), 454–55.

[94] Hooke (1705), 18; cf. Shapin and Schaffer (1985), 36; Jack (2009), 194.

[95] Oldroyd (1987), 151, Hooke (1705), 18.

The main difference between ordinary and philosophical histories is in their order. The latter aimed to provide material "so readily adapted, and rang'd for use, that in a moment, as 'twere, thousands of instances, serving for the illustration, determination or invention, of almost any inquiry, may be represented even to the sight."[96] Single experiments and observations are like letters of alphabet, "which seldom signify but when they are joyn'd and compounded in syllables or words." An appropriate collection of observations and experiments together is the needful basis for "reasoning and deducing from them." The philosophical work of discovery of causes, therefore, does not start once the descriptive work has ended. Experiments and observations are strictly linked to the quest for the causes through analogies, queries, and hypotheses.

The relevance of the instances' order in Hooke's plan for a philosophical history was consistent with his Baconian idea of induction, and with the role of analogies and comparisons.[97] Among the frailties of memory, Hooke listed the tendency to overestimate positive instances and reach easy conclusions. Human intellect often lacks the "patience to follow and prosecute the negative way of inquiry by rejection of the disagreeing natures."[98] Baconian induction was not a "simple enumeration," but a way "to separate a nature out by due rejections and exclusions, and then, after bringing enough negatives to bear, draw conclusions from affirmatives." According to Bacon, rejections and exclusions drive off as fumes all "volatile opinions" and avoid the premature, untimely conclusions drawn out of simple enumerations.[99] In Hooke's plan for a philosophical history, instances were ordered so as to further comparisons and rejections. Induction, in short, consisted in nothing else than "comparing the collected observations and ratiocinating from them."[100] In Hooke's hands, Bacon's induction did not follow natural history, rather it became part of it. It was essentially because of its comparative feature that Hooke included it in the new ministration to memory. Similarly, it was mainly for the presence of this kind of induction that Hooke's natural and experimental history could be described as a "philosophical history."

In Hooke's plan instances had also to be "interwoven here and there with some shorts hints of accidental remarks or theories, of corresponding or disagreeing received opinions, of doubts and queries and the like." Theoretical elements, such as hypotheses and queries, were also part of Hooke's philosophical history. Hooke distinguished hypotheses and suppositions already known from new queries that could help the understanding of observations and experiments. Being acquainted with "all sorts of hypotheses & theorys by which the phenomena of nature have been indeavourd to be solvd," the experimental philosopher will "be much sooner and better freed from prejudice; for by discovering experimentally the errors in this

---

[96] Hooke (1665), sig. d1r.

[97] Cf. Rossi (1986), 145; Perez-Ramos (1988), 266–9.

[98] Hooke (1705), 6; cf. Bacon (2004), 82–3.

[99] Bacon (2004), 110–1, 162–3, 254–5.

[100] Hooke (1705), 35.

or that hypothesis, 'twill be much easier taken off from adhering to any."[101] New hypotheses, on the other hand, are instrumental in the "discovery of the proprieties of bodies" and in the "finding out the nature of some general qualities." These hints were not seen as assertive statements, but rather instruments for a further investigation. Provisional explanations had to be verified by new experiments or observations. For Hooke, therefore, experiments were active interrogations of nature also because of their constant connection with theoretical elements. He considered an observation free from any theoretical elements as the likely source of "frivolous suppositions taken at random to solve one phenomenon."[102]

Hooke thought of queries and hypothesis "as hints to some further discovery, and not as axioms," or else "as doubtful problems and uncertain guesses and not as unquestionable conclusions, or matter of unconfutable science."[103] He planned to use them as instruments in the long quest for the invisible schematisms of matter:

> My meaning is not to make these quaerys extravangancy of phancy; but rather to make them consist of such as upon a serious consideration of the nature and conditions of the body to be examind seem those necessary to be unswerd before the constitution of that body can be known.[104]

The presence and relevance of theoretical elements did not affect the experimental character of the new philosophical history. "Nature it self," Hooke noted, "must give the last information, and tryalls must give the plainest evidence."[105] The human mind should direct the senses as a "lawful master, and not as a tyrant."[106] Hooke was aware that often experiments were realized and interpreted only in order to confirm some preconceived ideas. This awareness was shared by other seventeenth-century Baconians. William Petty, for instance, noted that experiments "would be more faithfully made and delivered, if they were not made to help out a theory, because that might prepossess and bias the experimenter."[107] Hooke, however, did not intend to separate experiments and theories. The new philosophical history was designed to introduce a much-needed virtuous interaction between these elements. In Hooke's view, hypotheses and queries were to be checked against experimental "touchstones." Like individuals, often sacrificed by Nature for the perpetuation of whole species, hypotheses were just instruments for the invention of causes, not the invention itself.[108]

For these reasons, Hooke introduced crucial experiments. Robert Boyle used this term for the first time with reference to Blaise Pascal's pneumatic experiments

---

[101] Ibid., 18–9, 28; Oldroyd (1987), 158.

[102] Hooke (1705), 357.

[103] Id. (1661), 26; Id. (1665), sig. b1r.

[104] Guildhall Library, London MS 1757.11, f. 103r.

[105] Royal Society Classified Papers, vol. XX, f. 181r.

[106] Hooke (1665), sig. b2r.

[107] Birch (1756–57), vol. IV, 8.

[108] Hooke (1661), 41–2; Birch (1756–57), vol. IV, 8.

of Puy de Dome.[109] In *Micrographia* Hooke employed it to present the experiment of refraction of light by Muscovy glasses. Since it showed, in Hooke's view, the insufficiency of the Cartesian theory of light and colours, the experiment was described "as one as our thrice excellent Verulam calls experimentum crucis, serving as a guide or land-mark, by which to direct our course in the search after the true cause of colours." Further references to Bacon's *instantia crucis* were included in some of Hooke's lectures delivered at the Royal Society. In a lecture on philosophical history, the "experimentum crucis as Lord Bacon calld it" was described as the way to determine controversies in natural philosophy. In a later paper, read in 1689, Hooke seemed to identify *instantia crucis* and *experimentum crucis*.[110] However, both Hooke and Boyle employed the expression only to refer to experiments. There are not crucial observations but only crucial experiments.[111] Although limited to mechanical alterations of the ordinary course of nature, Hooke's crucial experiments maintained a fundamental feature of Bacon's crucial instances. Linked to a Baconian form of induction, these experiments have a negative function; they are primarily "terminij, bound stones" showing the falsity of hypotheses. Crucial experiments, Hooke noted, inform "the traveller in naturall inquirys that here is the utmost limit, and that this way or that way is ne plus ultra, and therefore he must turne an other way."[112] Hooke's crucial experiments maintained the contingent character of Baconian crucial instances expressed by the analogy with land-marks. They did not end the research, but only direct it towards a new way.[113] The optical *experimentum crucis*, for instance, provided mainly a "negative information" about the limits of Cartesian theory. Its "affirmative and positive instruction" was just a new hypothesis that further experiments and observations could refine or reject. Describing his double prism experiment as an *experimentum crucis* of the new theory of light and colours, the young Newton thought perhaps of Hooke's crucial experiment rather than Baconian *instantia crucis*.[114] But he conferred, or rather aimed to confer, to his 1672 crucial experiment a deeply demonstrative character. In Newton's view, it did not produce another probable hypothesis, but it did demonstrate a certain theory.[115]

---

[109] Boyle 1999, vol. III, 50

[110] Hooke (1665), 54; Guildhall Library, London MS 1757.11, f. 104r; Royal Society Classified Papers, vol. XX, f. 178r.

[111] Anstey and Hunter (2008), 112; Pugliese (2004), 952.

[112] Royal Society Classified Papers, vol. XX, f. 178r.

[113] Hooke (1705), 35.

[114] Newton (1959–77), vol. I, 94.

[115] Hooke (1665), 54.

2.5   Shifting Boundaries

Hooke's project includes elements that can suggest the assumption of a Cartesian view of science, such as the stress on analogy and hypotheses, and the use of logical inferences in the quest for the invisible constituents of bodies. It is evident that a generally mechanical world-view influenced Hooke's ideas. But some epistemological elements entailed by any mechanical philosophy cannot be exclusively identified with Cartesianism. Different views of matter and knowledge, such as Gassendian atomism, also supported it. The reception of continental mechanism by British experimental philosophers was significantly influenced by their Baconian background. Moreover, it was not limited to Descartes, as the diffusion of Gassendi's Christianised version of atomism suggests. Cartesian corpuscular physics circulated widely among the members of the Royal Society, but the general acceptance of Cartesian philosophy was confined to marginal groups.[116] Fellows were generally aware of the importance of hypotheses and analogies in Descartes's physics. "'Tis upon billiars, and tennis-balls," Christopher Wren wrote, "upon the purling of sticks and tops, upon viol of water, or wedge of glass, that the great Descartes hath built the most refined and accurate theories that human wit ever reach'd to."[117] But this feature of Cartesian science was also the object of sharp criticism, for it was seen as linked to a wider neglect of experiments. In a letter to Samuel Hartlib dated 1649, for instance, William Petty strongly criticized Cartesian theories. Although Descartes is "to bee preferd before the Common schoole philosophers, who indeed have Nothing else but words," his principles "stand upon such narrow feet as those few Experiments mentioned in his Works are."[118] Similar scepticism was frequently expressed by the virtuosi.[119] Hooke read Descartes at Oxford, where he was part of the group gathered around Wilkins. The Baconian agenda of Wilkins and his associates likely influenced Hooke's reception of Cartesian ideas. As we have seen in Chap. 1, his readings on continental mechanism included also Gassendi's works. Like Wren, Hooke acknowledged the subtlety of the "most acute Descartes." Like Petty he often criticized the lack of experimental foundation in his physical theories. In *Micrographia*, Hooke lamented how little Descartes "applied himself experimentally."[120] In Hooke's eyes, Descartes made "that error, which the thrice

---

[116] Webster (1967), 168.

[117] Wren (1750), 224–7; Birch (1756–57), vol. I, 289.

[118] Petty to Hartlib, Hartlib Papers 7/123/1A; cf. McCormik (2009), 62–3; Boas Hall (1963), 86; Clarke (1989), 131–212.

[119] Henry (2013), 118–9; Laudan (1966a), 75; Rogers (1972), 238; Wilson (2008), 64–5. Which effective role experiments played in Descartes' work is a controversial question, on which see Buckwald (2008), 3; Osler (1994), 140–1; Clarke (1990), 106–8; Gaukroger (2002), 68; Baldassarri (2017), 117–8, 129. Suffice here to note that like many other natural philosophers, Descartes often did not practice the method he prescribed, cf. Osler (1994), 145–6.

[120] Hooke (1665), 46.

noble Verulam justly takes notes of, as such, and calls *philosophiae genus empiri-
cum, quod in paucorum experimentorum angustiis et obscuritate fundatum est*."[121]

Hooke did not maintain Descartes' confidence in a deductive science firmly
rooted in his metaphysics. In Hooke's Baconian and sylvatic view of nature, the
"mechanical operations" of the minute constituent particles could be quite different
"from those of bodies of greater bulks."[122] Hypotheses and analogies, albeit necessary, are valued less than the direct observation of phenomena. Hooke's methodological programme, therefore, presents fundamental non-Cartesian characteristics
that are ultimately due to a different image of nature. As Hooke's adoption of hypothetical elements did not entail the assumption of a Cartesian methodology, his criticism of Cartesian hypotheses did not entail the refusal of all Cartesian physics. On
the contrary, Hooke maintained important aspects of Cartesian mechanism despite
refusing their metaphysical foundation.

The view of Descartes as the main source of non-empiricist elements in the
methodological reflections of British experimental philosophers, such as Hooke and
Boyle, has often been related to an empiricist reading of early Baconianism.[123] The
Baconian elements of experimental philosophy have often been identified with a
supposedly empiricist conception of science. Bacon criticised the primacy of theories over experience in the traditional learning. At first glance this can seem condu-
cive to the banning of hypotheses. Bacon called for the definitive refusal of those
"wrong-headed philosophies which have put theses before hypotheses, led experience captive, and exulted over God's works." Instead of the volumes full of speculations used in the schools, the new philosophers should "read through with due
humility and reverence the volume of creatures, and dwell and reflect on it, and
purged by opinions, to study it with a pure and honest mind." Echoing this religious
intonation, Bacon described natural history as "the book of God's works" and
"another kind of Holy Writ." This book should contain "no more of antiquities, cita-
tions and differing opinions of authorities, or of squabbles and controversies, and,
in short, everything philological." Bacon knew that the pursuit of knowledge in the
schools was still largely based on ancient authorities, "a way of proceeding which is
altogether without foundation, and enveloped in opinions alone."[124] Dominated by
extravagant hypotheses and fanciful opinions, natural philosophy has thus become
similar to the fictional proceeding of astronomy.[125] The real knowledge consisted in
the "interpretation of nature" and not in "anticipations of the mind."

Bacon's scepticism on the traditional learning should not be mistaken for a general refusal of theories and hypotheses in natural philosophy. The case of astronomy
is significant in this respect.[126] Bacon's hostility towards the term 'hypothesis' was,

---

[121] Id. (1661), 41; Id. (1665), 28 (italics in the texts); cf. Bacon (2004), 100–1.

[122] Hooke (1678), 83, 94–5.

[123] Sargent (1987), 486; Id. (1995), 30–1.

[124] Bacon (2007), 11; Id. (2004), 131, 469, 457.

[125] Bacon (1996), 135; Id. (1857–74), vol. III, 603; Farrington (1964), 85 cf. Vickers (1985), 12.

[126] Bacon (2004), 59; Rossi (1984), 252; Urbach (1987), 31; Vickers (1992), 500.

perhaps, a consequence of its association with mathematical astronomy. In that tradition, the movements of celestial bodies were explained by means of conventional geometrical entities. Bacon's refusal of hypotheses in science was one and the same with the criticism of the preponderance of theories and opinions over experience in the traditional learning. His criticism of hypothetical astronomy and scholastic philosophy was directed to undermine the distinction between the two disciplines of astronomy and physics. Bacon called for a new non-hypothetical astronomy as an integral part of a new natural philosophy based on observations and experiments, such as the "crucial instance" called on to decide "whether the diurnal motion (by which to our eyes, the sun and stars rise and set) is a real motion of rotation in the heavens, or just apparent in the heavens but real in the earth."[127] Bacon's criticism of traditional learning, however, did not entail the empiricist conception of natural philosophy that scholars have often attributed to him.[128]

Baconian natural history did not consist of matters of facts rigidly separated from theoretical elements.[129] Describing the new natural history as "the primary matter of philosophy" or "the basic stuff and raw material of the true and legitimate induction," Bacon did not draw a boundary between experience and theory, between history and the interpretation of nature.[130] The whole *novum organum* aimed to elevate natural knowledge from the limited plans of experience or theory to that of a new integrated system of both theory and experience.[131] Bacon's criticism of the predominance of opinions and hypotheses in the traditional natural philosophy cannot be dissociated from his refusal of "the means of gaining experience" employed by the empirics. Bacon judged it "blind and stupid" to proceed from an individual case to another similar case without a systematic plan directed to the causes of the phenomena observed.[132] He defined this kind of experience an "unbound broom." "My route and plan," he wrote, "does not lead me to extract works from works, or experiments from experiments (as empirics do)." He often compared his idea of knowledge to a continuous scale from experience to axioms.[133] The well-known metaphor of ants, spiders and bees helps to clarify this point. Bacon's criticism was directed to both "dogmatists" and "empirics." The scope of his great instauration was "a chaste and lawful marriage between mind and nature."[134]

Bacon was not an empiricist, and he was not seen as such by many early Baconians. The space for hypotheses in the methodological programmes of Hooke and Boyle, for instance, can be considered an effect of the Baconian influence

---

[127] Bacon (2004), 325

[128] For instance Daston (1994), 152; Id. (2001), 746–50.

[129] Zagorin (2001), 382–3; Giglioni (2011), 46; Sargent (2012), 82–3.

[130] Bacon (1996), 104–5.

[131] Giglioni (2011), 101; Id. (2013), 59; Rusu (2012), 117; Stewart (2012), 88, 109.

[132] Bacon (2007), 110–1; cf. Hesse (1968), 115.

[133] Bacon (2004), 131, 175.

[134] Ibid., 111; Bacon (1857–74), vol. III 131, 583, 616; Farrington (1964)

combined with an eclectic approach to the mechanical philosophy.[135] Due to his corpuscular conception of matter, Hooke directed his "selective emphasis" to some aspects of Bacon's work.[136] On these he built a methodological programme wherein hypothetical elements play a significant role. Even outside the Royal Society, some peculiar features of the Baconian programme were not ignored.[137] Joshua Childrey, for instance, motivated the presence of theoretical elements in his natural history of Britain by appealing to the Baconian model. "I have here and there," Childrey wrote, "attempted to give the causes of the rarities I relate, having the example of my Lord Bacon for my authority." Like *Sylva silvarum*, his *Britannia baconica* contains "excursion ever and anon into Ætiology."[138]

As Bacon noted in the preface to the *Novum Organum*, his plan was "to establish degrees of certainty." In order to reduce the distance between mind and nature, data should be disposed in tables which introduce an external order to the chaos of experience. Inquiry does not end with an initial description of matters of fact; neither are these separate from the discovery of causes. "The new light of axioms educed from particulars" is an instrument to "point to and specify new particulars."[139] The interpretation of nature starts within natural history, by means of a series of clearly theoretical elements. Bacon lists some of these in the *Norma historiae presentis* of the *Historia naturalis et experimentalis*. The first consists in "topics of articles of inquiry." These are mainly questions that can originate new inquiries and lead the researcher towards new experiments and observations. They serve as "light to the present and stimulus to future inquiry," and Bacon describes them as "history in embryo." Hooke's and Newton's queries, albeit in different contexts, play a similar role. In order to "make the interpretation of nature readier," natural history contains "observations" on experiments, "speculations and, as it were, certain imperfect attempts at the interpretation of the causes." More than queries, these latter are proper theoretical elements. The Baconian interpretation of nature is a gradual process. It proceeds alongside the experimental work. Bacon describes it as opposed both to the empirics' limitation to experience and to the dogmatists' flying to the general axioms. Natural history should also include intermediate axioms between experience and the general causes of phenomena. These intermediate axioms are "provisional rules or imperfect axioms," which "are useful but not altogether true."[140] By introducing these theoretical elements in his natural history, Bacon distanced his project of instauration from the works of the "empirics." The distinction between "interpretation of nature" and "anticipation of mind," therefore, did not entail the exclusion of hypothetical elements from the Baconian programme.[141] As Antonio Pérez-Ramos has observed, the Baconian scale from experience to axioms

---

[135] Sargent (1987), 479; Anstey (2004), 254; Carey (1997), 253; Oldroyd (1972), 120–1.

[136] Hesse (1964), 265.

[137] Boas Hall (1983) 25.

[138] Childrey (1662), sig. B2r.

[139] Bacon (2004), 53, 158, 160–1.

[140] Bacon (2007), 14–7.

[141] Vickers (1992), 499; Urbach (1987), 33–4; Hesse (1968), 121–2; Laudan (1966b), 136; Lynch (2001), 9–11.

shows a "hypothetical-analogical character" and operates as a "strategy of gradual hypothetical inference."[142]

By now it should be evident that Hooke and other Restoration experimental philosophers were aware of the complex nature of the Baconian project, and selectively appropriated it. From Bacon, for instance, Hooke assumed the need to dispose of and regulate hypotheses to realise a new form of knowledge that joins human mind and nature. Hooke developed some non-empiricist features of Bacon's work which were consistent with his mechanic world-view. The nature of this programme contrasts with the traditional but still prevailing empiricist image of British experimental philosophy.[143] This view extends the traditional empiricist interpretation of Bacon to his early followers. Its main tenant is the association between experimental philosophy and empiricism, as both apparently eschewed hypotheses. It is significant that in this view Locke and Newton represent the main expressions of Baconianism.[144] Consequently, Hooke's call for a regulated use of hypotheses has often been considered inconsistent with Bacon's ideas.

A close look at the controversy between Hooke and Newton on the nature of light and colours in the early 1670s shows a different picture. Different ideas on the nature of matter were discussed during the controversy, and two alternative views of experimental philosophy emerged.[145] Methodological claims had a relevant role in the discussion.[146] Despite the fact that both denied a commitment to a specific theory of light, the atomist and the corpuscular philosophies were implicitly assumed by the contenders. Defending their respective ontological assumptions, Newton and Hooke engaged in a discussion on the value and use of hypotheses in natural philosophy. Aiming at a mathematical science of light, Newton criticized the hypothetical approach of those he called "naturalists," as opposed to the "mathematicians" engaged in the foundation of physical optics. The new theory submitted to the Royal Society in 1672, was for Newton "not an hypothesis but most rigid consequence" of experiments. It was "not conjectured by barely inferring 'tis thus because not otherwise or because it satisfies all phænomena (the philosophers universall topick) but evinced by ye mediation of experiments concluding directly & wthout any suspicion of doubt." Newton stressed the mathematical character of the new theory, which conferred on it the certainty that mechanical hypotheses could never attain.[147] In the new mathematical science of optics "reason and experiments" are opposed to "hypotheses."[148]

Hooke accepted the results of the Newtonian double prism experiments. But he refused to consider it as a decisive proof in favour of Newton's hypothesis. In his view the barrier established by Newton between experiments and theories was inappropriate to the rules of experimental philosophy as intended by Bacon:

---

[142] Pérez-Ramos (1988), 244, 254.

[143] Anstey (2005), 224–6.

[144] Anstey and Vanzo (2012), 500; Anstey (2011), 70–1, 73, 82.

[145] Guicciardini (2009), 21, 24–5.

[146] Zémplen (2011), 126–7, 130–1; Dear (1991), 136–7.

[147] Newton (1959–77), vol. I, 96–7, 174, 187–8, 198, 209.

[148] Id. (1779–85), vol. IV, 1.

I see noe reason why Mr. N. should make soe confident a conclusion that he to whome he writ did see how much it was besides the busness in hand to dispute about hypotheses. for I judge there is noething conduces so much to the advancement of Philosophy as the examining of hypotheses by experiments & the inquiry into Experiments by hypotheses. and I have the Authority of the Incomparable Verulam to warrant me.[149]

The reference to Bacon in defence of the use of hypotheses supports a revaluation of Newtonian Baconianism. As Alan Shapiro observed, the apparent continuity between Newton's experimental philosophy and the work of Bacon and the Royal Society's Baconians is "largely an illusion."[150] In his long war against Cartesian physics and mechanical philosophers, Newton appropriated the Baconians' idea of experimental philosophy and purified it of its hypothetical elements. Newton's experimental philosophy is more consistent with Locke's refusal of hypothesis rather than Bacon's use of them in natural history.[151] Hooke seems closer than Newton to representing the rich and manifold Baconian tradition of the Restoration. Assuming a Newtonian point of view, historians and philosophers of science have projected the post-newtonian distinction between speculative and experimental philosophy onto Bacon and the early Baconians. It is sufficient to shift away from Newton's focus to observe a more complex and historically accurate picture of Restoration science.

Bibliography

Anstey, Peter. 2004. The methodological origins of Newton's queries. *Studies in History and Philosophy of Science* 35: 247–269.

———. 2005. Experimental versus speculative natural philosophy. In *The science of nature in the seventeenth century: Patterns of change in early modern natural philosophy*, ed. Peter Anstey and John Schuster, 215–242. Dordrecht: Springer.

———. 2011. *John Locke and natural philosophy*. Oxford: Oxford University Press.

Anstey, Peter, and Michael Hunter. 2008. Robert Boyle's 'Design about Natural History'. *Early Science and Medicine* 13: 83–126.

Anstey, Peter, and Alberto Vanzo. 2012. The origins of experimental philosophy. *Intellectual History Review* 22: 499–518.

Applebaum, Wilbur. 1996. Keplerian astronomy after Kepler. *History of Science* 34: 451–504.

Bacon, Francis. 1857–74. *Works*, 7 vols., ed. Robert L. Ellis, James Spedding, Douglas D. Heath. London: Longman.

———. 1996. *Philosophical Studies c. 1611–c. 1619*, ed. Graham Rees. Oxford: Oxford University Press.

———. 2004. *The Instauratio magna, Part II: Novum organum*, ed. Graham Rees with Maria Wakely. Oxford: Oxford University Press.

———. 2007. *The Instauratio magna, Part III: Historia naturalis and Historia vitae*, ed. Graham Rees with Maria Wakely. Oxford: Oxford University Press.

---

[149] Id. (1959–77), vol. I, 110–11, 200.

[150] Shapiro (2004), 185.

[151] Downing (1997), 287–88.

Baldassarri, Fabrizio. 2017. 'Per experientiam scilicet, vel deductionem': Descartes' battle for scientia in the early 1630s. *Historia Philosophica* 15: 115–133.

Bennett, Jim. 1989. Magnetical philosophy and astronomy from Wilkins to Hooke. In *Planetary astronomy from the Renaissance to the rise of astrophysics, Part A: Tycho Brahe to Newton*, ed. R. Taton and C. Wilson, 222–230. Cambridge: Cambridge University Press.

———. 1997. Malpighi and the microscope. In *Marcello Malpighi anatomist and physician*, ed. Domenico Bertoloni Meli, 63–72. Florence: Olsckhi.

Bianchi, Massimo Luigi. 1987. *Signatura rerum: segni, magia e conoscenza da Paracelso a Leibniz*. Rome: Edizioni dell'Ateneo.

Birch, Thomas. 1756–57. *The History of the Royal Society of London*, 4 vols. London.

Boas Hall, Marie. 1963. Matter in seventeenth-century science. In *The concept of matter in modern philosophy*, ed. Ernan Mcmullin, 76–99. Notre Dame/London: University of Notre Dame Press.

———. 1983. Oldenburg, the philosophical transactions, and technology. In *The uses of science in the age of Newton*, ed. John Burke, 21–47. Berkeley/London: University of California Press.

Böhme, Helmut. 2005. The metaphysics of phenomena: Telescope and microscope in the works of Goethe, Leeuwenhoek and Hooke. In *Collection, laboratory, theatre. Scenes of knowledge in the 17th century*, ed. Helmar Schramm, Ludger Schwarte, and Jan Lazardzig, 355–393. Berlin/New York: Walter de Gruyter.

Boyle, Robert. 1999. *The Works of Robert Boyle*, 14 vols., ed. Michael Hunter and Edward Davies. London: Pickering and Chatto.

Bradbury, Saville. 1976. The quality of the image by the compound microscope: 1700–1840. In *Historical aspects of microscopy*, ed. Saville Bradbury and Gerard L'.E. Turner, 151–173. Cambridge: Royal Microscopical Society.

Bucciantini, Massimo. 2003. *Galileo e Keplero: filosofia, cosmologia e teologia nell'eta' della controriforma*. Turin: Einaudi.

Buchwald, Jed. 2008. Descartes' experimental journey past the prism and through the invisible world to the rainbow. *Annals of Science* 65: 1–46.

Buchwald, Jed, and Mordechai Feingold. 2013. *Newton and the origins of civilization*. Princeton / London: Princeton University Press.

Carey, Daniel. 1997. Compiling nature's history: Travellers and travel narratives in the early Royal Society. *Annals of Science* 54: 268–292.

Childrey, Joshua. 1662. *Britannia baconica*. London.

Clarke, Desmond. 1989. *Occult powers and hypotheses: Cartesian natural philosophy under Louis XIV*. Oxford: Clarendon.

———. 1990. The Discours and hypotheses. In *Descartes: il metodo e i saggi*, eds. Giulia Belgioioso, Guido Cimino, Pierre Costabel e Giovanni Papuli, 201–209. Rome: Istituto della Enciclopedia italiana.

Clucas, Stephen. 1994. In search of 'The True Logick': Methodological eclecticism among the 'Baconian reformers'. In *Samuel Hartlib and universal reformation: Studies in intellectual communication*, ed. Mark Greengrass, Michael Leslie, and Timothy Raylor, 51–74. Cambridge: Cambridge University Press.

Crosland, Maurice. 2005. Early laboratories c.1600–c1800 and the location of experimental science. *Annals of Science* 62: 233–253.

Da Costa Kaufnann, Thomas. 1993. *The mastery of nature: Aspects of art, science, and humanism in the Renaissance*. Princeton/London: Princeton University Press.

Daston, Lorraine. 1994. Baconian facts, academic civility, and the prehistory of objectivity. In *Rethinking objectivity*, ed. Allan Megill, 37–63. Durham/London: Duke University Press.

———. 2001. Perché i fatti sono brevi? *Quaderni Storici* 36: 745–770.

———. 2011. The empire of observation, 1600–1800. In *Histories of scientific observation*, ed. Loraine Daston and Elisabeth Lunbeck, 81–113. Chicago/London: The University of Chicago Press.

Dear, Peter. 1991. Narrative, anecdotes, and experiments: Turning experience into science in the seventeenth century. In *The literary structure of scientific argument: Historical studies*, ed. Peter Dear, 135–163. Philadelphia: University of Pennsylvania Press.

———. 1995. *Discipline & experience. The mathematical way in the Scientific Revolution*. Chicago/London: The University of Chicago Press.

———. 2006. The meaning of experience. In *The Cambridge history of science vol. 3 early modern science*, ed. Katharine Park and Lorraine Daston, 106–131. Cambridge: Cambridge University Press.

Downing, Lisa. 1997. Locke's Newtonianism and Lockean Newtonianism. *Perspectives on Science* 5: 285–310.

Espinasse, Margaret. 1974. The decline and fall of restoration science. In *The intellectual revolution of the seventeenth century*, ed. Charles Webster, 347–368. London: Routledge and Kegan Paul.

Farrington, Benjamin. 1964. *The philosophy of Francis Bacon*. Liverpool: Liverpool University Press.

Fournier, Marian. 2007. Personal styles in microscopy: Leeuwenhoek, Swammerdam and Huygens. In *From makers to users. Microscopes, markets, and scientific practices in the seventeenth and eighteenth centuries*, ed. Dario Generali and Marc Ratcliff, 211–230. Florence: Olschki.

Gabbey, Alan. 1985. The mechanical philosophy and its problems: Mechanical explanations, impenetrability, and perpetual motion. In *Change and progress in modern science*, ed. Joseph Pitt, 9–84. Dordrecht: Reidel.

———. 2001. Mechanical philosophies and their explanations. In *Late medieval and early modern corpuscular theories of matter*, ed. Christoph Lüthy, John Murdoch, and William Newman, 441–465. Leiden: Brill.

Gal, Ofer, and Raz Chen-Morris. 2010. Empiricism without the senses: How the instrument replaced the eye. In *The body as object and instrument of knowledge: Embodied empiricism in early modern science*, ed. Charles Wolfe and Ofer Gal, 121–147. Dordrecht: Springer.

Galilei, Galileo. 1890–1909. *Opere*, 20 vols, ed. Antonio Favaro. Florence: Barbera.

Garber, Daniel. 2010. Philosophia, historia, mathematica: Shifting sands in the disciplinary geography of the seventeenth century. In *Scientia in early modern natural philosophy: Seventeenth-century thinkers on demonstrative knowledge from first principles*, ed. Tom Sorell, Graham Allen Rogers, and Jill Kraye, 1–17. Dordrecht: Springer.

Gascoigne, John. 2013. Crossing the Pillars of Hercules, Francis Bacon, the scientific revolution and the new world. In *Science in the age of Baroque*, ed. Ofer Gal and Raz Chen-Morris, 215–237. Dordrecht: Springer.

Gaukroger, Stephen. 2001. *Francis Bacon and the transformation of early-modern philosophy*. Cambridge: Cambridge University Press.

———. 2002. *Descartes' system of natural philosophy*. Cambridge: Cambridge University Press.

———. 2006. *The emergence of a scientific culture: Science and the shaping of modernity 1210–1685*. Oxford: Clarendon Press.

———. 2010. The unity of natural philosophy and the end of Scientia. In *Scientia in early modern natural philosophy: Seventeenth-century thinkers on demonstrative knowledge from first principles*, ed. Tom Sorell, Graham Allen Rogers, and Jill Kraye, 19–33. Dordrecht: Springer.

Giglioni, Guido. 2011. *Francesco Bacone*. Rome: Carocci.

———. 2013. Francis Bacon. In *The Oxford handbook of British philosophy in the seventeenth century*, ed. Peter Anstey, 41–72. Oxford: Oxford University Press.

Glanvill, Joseph. 1665. *Scepsis scientifica*. London.

Grafton, Anthony. 2007. Renaissance histories of art and nature. In *The artificial and the natural: An evolving polarity*, ed. Barnadette Besaude-Vincent and William R. Newman, 185–210. Cambridge, MA: The Mit Press.

Grant, Edward. 2002. Medieval natural philosophy: Empiricism without observation. In *The dynamics of Aristotelian natural philosophy from antiquity to the seventeenth century*, ed. Cees Leijenhorst, Christoph Lüthy, and Johannes Thijssen, 141–168. Leyden: Brill.

Guicciardini, Niccolò. 2009. *Isaac Newton on mathematical certainty and method*. Cambridge, MA/ London: MIT Press.

Hamou, Philippe. 2001. *La mutation du visible: essai sur la portée épistémologique des instruments d'optique au XVIIe siècle, vol. 2. Microscopes et télescopes en Angleterre de Bacon à Hooke*. Villeneuve d'Ascq: Presses univiversitaires du Septentrion.

Hankins, Thomas, and Robert Silverman. 1995. *Instruments and the imagination*. Princeton/ London: Princeton University Press.

Harrison, Peter. 2011. Experimental religion and experimental science in early modern England. *Intellectual History Review* 21: 413–433.

Harvey, William. 1651. *Exercitationes de generatione animalium*. Amsterdam.

———. 1847. *The works of William Harvey*. Trans. Robert Willis. London: Sydenham Society.

Henry, John. 2013. The reception of Cartesianism. In *The Oxford handbook of British philosophy in the seventeenth century*, ed. Peter Anstey, 116–143. Oxford: Oxford University Press.

Hesse, Mary. 1964. Hooke's development of Bacon's method. In *Actes du dixième congrès international d'histoire des sciences*, 2 vols., ed. Henry Guerlac, vol. I: 265–268. Paris: Hermann.

———. 1968. Francis Bacon's philosophy of science. In *Essential articles for the Study of Francis Bacon*, ed. Brian Vickers, 114–139. London: Sidwick and Jackson.

Hooke, Robert. 1661. *An attempt for the explication of the phaenomena*. London.

———. 1665. *Micrographia*. London.

———. 1677. *Lampas*. London.

———. 1678. *Lectures and collections*. London.

———. 1705. *Posthumous works*, ed. Richard Waller. London.

———. 1726. *Philosophical experiments and observations*, ed. William Derham. London.

Hunter, Michael, and Paul Wood. 1986. Towards Solomon's house: Rival strategies for reforming the early Royal Society. *The British Journal for the History of Science* 24: 49–108.

Hutchinson, Keith. 1982. What happened to occult qualities in the scientific revolution? *Isis* 73: 233–253.

Jack, Jordynn. 2009. A Pedagogy of Sight: Microscopic Vision in Robert Hooke's Micrographia. *Quarterly Journal of Speech* 95: 192–202.

Laudan, Larry. 1966a. The clock metaphor and probabilism: The impact of Descartes on English methodological thought, 1650–65. *Annals of Science* 22: 73–104.

———. 1966b. *The idea of physical theory from Galileo to Newton: Studies in Seventeenth-Century Methodology*. PhD diss., Princeton University. Ann Arbor: University Microfilm International.

Lefèvre, Wolfgang. 2001. Galileo engineer: Art and modern science. In *Galileo in context*, ed. Jürgen Renn, 11–27. Cambridge: Cambridge University Press.

Lüthy, Christoph. 1996. Atomism, Lynceus, and the fate of seventeenth-century microscopy. *Early Science and Medicine* 1: 1–27.

Lynch, Michael. 2001. *Salomon's child: Method in the early Royal Society of London*. Stanford: Stanford University Press.

Mandelbaum, Maurice. 1964. *Philosophy, science and sense perception: Historical and critical studies*. Baltimore: Johns Hopkins University Press.

Manzo, Silvia. 2001. Experimentaión, instrumentos científicos y cuantificación en el método de Francis Bacon. *Man* 24: 48–84.

———. 2006. *Entre el atomismo y la alquimia: la teoria de la materia de Francis Bacon*. Buenos Aires: Editorial Biblos.

McCormick, Ted. 2009. *William Petty and the ambitions of political arithmetic*. Oxford/New York: Oxford University Press.

McGuire, J.E. 1970. Atoms and the analogy of nature. *History and Philosophy of Science* 1: 3–58.

Meinel, Christoph. 1998. Early seventeenth-century atomism: Theory, epistemology, and the insufficiency of experiment. *Isis* 79: 68–103.

Murdoch, John. 1982. The analytic character of late medieval learning: Natural philosophy without nature. In *Approaches to nature in the middle ages*, ed. Lawrence Roberts, 171–213. Brighton/New York: Center for Medieval and Early Renaissance Studies.

Newton, Isaac. 1779–85. *Opera quae extant omnia*, 5 vols, ed. Samuel Horsley. London.

———. 1959–77. *T he Correspondence of Isaac Newton*, 7 vols., ed. H. W. Turnbull, J. F. Scott, A. R. Hall and L. Tilling. Cambridge: Cambridge University Press.

Oldroyd, David. 1972. Robert Hooke's methodology of science as exemplified in his Discourse of Earthquakes. *The British Journal for the History of Science* 6: 109–130.

———. 1987. Some writings of Robert Hooke on procedures for the prosecution of scientific inquiry, including his 'Lectures of Things Requisite to a Ntral History'. *Notes and Records of the Royal Society of London* 41: 145–167.

Osler, Margaret. 1994. *Divine will and the mechanical philosophy: Gassendi and Descartes on contingency and necessity in the created world*. Cambridge: Cambridge University Press.

Palmieri, Paolo. 2008. *Reenacting Galileo's experiments: Rediscovering the techniques of seventeenth-century science*. Lewiston: Edwin Mellen Press.

Pérez-Ramos, Antonio. 1988. *Francis Bacon's idea of science and the maker's knowledge tradition*. Oxford: Clarendon.

Petty, William. 1647. *The advice of W. P. to Mr Samuel Hartlib for the advancement of some particular parts of learning*. London.

Principe, Lawrence. 1998. *The aspiring adept: Robert Boyle and his alchemical quest*. Princeton/London: Princeton University Press.

Pomata, Gianna. 2011. Observation rising: Birth of an epistemic genre, 1500–1650. In *Histories of scientific observation*, ed. Loraine Daston and Elisabeth Lunbeck, 45–80. Chicago/London: University of Chicago Press.

Pugliese, Patri. 2004. Robert Hooke. In *Oxford dictionary of national biography*, ed. H.C.G. Matthew and Brian Harrison, vol. 27, 951–958. Oxford: Oxford University Press.

Pyle, Andrew. 1995. *Atomism and its critics: Problem areas associated with the development of atomic theory of matter from Democritus to Newton*. Bristol: Thoemmes.

Rogers, Graham Allen. 1972. Descartes and the method of English science. *Annals of Science* 29: 237–255.

Rossi, Paolo. 1984. Ants, spiders, epistemologists. In *Francis Bacon: Terminologia e fortuna nel XVII secolo*, ed. Marta Fattori, 245–260. Roma: Edizioni dell'Ateneo.

———. 1986. *I ragni e le formiche: Apologia della storia della scienza*. Bononia: Il Mulino.

———. 1996. Bacon's idea of science. In *The Cambridge companion to Bacon*, ed. Markku Peltonen, 25–46. Cambridge: Cambridge University Press.

———. 2003. Sulla scienza e gli strumenti: cinque divagazioni baconiane. In *Musa musaei: Studies on scientific instruments and collections in honour of Mara Miniati*, ed. Marco Beretta, Paolo Galluzzi, and Paolo Triarico, 141–153. Florence: Olschki.

Ruestow, Edward. 1996. *The microscope in the Dutch republic. The shaping of discovery*. Cambridge: Cambridge University Press.

Rusu, Doina-Cristina. 2012. Francis Bacon: Constructing natural histories of the invisible. *Early Science and Medicine* 17: 112–133.

Sargent, Rose Mary. 1987. Robert Boyle's Baconian inheritance: A response to Laudan's Cartesian thesis. *Studies in History and Philosophy of Science* 17: 469–486.

Sargent, Rose-Mary. 1995. *The diffident naturalist: Robert Boyle and the philosophy of experiment*. Chicago/London: University of Chicago Press.

———. 2012. From Bacon to Banks: The visions and the realities of pursuing science for the common good. *Studies in History and Philosophy of Science* 43: 82–90.

Shapin, Steven, and Simon Schaffer. 1985. *Leviathan and the air pump: Hobbes, Boyle, and the experimental life*. Princeton/London: Princeton University Press.

Shapiro, Alan. 2004. Newton's "experimental philosophy". *Early Science and Medicine* 9: 185–217.

Stewart, Ian. 2012. Res, veluti per machinas, conficiatur: Natural history and the 'mechanical' reform of natural philosophy. *Early Science and Medicine* 17: 87–111.

Strano, Giorgio. 2012. Galileo's shopping list: An overlooked document about early telescope making. In *From Earth-bound to satellite: Telescopes, skills and networks*, ed. Alison D. Morrison-Low, Sven Dupré, Stephen Johnson, and Giorgio Strano, 1–19. Leiden: Brill.

Thomas, Jennifer. 2011. Compiling 'God's great book [of] universal nature': The Royal Society's collecting strategies. *Journal of the History of Collections* 23: 1–13.

Urbach, Peter. 1987. *Francis Bacon's philosophy of science*. La Salle: Open Court.

Valleriani, Matteo. 2010. *Galileo engineer*. Dordrecht: Springer.

Van Helden, Albert. 1983. The birth of modern scientific instrument, 1550–1700. In *The uses of science in the age of Newton*, ed. John Burke, 49–83. Berkeley/London: University of California Press.

Vertesi, Janet. 2010. Instrumental images: The visual rhetoric of self-presentation in Hevelius's Machina Coelestis. *The British Journal for the History of Science* 43: 209–243.

Vickers, Brian. 1984. Analogy versus identity: The rejection of occult symbolism 1580–1680. In *Occult and scientific mentalities in the Renaissance*, ed. Brian Vickers, 95–163. Cambridge: Cambridge University Press.

———. 1985. The Royal Society and English prose style: A reassessment. In *Rhetoric and the pursuit of truth: Language change in the seventeenth and eighteenth centuries*, ed. Brian Vickers and Nancy Struever, 6–76. Los Angeles: William Andrews Clark Memorial Library.

———. 1992. Francis Bacon and the progress of knowledge. *Journal of the History of Ideas* 53: 495–518.

Voltaire. 2004. *Philosophical dictionary*, trans. Theodore Besterman. London: Penguin.

Webster, Charles. 1967. Henry Power's experimental philosophy. *Ambix* 14: 150–178.

Wilson, Catherine. 1988. Visual surface and visual symbol: The microscope and the occult in early modern science. *Journal for the History of Ideas* 49: 85–108.

———. 1995. *The invisible world. Early modern philosophy and the invention of the microscope*. Princeton/London: Princeton University Press.

———. 2008. *Epicureanism and the origins of modernity*. Oxford/New York: Oxford University Press.

Wren, Stephen. 1750. *Parentalia or memories of the family of the Wrens*. London.

Zagorin, Perez. 2001. Francis Bacon's concept of objectivity and the idols of the mind. *The British Journal for the History of Science* 34: 379–393.

Zemplén, Gábor. 2011. The argumentative use of methodology: Lesson from a controversy following Newton's first optical paper. In *Controversies within the scientific revolution*, ed. Marcelo Dascal and Victor Boantza, 123–147. Amsterdam/Philadelphia: John Benjamins Publishing Company.

Chapter 3
Philosophical Algebra

3.1 Instauratio Incompleta

The observations of *Micrographia* aimed to provide some "meanest foundations" to eventually raise "superstructures" both in the study of nature and the reform of science. Like the rest of the book, the preface was intended by Hooke as a part of a more detailed methodological programme that was the subject of some Cutler lectures read at Gresham College in the second half of the 1660s. In 1664, Hooke only hinted at "another discourse" on the "manner of compiling a natural and artificial history" by means of "philosophical tables" that should be the "most useful for the raising of axioms and theories." But in the early Cutlerian lectures read after the publication of *Micrographia*, Hooke focused only on the first part of his methodological programme, namely a new philosophical history that would furnish the mind "with fit material to work on." He did not deal with the "rules and methods of proceeding or operating with this so collected and qualify'd supellex." This part of the new method was described in *Micrographia* as a "new kind of algebra, or analitik art." "By this, as by that art of algebra in geometry," Hooke claimed, "'twill be very easy to proceed in any natural inquiry, regularly and certainly." For this reason, it "might not improperly be call'd a philosophical algebra."[1] In the lectures read to the Gresham College and the Royal Society in the second half of the seventeenth century, Hooke did not go beyond the analogy between his "philosophical algebra" and the symbolic algebra of Viète and Descartes. In some late lectures read in the 1690s, he still referred to it as a project to complete.[2] Over more than thirty years, Hooke never fixed his views on the analytic component of his methodological programme. In the works and papers following *Micrographia*, there are contrasting references to it.

---

[1] Hooke (1665), sig. b1r–v, 93; Id. (1705), 7; cf. Hunter (1989), 299, 302–3; Id. (2003), 117.

[2] Pugliese (1982), 127–28.

© Springer Nature Switzerland AG 2020
F. G. Sacco, *Real, Mechanical, Experimental*, International Archives
of the History of Ideas Archives internationales d'histoire des idées 231,
https://doi.org/10.1007/978-3-030-44451-8_3

Stressing the relevance of some of these references over the rest, scholars have offered discordant accounts of what philosophical algebra could have been: a form of deductive logic combining hypotheses and experiments,[3] a real character,[4] or an operative procedure.[5] Additional confusion was due also to the fact that in Hooke's lectures "philosophical algebra" denotes both the "true method of building a solid philosophy" and the analytic part of it.[6]

In spite of the challenges of such incomplete and fragmentary evidence, Hooke's philosophical algebra is not an object of antiquarian curiosity but of historical understanding, for it is linked to the rest of his natural philosophy.[7] The young Hooke maintained a distinction between historical and philosophical work. The new history of nature was a "great work" and a philosophical enterprise "both sufficient and adapted for the perfecting [of] the knowledge of the works of nature." The discovery of the invisible and unknown nature of bodies, however, required "much deeper researches and ratiocinations, and very many vicissitudes of proceedings from axioms to experiments and from experiments to axioms." This was, in other words, the "business of the philosopher and not of the historian." The method to carry out this philosophical work was a *penus analytica* specifically designed to raise "axiomes and more general deductions from a sufficient stock of material" methodically collected. The new analytic instrument did not replace the theoretical elements of the philosophical history of nature, but rather built on them.[8] The absence of any significant description of this component of Hooke's methodological programme is consistent with the erasure of the already porous border between historical and philosophical domain. But even when this distinction had gradually disappeared, Hooke still maintained that one of his most relevant discoveries, the laws of elasticity, was mainly due to his new "art of invention or mechanical algebra."[9]

Hooke's quest for a deductive form of reasoning was consistent with his relocation of Bacon's induction within the new philosophical history. A new ministration to reason was needed, and Hooke looked for a logical instrument to insert in the new philosophical history that he had already elaborated. In Hooke's opinion, Baconian instauration was incomplete. "The extraordinary strength" of Bacon's experimental history of nature and art "carried him on almost to a miracle," but "he had the infelicity to have some prejudice for mathematicall learning" and did not understand the relevance of axiomatic systems and analysis for natural philosophy. "Several excellent hints to this purpose," Hooke added, "may be found scattered here and there in

---

[3] Hesse (1964), 266; Id. (1966b), 82; Laudan (1966), 148; Centore (1970), 36, 39; Kargon (1971), 78; Oldroyd (1972), 118; Id. (2000), 299–300; Hunter (2003), 123; Pugliese (2004), 952–53.
[4] Lewis (2001), 338; Salmon (1979), 192.
[5] Lynch (2001), 92–3, 150; Id. (2005), 198–9.
[6] Hooke (1705), 7.
[7] Hesse (1964), 265.
[8] Hooke (1705), 44, 61.
[9] Id. (1676), 26.

his *Novum Organum* and some other pieces of his writings; but they are but hints and doe yet need very much of thought and invention to produce them into a compleat art and method."[10] Despite these limits, "no man except the incomparable Verulam" had a clear idea of a new method for sciences. "He hath promoted it to a very good pitch, but there is yet somewhat more to be added, which he seem'd to want time to compleat."[11] Bacon's comparative induction remains useful for the methodical gathering and organization of the philosophical history of nature, but an effective ministration to reason can only be provided by the axiomatic and deductive proceeding of mathematics overlooked by Bacon. Ward expressed a similar view when he lamented that "it was a misfortune to the world that my Lord Bacon was not skilled in mathematicks, which made him jealous of their assistance in naturall enquires." When induction is no more useful in natural inquires, deductive forms of inference are needed to conclude the process of discovery.[12] Like Ward, Hooke shared a view of Bacon common to many natural philosophers in the mid-seventeenth century. According to Huygens, for instance, Bacon showed "very good ways" to make experiments and use them, but "he did not understand mathematics and lacked insight into physics."[13] Like many other "experimenters," Hooke noted, Bacon made "tryalls but out of some kind of aversenesse to geometricall and arithmeticall speculations made not calculations."[14] He understood that without "logick and mathematicks" natural philosophy is "lame and insignificant," but resolved to "make mathematicall learning to be but an eppendix or auxiliary to physicks, metaphysicks, mechanicks and magicks."[15]

The view of Bacon as distrustful of mathematics was a decisive component of early Baconianism.[16] But Hooke's criticism stressed also the consequences of Bacon's view of mathematics on his methodological programme and image of science. Bacon did not understand that the deductive logic employed in mathematics is "one of the first and most necessary of human inventions for exercising the faculty of reasoning."[17] For this reason, his instauration can be completed only by a new ministration to reason rooted in mathematics. The advantages offered by this latter to natural philosophy are twofold. Geometry and arithmetic are "the more demonstrative parts" of mathematics; they provide "numbers, weights, and measures to inquire into, examine and prove all things." Algebra is the "more inventive" part and "instructs and accustoms the mind to a more strict way of reasoning, to a more nice and exact way of examining, and to a much more accurate way of inquiring into the nature of things." Both in the "invention" and "demonstration" of knowledge,

---

[10] Leibniz (1991), 235.

[11] Hooke (1705), 6.

[12] Ward (1654), 25.

[13] Huygens (1888–1950), vol. X, 404.

[14] British Library, Sloane MS 1039, f.114 r.

[15] Royal Society Classified Papers, vol. XX, f.173 r.

[16] Domski (2013), 152.

[17] Royal Society Classified Papers, vol. XX, f.173 r.

mathematics is decisive for natural philosophy. Modelled on algebra, Hooke's new ministration will "show that even physical and natural enquires as well as mathematical and geometrical, will be capable also of demonstration."[18] Any experiment "fit to be made use of for any philosophical theories" should be carried out by means of quantitative standards of measure. The "certainty of quantity" is for Hooke a necessary component of his mechanical and experimental philosophy.[19] Modelled on Archimedes' work, Hooke's regulative ideal of science is "physicks geometrically handled."[20]

3.2 Mechanics' Philosophy

The myth of Bacon's and Baconians' neglect of quantification emerged in the mid-seventeenth century, and the achievements of physico-mathematics in the work of Newton proved decisive to its long resistance to historical deconstruction.[21] Less dismissive than Koyré in his assessment of Baconianism, Thomas Kuhn distinguished between experimental and mathematical contributions to modern science, yet this distinction further contributed to that myth.[22] In Kuhn's view, only Newton participated in the two separate traditions before they were joined in the nineteenth century by the complete mathematization of Baconian experimental sciences. Compared to the Newtonian synthesis, Hooke's contributions to physical sciences were seen by Kuhn as irremediably marked by a qualitative Baconian approach clearly distinct from Newton's physico-mathematics.[23] This view is significantly close to what Newton himself maintained of Hooke and other "naturalists."[24] A Galileian belief in the mathematical language of the book of nature informed Newton's approach to science.[25] Mechanics, for instance, after the *Principia* became the mathematical science of motion that is still studied in rational mechanics. In this field Newton did not leave room for the study of machines and the use of mechanical conjectures.[26] Newton positioned his physico-mathematic programme against both the Baconian and the Cartesian traditions. The principles of natural philosophy had to be deduced from experiments, and so analysis became the tool of mathematization of experimental science:[27]

---

[18] Hooke (1705), 7, 19.

[19] Id. (1726), 95.

[20] Id. (1705), 73, 85.

[21] See Rees (1985), 195–6; Manzo (2001), 51–2, 66–72; Pastorino (2011), 547–62; Jalobeanu (2015) 26–7; Id. (2016), 59–60, 65, 69; Mori (2017), 9, 20.

[22] Domski (2013), 145–6; cf. Koyré (1978), 39.

[23] Kuhn (1977), 46, 49–50.

[24] Newton (1959–77), vol. I, 96–7, 386, vol. II, 438; cf. Feingold (2001), 83.

[25] On Galileo see Palmerino (2006), 38–42; Id. (2016), 30–2.

[26] Guicciardini (2007), 169–70, 174, 181–3; Gabbey (1992), 308, 319.

[27] Guicciardini (2009), 24–6; Shapiro (1989), 226; Borghero (2005), 445–6.

As in mathematicks, so in natural philosophy the investigation of difficult things by the method of analysis ought to precede the method of composition. This analysis consists in making experiments and observations, in drawing general conclusions from them by induction, and admitting of no objection against the conclusions but such as are taken from experiments, or other certain truths. For hypotheses are not to be regarded in experimental philosophy.[28]

Considered against this successful model of knowledge, Hooke's "physics geometrically handed" has been confined to the realm of qualitative and pre-paradigmatic science. His plan to incorporate analysis into experimental philosophy via "philosophical algebra" has also been considered an "interesting," but definitively inferior, "parallel" to Newton's.[29] A reappraisal of Hooke's view, therefore, would be possible only outside what has been called the Koyré-Kuhn distinction between experimental and mathematical traditions in early modern science.[30] A new image of Renaissance learning has proved this dichotomy too rigid to depict adequately the complex philosophical landscape in which modern science emerged.[31] Although it does not fit into the Koyré-Kuhn dichotomy, a Vitruvian and Archimedean approach to mathematics, mechanics, and nature was widely diffused among early modern artisans, engineers, and natural philosophers.[32] The introduction of mathematics into the domain of natural philosophy took place in different ways. An intellectual and contemplative road was represented by the Platonic belief in the mathematical nature of the universe. A more practical way was offered by mechanics and mixed mathematics.[33] In this latter, measurement was not rooted in a mathematic ontology, and quantification did not have causal value, but only an experimental function.[34] Renaissance mechanics aimed at a theoretical understanding of nature by means of the study of machines and mechanical devises.[35] This knowledge, as Guidoubaldo Del Monte noted, was never to be "considered apart from either geometrical demonstrations or actual motion."[36] As a mixed science, mechanics was an art that declines "from the purity, simplicities, and immateriality of our principall science of magnitude," that is from pure geometry. It "nevertheless," John Dee added, "use[s] the great ayde, direction, and method of the sayd principall science."[37] Since it "teacheth the drawing, measuring and proportion of figures," geometry is in Robert Recorde's view a fundamental component of the work of "carpenters, karvers, joiners." As Plato claimed, it is also fundamental for the "learned professions in logike

---

[28] Newton (1779–85), vol. IV, 263.

[29] Hintikka and Remes (1974), 107, 115 n.11.

[30] Hall (1987), 494.

[31] Smith (2009), 362; Id. (2004), 18; Dear (1998), 175.

[32] Hoyrup (1992), 81–2, 85; Bennett (1998), 197–8; Id. (2002), 215–8.

[33] Dear (2007), 431–2.

[34] Henry (2011), 704.

[35] Dear (2011), 149, 155; Clucas (1999), 146–7, 150.

[36] Del Monte (1575), sig. B1v; Drake and Drabkin (1966), 245.

[37] Dee (1570), sig. 1b; cf. Johnston (2012), 471–5.

and rhetorike and partes of philosophy."[38] Even Henry Saville, who was influenced by Proclus's commentary on Euclid, acknowledged the relevance of mechanics and its fundamental link to speculative mathematics. The statutes of the Savilian chairs of mathematics at Oxford reflected Savile's late ideas.[39] These two ways, however, were not always alternative and separate. They often combined in the works of learned craftsmen, engineers, and experimenters. Contrary to Platonic mathematism, these learned practitioners often maintained that nature was a big machine, and God was a mechanic as well. In their work, an operative approach to mathematics favoured the primacy of experience and a mechanical notion of physical causes.[40]

Of this mechanics' philosophy, Hooke provided perhaps one of the most refined expressions.[41] His plan for a "real, mechanical, experimental" philosophy was influenced by Wilkins' Vitruvian approach to mathematics, and the education that he received at Oxford from the Savilian professor of mathematics Seth Ward.[42] His association with Gresham College, largely due to Wilkins, strengthened Hooke's links with the cultural tradition embodied in that institution.[43] Hooke's plan for a "physics geometrically handled" was rooted in practical mechanics. Physics and mechanics, like any other part of "mixed mathematicks," are "all indeed but sproutes shooting forth and growing from the stock and root of geometry."[44] As a mixed science, mechanics shows "a way and method of applying" mathematics to the study of nature. This application takes place at two levels. Mechanics "enters the mind into a method of accurate and demonstrative inquiry." In Hooke's methodological programme, philosophical algebra is probably the logical instrument that best reflects this philosophical contribution of mechanics. This latter also provides a model of physical operations and "gives a scheme of the laws and rules of motion."[45] Hooke's "physics geometrically handled" aims to establish quantitative principles and mathematical laws. Nature is not for Hooke a book written in mathematical characters, as Galileo claimed in *The Assayer*.[46] It is rather a Baconian *sylva*, whose elementary components are geometrically structured in various and impervious textures and schematisms. By means of microscopes, it is possible to disclose some of these "geometrical figures." But this "mathematicks of nature" cannot be known a priori.[47] The operations of nature are more "secret and abstruse and hid from our discerning" than those of the "more gross and obvious ones of engines." Like machines designed and built by engineers and craftsmen, natural bodies have a

---

[38] Recorde (1551), sig. tv, t1v, A1r.

[39] Savile (1621), 20, 27; Goulding (1999), 138–9, 144.

[40] Bennett (1986), 5, 11; Rossi (2002), 7.

[41] Bennett (1986), 23–4; Bennett (1998), 220.

[42] Id. (1982), 12.

[43] Feingold (1999), 174–5, 180.

[44] Royal Society Classified Papers, vol. XX, f. 173r.

[45] Hooke (1705), 19–20.

[46] Galileo (1890–1909), vol. VI, 232.

[47] Hooke (1665), 87.

mechanical structure that follows geometrical principles. Some of these, Hooke claims, "may be as capable of demonstration and reduction to a certain rule as the operations of mechanicks or art." Nature "always act[s] in a regular and uniform" manner, which can be discovered and "reduced under certain rules." The order of nature is "geometrical or mechanical." As in mechanics, "mathematical exactness" is not achievable in "the science of physics" or in "natural and experimental philosophy." The geometrical proprieties of natural bodies are, like artificial engines, affected by their material dimension. Thus, "nature itself does not so exactly determine its operations, but allows a latitude almost to all of its workings." This "latitude," though, "seems to be restrain'd within certain limits, and beyond those is neither excessive on the one hand or defective on the other." In contrast to physico-mathematics, Hooke's "physics geometrically handled" focuses on the material components of natural phenomena and is considered "sufficient to come as near to the truth as the matter is capable of."[48]

This evidently appeared in Hooke's approach to the law of free fall. Like Marine Mersenne, Hooke accepted the law but questioned the relationship between experiment and theory established by Galileo.[49] According to the law, the speed of bodies moving from rest to the centre of the earth increases in direct proportion to the square of the times of their descent. Galileo claimed that the law "must be assumed to be verified whenever the accidental and external impediments are removed," and acknowledged that "we cannot remove" the resistance of the medium.[50] Hooke found that Galileo's law "would hold very near" only in absence of any "gross fluid," for in "a medium wherein there was a resisting fluid body, it would not hold in any wise." Through a series of experiments on different mediums, Hooke observed that the more rarefied the medium, the closer the behaviour of falling bodies to the ratio found by Galileo. Even through the thinnest medium, though, the ratio was never "mathematically true." The Galilean law, Hooke concluded, has "been made upon a theory, and not upon experiments."[51]

The use of different mediums was part of an experimental programme focusing on elastic bodies moving through material fluids. To these bodies, Hooke found that a different and more general principle than Galileo's law applied. Nonetheless, in his argument Hooke followed the *filosofo geometra* Galileo in "deduct[ing] the material hindrances" in order to "recognize in concrete the effects which he has proved in the abstract."[52] He admitted that "in vacuo the descent of all bodies was equally swift, increasing continually its velocity by a duplicate proportion to the time of continuance." Experiments, however, proved that "in all gravitating mediums somewhat of that proportion is impeded."[53] For this reason, Galileo's law was

---

[48] Id. (1705), 20, 38, 172.

[49] Palmerino (2010), 52–3.

[50] Galileo (1890–1909), vol. VIII, 118–9, 209; Id. (1974), 77–8, 166; see also Palmerino (2006), 16.

[51] Birch (1756–57), vol. I, 486; vol. III, 398.

[52] Galileo (1890–1909), vol. VII, 234; Id. (1953), 207.

[53] Birch (1756–57), vol. III, 309.

in Hooke's eyes a theoretical ratio that could not apply to any physical motion, for physical motions always took place through mediums of varying density. The mathematical ratio could never be observed. It was a theoretical mathematical law, not the mechanical physical law that Hooke was looking for by means of his experiments on motion through different mediums. Hooke did not accept Galileo's opinion that once the "imperfection of matter capable of contaminating the purest mathematical demonstration" was removed, concrete machines would operate according to the same principle as their "abstract ideal counterpart," and "purely mathematical demonstrations [would] be produced" about any natural body.[54] For Hooke, on the contrary, "speculative geometry" dealt with entities that are abstract and only logically "possible." Their properties might well apply to natural and artificial concrete bodies. By drawing lines, practical geometers might "experimentally" verify the "sensible truth" of ideal geometrical figures, but could not demonstrate them because of an ontological divide between what is logically possible and what is real. Similarly, mechanics "doth find many difficulties in the actual performance" of ideal machines.[55]

In spite of this ontological divide, speculative mathematics remained an indispensable component of both natural philosophy and practical mechanics. Experiments and inquires without measure and quantification "will be of little significancy, and their productions will prove but like seed springing up from stony ground which will soon wither and dry."[56] Geometry provided the foundation on which the "conjunction of physical and philosophical with mechanical and experimental knowledge" could be built. This conjunction was necessary both for the increase of natural knowledge and the development of technology.[57] The operative dimension of Hooke's "physics geometrically handled" exceeded the speculative one, for he seemed to believe with Bacon that "the discovery of fruits and works as it were guarantees and underwrites the truth of philosophies."[58]

Instruments were often a bridge between mathematical sciences, such as astronomy, and the work of mechanics and experimenters.[59] In Hooke's physics, mechanical instruments were the main way to measure and quantify. Only by means of instruments could the "quality or degree of the proprieties, powers, and affection of bodies" be measured and quantitatively determined. Like microscopes, instruments of measure showed some features of natural phenomena that our bare senses cannot reach. While the former "help the eye to make invisible bodies and parts visible," the latter "help the hand to make intractable bodies tractable and ponderable, and comparable." The hidden and inaccessible geometrical properties of matter are not accessed by abstraction from material contingencies. Hooke, instead, followed a

---

[54] Galileo (1890–1909), vol. VIII, 51; Id. (1974), 12–3.

[55] Hooke (1705), 19–20, 523, 525, 532.

[56] Royal Society Classified Papers, vol. XX, f. 173r.

[57] Hooke (1674), 16.

[58] Bacon (2004), 116–7; cf. Lynch (2001), 83, 91; Perez-Ramos (1988), 173 n. 9.

[59] Kuhn (1977), 49; Van Helden (1983), 60; Bennett (2002), 218.

mechanical path to what he described as "the matematicks of nature," which was built on the isomorphism between natural and artificial bodies.[60] The very existence of philosophical instruments depended on the mechanical nature of the bodies to which they applied. Thus, the mechanical principles on which instruments were built provided models of understanding of what was observed.[61] Since the "geometrical mechanisme of nature" observed through the microscope was not different from the mechanical and geometrical structure of the instrument itself, the knowledge of the properties of physical bodies could provide "admirable advantages towards the increase of the operative and the mechanick knowledge."[62]

3.3   The Most Perfect Idea of Algebra

The *penus analaytica* referred to in 1664 was the ministration to intellect that was needed to complete Hooke's Baconian set of artificial instruments for the human senses.[63] Inspired by the Renaissance revival of ancient geometry, Hooke intended analysis as a tool for discovery and the expansion of knowledge.[64] "The methods of attaining this end," Hooke noted, "may be two, either the analytick, or the synthetic. The first is proceeding from the causes to the effects. The second from the effects to causes." The "synthetic method" is "more proper for experimental inquiry." Like a gardener "who prepares his ground and sows his seeds, and diligently cherishes the growing vegetable, supplying it continually with fitting moisture, food, shelter, etc.," the synthetic method is the safest way to great "discoverys in physicks," but it is not "to be expected that a production of such perfection as this is designed, should in an instant be brought to its compleat ripeness and perfection." It is "an imperfection of our nature that we are quickly weary of this more tedious and sure way" and allow the brain and fancy to dominate over the rest of human faculties. It is necessary "to accustom the mind to attention and circumspection," because without the needed examinations "we are too apt to run away with a thing, and think we know it and see it clearly before we are sure we do."[65] Long and apparently redundant observations and inquires, though, can produce the opposite effect. By "being too scrupulous and exact about every circumstance," we might "confine and straighten it too much." Hooke's ministrations aim to establish a balance between sense and understanding. Analysis is needed to turn experimental inquiries into efficient and productive researches into the true nature of bodies. "In truth," Hooke notes, "the

---

[60] Hooke (1665), 87; Id. (1705), 36; Id. (1726), 114–5.

[61] Bennett (2003), 67–8; Id. (2006), 67–8; Fournier (1996), 52; Guicciardini (2011), 75; Henderson (2019), 416.

[62] Hooke (1665), sig. A2v, 91.

[63] Id. (1705), 61.

[64] Hintikka and Remes (1974), 7.

[65] Hooke (1705), 70, 330; Guildhall Library, London MS 1757.11, f. 105r.

synthetick way by experiments observations etc. will be very slow if it be not assisted by the analytick, which proves of excellent use." This method assumes universal causes as known and "branches itself out into the more particular and subordinate." In Hooke's methodological programme, it is complementary but not alternative to synthesis and induction, which provide the indispensable foundation of any experimental knowledge.[66] As a form of analysis, the algebra of Viète, Harriot, and Descartes "supposes the thing to be already done and known, which is the thing sought and to be found out." In the inference from the known to the unknown, algebra shows "the very grounds and proceedings of ratiocinations":

> For in this we see, as it were, how from the most obvious and sensible object we are carried to the highest pitch of ratiocination upon what ground axioms are built and how they are all derived from the most plain and obvious information of the senses, and that they are not innate in us or infused, but acquired habits.[67]

In Hooke's eyes, modern algebra is a model of analysis consistent with an experimental form of knowledge. In contrast to Cartesian philosophy, Hooke's use of algebra as analytical tool does not aim to emancipate human knowledge from senses and empirical experience.[68] It does not lead to an axiomatic form of physics,[69] but is consistent with experimental philosophy. Hooke's call for a philosophical algebra echoes the view of natural philosophy drawn by Wren, i.e., a form of knowledge "order'd into a geometrical way of reasoning from ocular experiment[s]" that leads to a "real science of nature, not an hypothesis of what nature might be."[70]

Mathematics, according to Wren, provided the logic that was necessary in all sciences.[71] As William Petty noted, algebra is a "kind of logick" that applies not only to numbers and quantities, but to "several species of things."[72] If applied beyond mathematics, according to Hooke, "this method" could improve natural philosophy because the "progress of the mind or reason is the same in all kinds of ratiocinations."[73] Was the new symbolic algebra a model for Hooke's third ministration or was it the ministration itself?[74] Some lectures on algebra read in 1665 suggest that Hooke's philosophical algebra was far from a simple and direct application of symbolic algebra to natural philosophy. This new mathematical tool had been "in great part perfected" and "very far promoted" since Viète, although it is "yet in very particulars deficient." Hooke's aim was to contribute to the perfection of modern algebra, "making it useful and applicable for the invention and finding out of any

---

[66] Hooke (1665), sig. b2r; Id. (1705), 330–1.

[67] Royal Society Classified Papers, vol. XX, f. 65r.

[68] Cf. Descartes (1964–74), vol. VII, 20, 35; see also Timmermans (1995), 112–4; Id. (1999), 444–6; Schuster (2013), 91–2.

[69] As described by Bertoloni Meli (2010), 24, 38.

[70] Wren (1750), 204.

[71] Ward (1740), 31.

[72] Petty (1927), vol. II, 10.

[73] Guildhall Library, London MS 1757.11, f. 105r.

[74] Pugliese (1982), 69; Mulligan (1992), 152.

other inquiry." Hooke saw the different forms and uses of algebra in early modern mathematics as an obstacle to his methodological programme. It was necessary, therefore, to identify "which idea of algebra may be the most perfect" so as to find out "which of all the methods thus far received most closely approaches it."[75] Hooke's approach to symbolic algebra, in short, was subordinated to his philosophical interest. For this reason, like Wallis, Hooke considered the work of Viète as a watershed in the history of algebra.[76] The neglect for the contribution of Italian and German mathematicians before Viète was linked to Hooke's view of algebra not only as a tool for the resolution of geometrical problems, but mainly as the form of analysis closer to the way the human intellect proceeds, or rather should proceed, along the right path to truth. Viète's "method of proceeding by species" is better than "the cossick algebra or numbring algebra" because symbols help the mind in dealing with relationships between objects better than numbers.[77] The French mathematician defined analysis as the best tool for discovery in mathematics, which was to be carried out by means of "species" and not numbers.[78] The progress offered by Viète's *logistica speciosa* over ancient mathematics was in Hooke's opinion as great as that guaranteed by the contributions of Harriot. Symbols were a significant component of Harriot's mathematical and philosophical work.[79] His new notation was easier to use and more transparent than Viète's, for it revealed the structure of algebraic procedures better than the verbal description still largely employed by Viète. Despite this, Viète made it possible to denote geometrical quantities by means of algebraic symbols, thus leading to the introduction of algebra in geometry.[80] Descartes completed this process, for unlike Viète he gave a non-dimensional interpretation of algebraic operations.[81] In *Géométrie*, the product of two segments is still a segment, not a square.[82] But in Hooke's eyes, Descartes' contribution consisted mainly in the improvement of the symbolism along the lines set by Harriot.[83] Even William Oughtred improved Harriot's symbolism. The "short and significant characters" introduced in his *Elementi decimi Euclidis Declaratio* are better than Harriot's. Hooke found disappointing "the method of proceeding in the inventions of problems" in Oughtred's *Clavis mathematicae*,[84] which Ward and Willis edited and used as a textbook at Oxford while Hooke was a student there.[85] Despite his

---

[75] Guildhall Library, London MS 1757.11, f. 117r; Royal Society Classified Papers, vol. XX, f. 65v.

[76] Wallis (1685), 14, 66; cf. Serfati (1998), 239; Maieru (2007), 174, 476–7.

[77] Royal Society Classified Papers, vol. XX, f. 66r.

[78] Viète (1646), 4, 10; see also Cajori (1928), vol. I, 183; Freguglia (1988), 67; Id. (1999), 126; Charbonneau (2005), 66, 73, Panza (2007), 402.

[79] Stedall (2007), 383.

[80] Id. (2002), 50, 94; Seltman (2000), 184.

[81] Bos (2001), 264; Freguglia (1999), 116, 168, 222.

[82] Descartes (1964–74), vol. VI, 371.

[83] Cf. Stedall (2002), 112.

[84] Royal Society Classified Papers, vol. XX, f. 66r.

[85] Stedall (2002), 63.

criticism, the *Clavis* was probably an early source for Hooke's philosophical algebra. Oughtred described the analytic art as the "Ariadne's thread" of the secrets of mathematics, and introduced the idea of a *penus analytica* to extend the use of algebra beyond mathematics.[86]

Hooke's lectures on algebra show a preponderant interest in symbolism. "The first thing therefore to be lookd after in Algebra," he wrote, "is a most plain, simple, short and most significant character, whereby both the quantity and operation and effects may be most distinctly, plainly, briefly and significantly exprest". For this reason, Hooke considered the contribution of the French mathematician Pierre Hérigone superior to those of Descartes and Harriot.[87] Unlike arithmetic, Viète's *algebra speciosa* was not for Hérigone limited to mathematical problems, but could be employed for the discovery and demonstration of different forms of knowledge.[88] In Hérigone's *Cursus mathematicus*, the verbal description still widely employed by Viète to explain the process of demonstration was replaced by symbols and a tripartite graphic structure.[89] This aspect of Hérigone's work was emphasised by Hooke, because it provided space to "set down the reason of each deduction." But the very "idea of algebra" was provided by John Pell. Hooke described Pell's contribution as superior to any other because of "the significancy of the characters and the plainnes and regularity of his method of proceding and deducing."[90] Hooke was here referring to the manuscript of Johann Heinrich Rahn's *An introduction to Algebra*. Pell revised the work of his former Swiss pupil after he returned to London, and published it "much altered and augmented" three years later. As Pell explained in the preface, the translation of the original manuscript was completed by May 18, 1665.[91] Hooke perhaps had the opportunity to see the manuscript and discuss its edition with Pell before he read his lectures in June, 1665. As he wrote in a letter to Theodore Haak, Pell knew Hooke already in 1664, and considered him as "a man very industrious, dexterous and ingenious."[92] Along with Haak, Pell was involved in the projects of reform of the institutions of learning by Samual Hartlib and his circle.[93] In *An idea of Mathematics*, Pell proposed a reform of mathematical education and introduced his views on the relationship between algebra and natural knowledge. Even though their friendship was abruptly broken off by Hooke in the late 1670s, he re-published Pell's Latin *Idea matheseos* in the *Philosophical Collections* of 1682, along with letters of Descartes and Mersenne.[94] The adoption of Hérigone's

[86] Oughtred 1652: Sig A2, 63.

[87] Royal Society Classified Papers, vol. XX, f. 66r.

[88] Hérigone (1634), vol. II, 2.

[89] Massa Esteve (2008), 291, 296.

[90] Royal Society Classified Papers, vol. XX, f. 66r.

[91] Rahn (1668), sig. A2.

[92] Pell to Haak, 4 February 1664, British Library MS 4365, ff. 9r, 13r.

[93] Stedall (2002), 128–31; Barnett (1957), 206; Barnett (1962), 76–8, 98–113.

[94] Pell (1680); Malcom 2000: 280; cf. Moxham (2016), 489–90.

tripartite structure in 1640 by Pell was part of a wider project of an "algebra of knowledge" linked to a universal language.[95]

3.4 An Algebra of Algebras

The symbolism developed by modern mathematicians, especially Oughtred and Hérigone, represented for Hooke a model for the creation and establishment of any artificial language "whenever the occasion calls for it." Its characters should be the "briefest that can be made, and also easy for the memory and intellect." In the algebraic lectures, Hooke undertook a system of "shorthand symbols," while in a later lecture he proposed a "mathematicall language."[96] The "rule" adopted for algebra applies to the language of science as well. As Hooke noted in some early lectures at Gresham College, "the philosophical words of all languages yet known in the world, seem to be for the most part very improper marks set on confused and complicated notions."[97] Algebraic symbolism was a model for a more general reform of scientific language and knowledge. "Names therefore or characters ought to be adequate to the thoughts or notions of things to be communicated."[98] Since the mid-1660s, Hooke undertook a project of classification of natural objects propaedeutic to the reform of scientific language.[99] His classification was clearly influenced by Wilkins.[100] As he wrote to Leibniz, Hooke "spent some thoughts" upon Wilkins's project.[101] Like Wilkins's real character, Hooke's philosophical language was based on the "enumeration and description of such things or notions as are to have marks or names assigned to them."[102] If the groundwork of the classification was modelled on Wilkins, Hooke's grammar seemed different. In the draft lecture on "mathematical language," Hooke considered both the use of letters and new symbols. The former were widely used in algebra as a universal system of signs independent of the nature of the objects to which they referred. Letters, wrote Frans Van Schooten, are "very simple" characters that can be used to denote any quantity "universally and in abstract."[103] As Hooke knew, besides having the virtue of brevity, shorthand helps to perform operations almost in a mechanical way.[104] This particular aspect of algebraic symbolism was stressed by Hooke.

---

[95] Malcolm and Stedall (2005), 265–8.

[96] Guildhall Library, London MS 01757. 12.3, ff. 118v, 119r, 121r–v.

[97] Hooke (1705), 10.

[98] Royal Society Classified Papers, vol. XX, f. 160r.

[99] Hooke (1705), 20–1.

[100] Slaughter (1982), 159–60; Pugliese (1982), 113; Lynch (2001), 116–7.

[101] Leibniz (1991), 234.

[102] Wilkins (1668), 20.

[103] Van Schooten (1651), sig. ∗∗3 r–v, 1–2.

[104] Bashmakova and Smirnova (2002), 94.

The draft on "mathematical language" might suggest that Hooke's philosophical algebra was a form of real character and universal language.[105] Hooke contributed to Wilkins's project and was also involved in the unsuccessful revision of it undertaken by Aubrey and other members of the Royal Society.[106] "I had the happiness," Hooke wrote to Leibniz in 1680, "to be with him a great part of the time he was compiling of it and have had many debates with him of that partycular thought."[107] Like other fellows of the Royal Society, he used Wilkins' artificial language.[108] Wilkins was aware "of the sundry defects" of his work, and since its publication asked the Society to select some fellows to "offer their thoughts concerning what they judge fit to be amended in it."[109] But the difference of opinions among the members of the committee nominated by the Society prevented them from carrying out any reform.[110] In Hooke's opinion, Wilkins's "universal and real character" was "as well for philosophical as for common and constant use."[111] Francis Lodwick, a member of the committee and friend of Hooke, distinguished the "vocall characters" that apply to sounds and mainly consist of alphabetical letters, from the "real characters" that directly express the nature of things themselves. Like Wilkins, Lodwick believed that real characters were "also expressable in a distinct language" that could become universal.[112] But these two functions were often distinguished. The real character was seen as a philosophical language, whereas the universal language was intended as a new lingua franca or substitute for Latin.[113] For Andrew Paschall, also a member of the Society committee, this distinction applied to the projects of Wilkins and Ward. The exchange of opinions between Paschall and Hooke on this question suggests that Hooke's consideration of, and debt to, Wilkins' *Essay* was separate from his view of symbolic language.[114] Since Wilkins built on "an orderly enumeration of the severall objects of our thoughts" and assigned to each of them a "distinct marke," his project produced a universal language. Ward, on the contrary, aimed to "contract the great number of marks of those objects" to obtain a symbolic language "not for the use of common discourse but for the service of severe and strict reasoning."[115] Hooke agreed that Ward's language was "only designed for philosophy," but maintained that Wilkins aimed to fulfil both uses.[116] As he wrote to Leibniz, "a language somewhat like that of the bishop of Chester [Wilkins] would

---

[105] Slaughter (1982), 182–83.

[106] Poole (2012), 152.

[107] Leibniz (1991), 234.

[108] Illife (1992), 35–8; Andrade (1936), 12; Turner (1994), 306–7; Poole (2018), 8, 13.

[109] Wilkins (1668), sig. A1v.

[110] Lewis (2001), 332, 339; Poole (2010), 58–61.

[111] Hooke (1676),150–1.

[112] Lodwick (2011), 148–9.

[113] Lewis (2007), 108–9; Salmon (1979), 188; Slaughter (1982), 126; cf. Clauss (1982), 545–6.

[114] Lewis (2001), 352–3, 354 n. 39.

[115] Paschall to Hooke, 21 February 1680. Royal Society Early Letters P1 57, f. 96r.

[116] Hooke to Paschall, 1 March 1679. Royal Society Early Letters H3 61.

be of use for promotion of science." A universal language, however, was not what Hooke was interested in. "My aymes," he added, "have always been much higher."[117] Like Ward, Hooke intended the new language as a set of symbols modelled on the most advanced algebraic notation.[118] Assigned to a limited number of simple notions, symbols could be associated, in Ward's view, so as to "represent to the very eye all the elements of their composition and so deliver the natures of things." As in the "specious Analytics," by means of this philosophical language "exact discourses may be made demonstratively."[119] Even Hooke aimed to extend to natural philosophy that "methodical process of ratiocination" expressed in symbolic algebra.[120] His aim was to devise an instrument "expressing and remembering things and notions," but also able to "direct and regulate assist and necessitate and even compel the mind to find out and comprehend whatsoever is knowable." Rather than a new universal language, Hooke's *penus analytica* was intended as the "algebra of algebras or the science of methods."[121]

The new ministration was expected to transform the deduction from experiments and observations into a mechanical process, though not a calculating machine similar to that devised by Leibniz and quickly reproduced by Hooke at the Royal Society in 1673. In a letter to Hooke, Leibniz explained his plan for a universal and symbolic language (*scripturam quandam universalem*) by means of which the processes of discovery and demonstration could be reduced to calculus. Like a telescope, Leibniz's symbolic language was expected to emend and magnify human intellectual faculty.[122] In *Micrographia*, Hooke compared the method he employed for the solution of mechanical problems to geometrical algebra. Both were "applicable to physical inquires."[123] But he proved dismissive of arithmetic machines, preferring calculations by means of pen and paper.[124] His emphasis on the symbols and the graphic structure introduced by Hérigone and Pell in algebra is consistent with the importance that Hooke attributed to the visual components of the deduction that these elements made possible. Hérigone and Pell provided the closest model to the "most perfect idea" of algebra, wherein "the ratiocination does most distinctly and plainly appear."[125] Like the two mathematicians, Hooke considered the use of few and plain symbols as relevant as the synoptic view offered by the new algebra.[126] The tripartite graphic introduced by Hérigone and Pell was closer to a synoptic

---

[117] Leibniz (1991), 391.

[118] Hesse (1966a), 79; see also Rossi (2006), 151.

[119] Ward (1654), 21.

[120] Royal Society Classified Papers, vol. XX, f. 65r–v.

[121] Leibniz (1991), 234–35.

[122] Leibniz (1991), 314; see also Leibniz (1903), 157, 334, 337–8; cf. Maat (2004), 301, 308, 382–3; Pasini (1997), 39; Rutherford (1995), 230–1.

[123] Hooke (1665), sig. d2r.

[124] Jones (2016), 63–5.

[125] Royal Society Classified Papers, vol. XX, f. 65v.

[126] See Russo (1959), 204.

view, allowing a view "by the same cast of the eye as it were the deduction itself and the cause of that deduction" in an uninterrupted sequence on the page. What made algebra a model for the new ministration to reason was, for Hooke, its ability to show, "evidently spread before our eyes, how invention is prosecuted and carried on in the braine." By means of this instrument, "the eye, the hand, nay the pen almost is made to doe greater things, than can be done by the braine, reason or the very soule of man." The certainty of deductions and the demonstrative nature of any conclusion reached by algebra are due to the use of symbols, and a new graphic structure by means of which "a whole series of ratiocinations" is displayed in a restricted space. Through a "small cast of an eye as it were, and in an instant almost, one is enabled to examin, and compare and change, and transpose and order any part of it, as he pleases, with very litle trouble and the greatest certainty."[127] This was the essence of Hooke's methodological programme: a set of mechanical and logical instruments that corrected, assisted, and improved human faculties made deficient after the Fall. The symbols and the synoptic view developed by Hérigone and Pell rendered the new algebra a congenial model for this project.

In the 1660s, Hooke's project seemed closer to a graphic form and symbolic notation for the new philosophical history.[128] Written in real character and with an algebraic structure, this latter was for Hooke the best Baconian way to recover the direct and true knowledge of things themselves lost after the Fall.[129] Philosophical algebra was also a means to improve the ability to collect, preserve and arrange experiments and observations to facilitate the process of deduction.[130] This was also a virtue of algebra, for it "serves as much to aid the memory as to cultivate intelligence and reasoning."[131] The method of Hérigone and Pell perfected this feature. Since it displayed the whole series of deductions and axioms involved in it, both memory and ratiocination were not disturbed by the need to verify each phase of the inference.[132] In the early Cutlerian lectures, a similar arrangement was also part of Hooke's new philosophical history. By means of shorthand and abbreviations, the "whole history may be contracted into as little space as is possible," and expressed in as "few letters or characters as it has considerable circumstances." The graphic structure of the new history was intended to be very different from common ones. Space was reserved for schemes, drawings, illustrations, and "schedules." After the positive reception of *Micrographia*, images and illustrations could not be excluded from Hooke's project. Like symbols, images could express the nature of things observed better than any verbal description, which would also inevitably be longer. As in the algebra, theoretical elements and empirical operations were planned to be integrated into the new philosophical history.[133]

---

[127] Royal Society Classified Papers, vol. XX, f. 65r–v.

[128] Lewis (2007), 153–4; Johns (1998), 433; Yeo (2014), 240, 249–52.

[129] Cf. Formigari (1988), 12; Harrison (2007), 152, 172–3.

[130] Hooke (1705), 18.

[131] Guildhall Library, London 1757.12., f. 171r.

[132] Royal Society Classified Papers, vol. XX, f. 65v.

[133] Hooke (1705), 64–5; cf. Fletcher (2014), 64–5; Henderson (2019), 420.

In *A Method for making a History of the Weather*, published by Thomas Sprat in 1667, Hooke perhaps followed some of the principles of the philosophical history that he had been proposing in some coeval lectures. The "scheme at one view representing to the eye the observations of the weather for a month" lacked algebraic symbols or shorthand. Nonetheless, it displayed observations, measurements, and "general deductions" on the same page.[134] Illustrations were the only graphic component of Hooke's early thoughts on philosophical algebra that remained in his works after 1667. Hooke planned to "fully explain" how symbols and new graphic structures could be introduced in a "second part" of his lectures, but he probably never wrote it.[135] Instead, he just referred here and there in his lectures to the need to integrate analysis and synthesis into natural philosophy. Thus, what his philosophical algebra could have been remains a question impossible to solve by means of the historical evidence that we currently have. Despite the absence of an adequate theorization of the *penus analytica*, as a set of ministrations to human faculties Hooke's philosophical algebra took the form of a methodological programme aiming at the integration of hypotheses and experiments, queries and observations, analyses and syntheses, induction and deduction.[136] Even though the clear distinction between historical and deductive components of natural philosophy gradually disappeared in his work,[137] an innovative combination of hypotheses and experiments remained a constant of both Hooke's methodological claims and his scientific praxis.[138]

Bibliography

Andrade, Edward. 1936. The real character of Bishop Wilkins. *Annals of Science* 1: 4–12.
———. 1960. Robert Hooke, F. R. S. (1635–1703). *Notes and Records of the Royal Society of London* 15: 137–145.
Bacon, Francis. 2004. *The Instauratio magna, Part II: Novum organum*, ed. Graham Rees with Maria Wakely. Oxford: Oxford University Press.
Barnett, Pamela. 1957. Theodore Haak and the early years of the Royal Society. *Annals of Science* 13: 205–218.
———. 1962. *Theodore Haak F. R. S. (1605–1690)*. The Hague: Mouton and Co.
Bashmakova, I.G., and G.S. Smirnova. 2002. The evolution of literal algebra. In *Mathematical evolutions*, ed. Abe Shenitzer and John Stillwell, 83–97. Washington, DC: The Mathematical Association of America.
Bennett, Jim. 1982. *The mathematical science of Christopher Wren*. Cambridge: Cambridge University Press.
———. 1986. The mechanics' philosophy and the mechanical philosophy. *History of Science* 24: 1–28.

---

[134] Sprat (1667), 179.

[135] Hooke (1705), 64.

[136] Ibid., 173, 330–1, 553; Royal Society Classified Papers, vol. XX, ff. 65r, 169r.

[137] Mulligan (1992), 158.

[138] Hooke (1665), 169; see also Andrade (1960), 138; Sciuto (1983), 54.

———. 1998. Practical geometry and operative knowledge. *Configurations* 6: 195–222.

———. 2002. Geometry in context in the sixteenth century: The view from the museum. *Early Science and Medicine* 7: 214–230.

———. 2003. Hooke's instruments. In *London's Leonardo. The life and work of Robert Hooke*, ed. Jim Bennett, Michael Cooper, Michael Hunter, and Lisa Jardine, 63–104. Oxford: Oxford University Press.

———. 2006. Instruments and ingenuity. In *Robert Hooke: Tercentennial studies*, ed. Michael Cooper and Michael Hunter, 65–76. Aldershot: Ashgate.

Bertoloni Meli, Domenico. 2010. The axiomatic tradition in seventeenth-century mechanics. In *Discourse on a new method: Reinvigorating the marriage of history and philosophy of science*, ed. Mary Domski and Michael Dickson, 23–41. La Salle: Open Court.

Birch, Thomas. 1756–57. *The history of the Royal Society of London*, 4 vols. London.

Borghero, Carlo. 2005. L'analisi da Descartes a Kant. *Giornale critico della filosofia italiana* 84: 433–469.

Bos, Henk. 2001. *Redefining geometrical exactness: Descartes' transformation of the early modern concept of construction*. Dordrecht: Springer.

Cajori, Florian. 1928. *A history of mathematical notations*, 2 vols. London: Open Court Company.

Charbonneau, Louis. 2005. L'algèbre au cœur du programme analytique. In *François Viète: un mathématicien sous la Renaissance*, ed. Évelyne Barbin and Anne Boyé, 53–73. Paris: Vuibert.

Centore, Floyd. 1970. *Robert Hooke's contributions to mechanics*. The Hague: Martinuus Nijhoff.

Clauss, Sidonie. 1982. John Wilkins' Essay toward a real character: Its place in seventeenth-century episteme. *Journal of the History of Ideas* 43: 531–553.

Clucas, Stephen. 1999. 'No small force': Natural philosophy and mathematics in Thomas Gresham's London. In *Sir Thomas Gresham and the Gresham College: Studies in the intellectual history of London in the sixteenth and seventeenth centuries*, ed. Francis Ames-Lewis, 146–173. Aldershot: Ashgate.

Dear, Peter. 1998. The mathematical principles of natural philosophy: Towards a heuristic narrative for the scientific revolution. *Configurations* 6: 173–193.

———. 2007. Towards a genealogy of modern science. In *The mindful hand: Inquiry and invention from the late Renaissance to early industrialization*, ed. Lissa Roberts, Simon Schaffer, and Peter Dear, 431–441. Amsterdam: Koninklijke Nederlandse Akademie va Wetenschappen.

———. 2011. Mixed mathematics. In *Wrestling with nature. From omens to science*, ed. Peter Harrison, Ronald Numbers, and Michael Shank, 149–172. Chicago/London: University of Chicago Press.

Dee, John. 1570. Mathematicall preface. In *The elements of geometrie of the most ancient philosopher Euclide of Megara*. Trans. Henry Billingsley. London.

Del Monte, Guido Ubaldo. 1575. *Mechanicorum libri*. Pesaro.

Descartes, René. 1964–74. *Oeuvres*, 12 vols., ed. Charles Adam and Paul Tannery. Paris: Vrin.

Domski, Mary. 2013. Observation and mathematics. In *The Oxford handbook of British philosophy in the seventeenth century*, ed. Peter Anstey, 144–168. Oxford: Oxford University Press.

Drake, Stillman, and Israel Drabkin, eds. 1966. *Mechanics in sixteenth-century Italy: Selections from Tartaglia, Benedetti, Guido Ubaldo and Galileo*. Madison/London: University of Wisconsin Press.

Feingold, Mordechai. 1999. Gresham College and London practitioners: The nature of the English mathematical community. In *Sir Thomas Gresham and the Gresham College: Studies in the intellectual history of London in the sixteenth and seventeenth centuries*, ed. Francis Ames-Lewis, 174–188. Aldershot: Ashgate.

———. 2001. Mathematicians and naturalists: Sir Isaac Newton and the Royal Society. In *Isaac Newton's natural philosophy*, ed. I. Bernard Cohen and Jed Buchwald, 77–102. Cambridge, MA: The MIT Press.

Fletcher, Puck Francis. 2014. *Space, spatiality, and epistemology in Hooke, Boyle, Newton, and Milton*. DPhil disseration, University of Sussex.

Formigari, Lia. 1988. *Language and experience in 17th-century British philosophy*. Amsterdam/Philadelphia: Benjamins Publishing.

Fournier, Marian. 1996. *The fabric of life: Microscopy in the seventeenth century*. Baltimore: Johns Hopkins University Press.

Freguglia, Paolo. 1988. *Ars analytica: matematica e methodus nella seconda metà del Cinquecento*. Busto Arsizio: Bramante.

———. 1999. *La geometria fra tradizione e innovazione: temi e metodi geometrici nell'età della rivoluzione scientifica 1550–1650*. Turin: Bollati Boringhieri.

Gabbey, Alan. 1992. Newton's Mathematical principles of natural philosophy: A treatise on 'mechanics'? In *The investigation of difficult things: Essays on Newton and the history of exact sciences in honour of Derek T. Whiteside*, ed. P.M. Harman and Alan Shapiro, 305–322. Cambridge: Cambridge University Press.

Galilei, Galileo. 1890–1909. *Opere*, 20 vols, ed. Antonio Favaro. Florence: Barbera.

———. 1974. *Two new sciences, including Centres of gravity and Force of percussion*. Trans. Stillman Drake. Madison/London: University of Wisconsin Press.

Goulding, Robert. 1999. Testimonia humanitatis: The early lectures of Henry Savile. In *Sir Thomas Gresham and the Gresham College: Studies in the intellectual history of London in the sixteenth and seventeenth centuries*, ed. Francis Ames-Lewis, 125–145. Aldershot: Ashgate.

Guicciardini, Niccolò. 2007. "Mechanica rationalis" and "philosophia naturalis" in the auctori praefatio to Newton's Principia. In *Mechanic and cosmology in the medieval and early modern period*, ed. Massimo Bucciantini, Michele Camerota, and Sophie Roux, 169–186. Florence: Olschki.

———. 2009. *Isaac Newton on mathematical certainty and method*. Cambridge, MA/London: MIT Press.

———. 2011. *Newton*. Rome: Carocci.

Hall, Alfred Rupert. 1987. Alexandre Koyré and the scientific revolution. *History and Technology* 4: 285–295.

Harrison, Peter. 2007. *The fall of man and the foundation of science*. Cambridge: Cambridge University Press.

Henderson, Felicity. 2019. Robert Hooke and the visual world of the Early Royal Society. *Perspectives on Science* 27: 395–443.

Henry, John. 2011. The origins of the experimental method: Mathematics or magic? In *Departure for modern Europe: A handbook of early modern philosophy (1400–1700)*, ed. Hubertus Busche, 702–714. Hamburg: Felix Meiner Verlag.

Hérigone, Pierre. 1634. *Cursus mathematicus*, 4 vols. Paris.

Hesse, Mary. 1964. Hooke's development of Bacon's method. In *Actes du dixième congrès international d'histoire des sciences*, 2 vols., ed. Henry Guerlac, vol. I, 265–268. Paris: Hermann.

———. 1966a. Hooke's vibration theory and the isochrony of springs. *Isis* 57: 433–441.

———. 1966b. Hooke's philosophical algebra. *Isis* 57: 67–83.

Hintikka, Jaakko, and Unto Remes. 1974. *The method of analysis: Its geometrical origin and its general significance*. Dordrecht: Reidel.

Hooke, Robert. 1665. *Micrographia*. London.

———. 1674. *An attempt to prove the motion of the Earth by observations*. London.

———. 1676. *A description of helioscopes*. London.

———. 1705. *Posthumous works*, ed. Richard Waller. London.

———. 1726. *Philosophical experiments and observations*, ed. William Derham. London.

Høyrup, Jens. 1992. Archimedism, not Platonism: On a malleable ideology of Renaissance mathematics (1400 to 1600), and on its role in the formation of seventeenth-century philosophies of science. In *Archimede: mito, traditione, scienza*, ed. Corrado Dollo, 81–110. Florence: Olschki.

Hunter, Michael. 1989. *Establishing the new science: The experience of the early Royal Society*. Woodbridge: Boydell Press.

———. 2003. Hooke the natural philosopher. In *London's Leonardo. The life and work of Robert Hooke*, ed. Jim Bennett, Michael Cooper, Michael Hunter, and Lisa Jardine, 105–162. Oxford: Oxford University Press.

Huygens, Christiaan. 1888–1950. *Oeuvres completes*, 22 vols., ed. Société hollandaise des sciences. The Hague: Martinus Nijhoff.

Illife, Rob. 1992. 'In the Warehouse': Privacy, property and priority in the early Royal Society. *History of Science* 30: 29–68.

Jalobeanu, Dana. 2015. *The art of experimental natural history: Francis Bacon in context.* Bucharest: Zeta Books.

———. 2016. Disciplining experience: Francis Bacon's experimental series and the art of experimenting. *Perspectives on Science* 24: 324–342.

Johns, Adrian. 1998. *The nature of the book: Print and knowledge in the making.* Chicago/London: Chicago University Press.

Johnston, Stephen. 2012. John Dee on geometry: Text, teaching, and the Euclidean tradition. *Studies in History and Philosophy of Science* 43: 470–479.

Jones, Matthew. 2016. *Reckoning with matter: Calculating machines, innovation, and thinking about thinking from Pascal to Babbage.* Chicago/London: University of Chicago Press.

Kargon, Robert. 1971. The testimony of nature: Boyle, Hooke and experimental philosophy. *Albion* 3: 72–81.

Koyré, Alexandre. 1978. *Galileo studies.* Atlantic Highlands: Humanities Press.

Kuhn, Thomas. 1977. *The essential tension: Selected studies in scientific tradition and change.* Chicago/London: University of Chicago Press.

Laudan, Larry. 1966. *The idea of physical theory from Galileo to Newton: Studies in Seventeenth-Century Methodology.* PhD diss., Princeton University. Ann Arbor: University Microfilm International.

Leibniz, Gottfried Wilhelm. 1903. *Opuscules et fragments inédits de Leibniz,* ed. Louis Couturat. Paris: Alcan.

———. 1991. *Mathematischer naturwissenschaftlicher und technischer Briefwechsel. Dritter Band 1680–Juni 1683.* Berlin: Akademie Verlag.

Lewis, Rhodri. 2001. The effort of the Aubrey correspondence group to revise John Wilkins' Essay (1668) and their context. *Historiographia Linguistica* 28: 331–364.

———. 2007. *Language, mind and nature: Artificial languages in England from Bacon to Locke.* Cambridge: Cambridge University Press.

Lodwick, Francis. 2011. *On language, theology, and utopia,* ed. Felicity Henderson and William Poole. Oxford: Clarendon Press.

Lynch, Michael. 2001. *Salomon's child: Method in the early Royal Society of London.* Stanford: Stanford University Press.

———. 2005. A society of Baconians? The collective development of Bacon's method in the Royal Society of London. In *Francis Bacon and the refiguring of early modern thought: Essays to commemorate The Advancement of Learning (1605–2005),* ed. Julie Robin Salomon and Chaterine Gimelli Martin, 173–202. Aldershot: Ashgate.

Maat, Jaap. 2004. *Philosophical languages in the seventeenth century: Dalgarno, Wilkins, Leibniz.* Dordrecht: Kluwer.

Maierú, Luigi. 2007. *John Wallis: una vita per un progetto.* Soveria Mannelli: Rubbettino.

Malcolm, Noel. 2000. The publications of John Pell, F. R. S. (1611–1685): Some new light and some old confusions. *Notes and Records of the Royal Society of London* 54: 275–292.

Malcolm, Noel, and Jacqueline Stedall. 2005. *John Pell (1611–1685) and his correspondence with Sir Charles Cavendish: "The mental world of an early modern mathematician".* Oxford: Oxford University Press.

Manzo, Silvia. 2001. Experimentaión, instrumentos científicos y cuantificación en el método de Francis Bacon. *Manuscrito* 24: 48–84.

Massa Esteve, Maria Rosa. 2008. Symbolic language in early modern mathematics: The Algebra of Pierre Hérigone (1580–1643). *Historia Mathematica* 35: 285–301.

Mori, Giuliano. 2017. Mathematical subtleties and scientific knowledge: Francis Bacon and mathematics, at the crossing of two traditions. *The British Journal for the History of Science* 50: 1–21.

Moxham, Noah. 2016. Authors, editors and newsmongers: Form and genre in the "Philosophical Transactions" under Henry Oldenburg. In *News networks in early modern Europe*, ed. Joad Raymond and Noah Moxham, 465–492. Leiden: Brill.

Mulligan, Lotte. 1992. Robert Hooke and certain knowledge. *The Seventeenth Century* 7: 151–169.

Newton, Isaac. 1779–85. *Opera quae extant omnia*, 5 vols., ed. Samuel Horsley. London.

———. 1959–77. *The correspondence of Isaac Newton*, 7 vols., ed. H. W. Turnbull, J. F. Scott, A. R. Hall and L. Tilling. Cambridge: Cambridge University Press.

Oldroyd, David. 1972. Robert Hooke's methodology of science as exemplified in his 'Discourse of Earthquakes'. *The British Journal for the History of Science* 6: 109–130.

———. 2000. Robert Hooke. In *Encyclopaedia of the scientific revolution: From Copernicus to Newton*, ed. Wilbur Applebaum, 299–301. New York: Garland Publishing.

Oughtred, William. 1652. *Clavis mathematicae denuo limata ... editio tertia*. Oxford.

Palmerino, Carla Rita. 2006. The mathematical characters of Galileo's book of nature. In *The book of nature in early modern and modern history*, ed. Klaas van Berkel and Arjo Vanderjagt, 27–44. Leuven: Peeters.

———. 2010. Experiments, mathematics, physical causes: How Mersenne came to doubt the validity of Galileo's law of free fall. *Perspectives on Science* 18: 50–76.

———. 2016. Reading the book of nature: The ontological and epistemological underpinnings of Galileo's mathematical realism. In *The language of nature: Reassessing the mathematization of natural philosophy in the seventeenth century*, ed. Geoffrey Gorham, Benjamin Hill, Edward Slowik, and Kenneth Waters, 29–50. Minneapolis: University of Minnesota Press.

Panza, Marco. 2007. Classical sources for the concept of analysis and synthesis. In *Analysis and synthesis in mathematics: History and philosophy*, ed. Marco Panza and Michael Otte, 365–414. Dordrecht: Kluwer.

Pasini, Enrico. 1997. Arcanum artis inveniendi: Leibniz and analysis. In *Analysis and synthesis in mathematics: History and philosophy*, ed. Michael Otte and Marco Panza, 35–46. Dordrecht: Kluwer.

Pastorino, Cesare. 2011. Weighing experience: Experimental histories and Francis Bacon's quantitative program. *Early Science and Medicine* 16: 542–570.

Pell, John. 1680. Idea matheseos. *Philosophical Collections* 5: 127–145.

Pérez-Ramos, Antonio. 1988. *Francis Bacon's idea of science and the maker's knowledge tradition*. Oxford: Clarendon.

Petty, William. 1927. *The Petty papers*, 2 vols., ed. Henry 6th Marquis of Lansdowne. London: Constable and Company.

Poole, William. 2010. *The world makers: Scientists of the restoration and the search for the origins of the Earth*. Oxford: Peter Lang.

———. 2012. Vossius, Hooke, and the early Royal Society's use of sinology. In *The intellectual consequences of religious heterodoxy 1600–1750*, ed. John Robertson and Sarah Mortimer, 135–153. Leiden: Brill.

———. 2018. Seventeenth-century 'double writing' schemes, and a 1676 letter in the phonetic script and real character of John Wilkins. *Notes and Records of the Royal Society of London* 72: 7–23.

Pugliese, Patri. 1982. *The scientific achievement of Robert Hooke*. PhD dissertation, Harvard University. Ann Arbor: University Microfilm International.

———. 2004. Robert Hooke. In *Oxford dictionary of national biography, vol. 27*, ed. H.C.G. Matthew and Brian Harrison, 951–958. Oxford: Oxford University Press.

Rahn, Johann Heinrich. 1668. *An introduction to algebra ... much altered and augmented by D. P.* London.

Recorde, Robert. 1551. *The pathway to knowledge*. London.

Rees, Graham. 1985. Quantitative reasoning in Francis Bacon's natural philosophy. *Nouvelles de la République des lettres* 1: 27–48.

Rossi, Paolo. 2002. I filosofi e le machine. In *L'uomo e le macchine*, ed. Mimma Bresciani Califano, 3–26. Florence: Olschki.

———. 2006. *Logic and the art of memory: The quest for a universal language.* Trans. Stephen Clucas. London/New York: Continuum.

Russo, Filippo. 1959. La construction de l'algèbre au XVIe siècle. Étude de la structure d'une evolution. *Revue d'histoire des sciences* 12: 193–208.

Rutherford, Donald. 1995. Philosophy and language in Leibniz. In *The Cambridge companion to Leibniz*, ed. Nicholas Jolley, 224–269. Cambridge: Cambridge University Press.

Salmon, Vivian. 1979. *The study of language in 17th-century England.* Amsterdam: John Benjamins Publishing.

Savile, Henry. 1621. *Praelectiones tresdecim in principium elementorum Euclidis.* Oxford.

Schuster, John. 2013. *Descrates-agonistes: Physico-mathematics, method and corpuscular-mechanism, 1618–33.* Dordrecht: Springer.

Sciuto, Maurizio. 1983. *Induzione e meccanicismo: il programma metodologico e scientifico di Robert Hooke.* MA dissertation, University of Florence.

Shapiro, Alan. 1989. Huygens' Traité and Newton's Opticks: Pursuing and eschewing hypotheses. *Notes and Records of the Royal Society of London* 43: 223–247.

Serfati, Michel. 1998. Descartes et la constitution de l'écriture symbolique mathématique. *Revue d'histoire des sciences* 51: 237–290.

Seltman, Muriel. 2000. Harriot's algebra: Reputation and reality. In *Thomas Harriot: An Elizabethan man of science*, ed. Robert Fox, 153–185. Aldershot: Ashgate.

Slaughter, M.M. 1982. *Universal language and scientific taxonomy in the seventeenth century.* Cambridge: Cambridge University Press.

Smith, Pamela. 2004. *The body of the artisan: Art and experience in the scientific revolution.* Chicago/London: University of Chicago Press.

———. 2009. Science on the move: Recent trends in the history of early modern science. *Renaissance Quarterly* 62: 345–375.

Sprat, Thomas. 1667. *History of the Royal Society.* London.

Stedall, Jacqueline. 2002. *A discourse concerning algebra: English algebra to 1685.* Oxford: Oxford University Press.

———. 2007. Symbolism, combinations, and visual imagery in the mathematics of Thomas Harriot. *Historia Mathematica* 34: 380–401.

Timmermans, Benoît. 1995. *La résolution des problèmes de Descartes à Kant: l'analyse à l'âge de la révolution scientifique.* Paris: Presses Universitaires de France.

———. 1999. The originality of Descartes' conception of analysis. *Journal of the History of Ideas* 60:433–447.

Turner, Anthony J. 1994. Learning and language in the Somerset levels: Andrew Paschall of Chedsey. In *Learning, language and invention: Essays presented to Francis Maddison*, ed. Willem D. Hackmann and Anthony J. Turner, 297–308. Ashgate: Variorum.

Van Helden, Albert. 1983. The birth of modern scientific instrument, 1550–1700. In *The uses of science in the age of Newton*, ed. John Burke, 49–83. Los Angeles/London: University of California Press.

Van Schooten, Frans. 1651. *Principia matheseos universalis seu introductio ad geometriae methodum Renati Des Cartes, edita ab Er. Bartholino.* Leyden.

Viète, François. 1646. *Opera mathematica.* Leiden.

Wallis, John. 1685. *A treatise of algebra both historical and practical.* London.

Ward, John. 1740. *The lives of the professors of Gresham College.* London.

Ward, Seth. 1654. *Vindiciae academiarum.* Oxford.

Wilkins, John. 1668. *An essay towards a real character and a philosophical language.* London.

Wren, Stephen. 1750. *Parentalia or memories of the family of the Wrens.* London.

Yeo, Richard. 2014. *Notebooks, English virtuosi, and early modern science.* Chicago/London: University of Chicago Press.

Chapter 4
A Larger Scheme

4.1 A Large New Prospect into Nature

Hooke did not write a treatise on the origins of the physical qualities of bodies; rather he contributed theories and hypotheses to what we consider today as different, sometimes distant, disciplines. In these contributions, however, his ideas on the fundamental constitution of matter emerged.[1] Although far from being systematic and consistent, these ideas nonetheless led to a theory of matter that sometimes played the role of premise to his experimental and speculative work, but was often reshaped by new evidence which emerged from his manifold researches. Matter theory, in short, was part of Hooke's wider image of nature.

The very nature of Hooke's theory of matter, however, has hitherto been a controversial topic. Following Marie Boas, many scholars have described Hooke's ideas on matter as "essentially Cartesian."[2] He seemed sceptical about the existence of a physical void, and largely employed elastic ethereal fluids to account for phenomena. Principles such as congruity, in this perspective, confirm rather than question the mechanical nature of Hooke's theory, for they replace the magical principles of sympathy and antipathy by a mechanical explanation.[3] In contrast to this view, Penelope Gouk noted that the musical analogies often employed to explain this new principles prove the influence of natural magic on Hooke's view of the universe.[4]

By describing Hooke as an "incongruous mechanist," John Henry has provided a more balanced view. Congruity is an active principle that coexists with mechanical

---

[1] Chapman (2005), 44.

[2] Boas (1952), 453–5; cf. Pumfrey (1987), 11; Centore (1970), 63–7.

[3] Ehrlich (1992), 50, 67; Id. (1995), 127, 130, 135–7.

[4] Gouk (1980), 575, 577, 585; Id. (1999), 194, 204, 214, 222–3; cf. Kassler and Oldroyd (1983), 588–9.

© Springer Nature Switzerland AG 2020
F. G. Sacco, *Real, Mechanical, Experimental*, International Archives
of the History of Ideas Archives internationales d'histoire des idées 231,
https://doi.org/10.1007/978-3-030-44451-8_4

ideas in Hooke's natural philosophy.[5] The apparent inconsistency between these two elements is mainly due to the prevalence of a strictly Cartesian view of mechanical philosophy, which misrepresents the wider philosophical landscape of the seventeenth century. What was a mechanical philosophy? Who could be considered a mechanical philosopher? These questions could be answered in very different ways. A mechanical philosophy usually entailed a reductionist approach to scientific explanation and an emphasis on motion and the laws of motion as the causes of all physical processes.[6] These elements, however, could lead to many different forms of natural philosophy. Cartesian mechanism was only one of these. As Gassendi showed, active principles such as the connatural motion of atoms were still part of many mechanical philosophies.[7]

Long before Newton's dynamic version of the new philosophy emerged, English natural philosophers paved the way by introducing active principles, and assimilating chymical qualities.[8] Hooke first read the continental mechanical philosophers as a student at Christ Church College in Oxford where, as Ward claimed, there was "scarce any hypothesis, which hath been formerly or lately entertained by judicious men, and seemes to have in it any clearnesse or consistency, but hath here its strenuous assertours."[9] Atomism, chymistry, and magnetical philosophy were especially discussed among the group of Baconian fellows gathered around the warden of the Wadham College, Wilkins. Hooke worked as assistant to a member of this group, Willis, while the latter was developing his ideas on fermentation and fevers. Willis' hypotheses were rooted in a corpuscularian form of chymical philosophy.[10] To the three chymical principles, Willis added water and earth. All of them were described as "the simplest and most elementary" portions in which sensible bodies could be divided. The physical properties of compounded bodies originated from the "affinities" among the elements.[11]

Along with Willis' eclectic natural philosophy, Bacon's work influenced the reception of continental mechanical philosophies by Hooke and many others of his generation.[12] Bacon's influence, in fact, was not limited to methodology.[13] His ideas on matter played a very relevant role in shaping Hooke's approach towards mechanism. Bacon's emphasis on the dynamic components of nature can be considered a source of Hooke's mature matter theory. Hooke scholars have hitherto focused only on mechanical philosophers such as Descartes and Hobbes because of the supposed

---

[5] Henry (1986), 348; Id. (1989), 150, 156–7; cf. Lynch (2005), 191.

[6] Mclaughlin (2006), 97–8; Pyle (1995), 506–8; Machamer et al. (2012), 372; Gabbey (1993), 134–6.

[7] Clericuzio (2000), 1, 7, 63–5; Massignat (2000), 192; Lolordo (2007), 134, 140–4.

[8] Schaffer (1987), 56; Henry (1986), 336–7, 342; Giudice (2006), 40–1, 55; Rogers (1997), 214–5; Clericuzio (1993), 318, 326–7; Emerton (1984), 110–1, 116. On the use of the terms 'chymistry' and 'chymical' see Newman and Principe (1998), 41.

[9] Ward (1654), 2.

[10] Clericuzio (1993), 318.

[11] Willis (1659), 3–4, 13–4.

[12] Henry (2013), 116–35.

[13] For a different opinion see Rees (1977), 29.

inconsistency between Bacon's "speculative philosophy," as Graham Rees described it,[14] and mechanical philosophy.[15] But Bacon's speculative philosophy included principles derived from Renaissance vitalism, chymistry, and ancient atomism. Following Democritus, Bacon aimed to reach the fundamental components of matter, latent processes and schematisms. In Bacon's matter theory, the appetites or ultimate motions operate on *minima* to produce bodies with different structures and qualities.[16] This eclectic system was interpreted in different and contrasting ways in the seventeenth century.[17] Schematisms and textures could be seen as corpuscular structures of bodies. The explanation of heat in the second book of the *Novum Organum* could be used to support a mechanical worldview. Bacon described a universe in constant motion, in which the smallest and invisible portions of matter are never at rest.[18] As the rest of this chapter will show, this image was appropriated by Hooke. Along with the project of a new experimental history of nature, Bacon's speculative philosophy influenced Hooke's selective reading of Descartes and other continental philosophers and his development of a new form of mechanical philosophy.

In Hooke's work, the *Lectures de potentia restitutiva* are the only place where a consistent matter theory can be found. Published for the first time in 1678, the lectures included a detailed treatment of elasticity and Hooke's law. Hooke claimed that he knew the law long before. "It is now," he wrote in 1678, "eighteen years since I have found it out, but designing to apply it to some particular use, I omitted publishing thereof." Some hints were included in the *Attempt* published in 1661. The *Lectures* were presented by Hooke as a publication of ideas on springs and the properties of matter that he had maintained since he published his first scientific treatise.[19] This narrative suggests that Hooke's matter theory had been almost unaltered during the two most productive decades of his career.[20] Historical evidence, however, indicates otherwise. Only in 1664, for instance, was he assigned experiments on the elasticity of wood and springs by the Royal Society. When, in 1673, Petty called for a "good theory of the springiness in bodies," Hooke stated "that formerly he had explained it, in a discourse of his, brought in upon the occasion of the odd phaenomenon of the pipe of mercury standing top-full far above ordinary station." This discourse was read at the Royal Society on November 9, 1672.[21] The links between the theory of elasticity and phenomena such as capillarity and the so-called anomalous suspension are very significant. Hooke's ideas on matter were not kept in a philosophical vacuum in the two decades when he engaged with an

---

[14] Rees (1980), 553.

[15] cf. Hesse (1966), 434; Ehrlich (1992), 78.

[16] Bacon (2004), 88–9, 210–1, cf. Giglioni (2011), 19–22, 61–2; Id. (2013a), 45.

[17] Clucas (1997), 257; Giglioni (2013b), 38–51.

[18] Giglioni (2011), 65, 73.

[19] Hooke (1678), 1, 6.

[20] Hesse (1966), 433; Gal (2002), 131–2.

[21] Birch (1756–57), vol. I, 384; vol. II, 58–61; vol. III: 109.

impressive series of scientific questions. During this time Hooke's experimental agenda widely expanded, and relevant differences can be traced between the ideas expressed in the *Attempt* of 1661, *Micrographia*, and the *Lectures* of 1678. Many of the new phaenomena investigated by Hooke raised questions about strictly mechanical explanations, such as Descartes'. Engaged in the construction of a new experimental history of nature, Hooke was increasingly aware that traditional mechanical philosophy could not offer an adequate explanation for a great number of the natural phaenomena that emerged in new experiments. Some of the operations of nature are "secret" because "the manner of proceeding is more obscure and difficult to be found, and not yet discoverable by the senses, or any known way." Attractions, "sympathetical and antipathetical" effects, and the "operations of bodies at a much greater distance" demanded new philosophical answers. By means of these "we give ourselves a new or sixth sense, which will open us a large new prospect into nature that we dreamt not of before."[22]

4.2 Congruity

The first of the puzzling phenomena faced by Hooke was capillarity. In one of his *New Experiments* Boyle aimed "to try whether or no the pressure of the Air might reasonably be suppos'd to have either the principal, or at least a considerable Interest in the raising" of water and other liquids in thin glass pipes. Boyle did not hide his surprise when he observed that the vacuum of the pump did not affect the phenomena. He also noted that the level reached by liquids is higher in thinner tubes, the rising quicker in wet tubes, and the surface of the liquids within the tubes is convex. "The cause of this ascension of the Water," he noted, "appear'd to all that were present so difficult, that I must not stay to enumerate the various conjectures that were made at it, much less examine them." Only the opinion of "an ingenious Man" was considered worth being reported. According to "his ingenious conjecture," the cause of the phenomena is the "greater pressure made upon the Water by the Air without the Pipe, then by that within it." This conjecture was based on the fact that water consisted "of corpuscules more pliant to the internal surface of the glasse."[23] The "ingenious man" was probably Hooke, who one year later published a small tract to support the "ingenious conjecture" reported by Boyle.

Hooke was aware that the absence of air in the pump demanded a more elaborate explanation. He maintained that the glass of the pump did not stop the ether from entering the receiver; only atmospheric air needed for respiration and combustion was removed.[24] Even if there was no air in the pump, an "unequal pressure" on the water internal and external to the thin pipes was produced by the "much greater

---

[22] Hooke (1705), 43, 46; cf. Boyle (1999), vol. VIII, 136; Wren (1750), 221.

[23] Boyle (1999), vol. I, 350–3.

[24] Hooke (1665), 14, 32, 99–100.

inconformity or incongruity (call it what you please) of Air to Glass, and some other bodies, than there is of water to the same." Water is pushed up within the thin tubes by an attractive force operating between the liquid and the glass. In this perspective, the thinner the tubes, the bigger the surface of contact between water and glass, the stronger the effect of their congruity and the higher the level reached by the liquid. Congruity rather than atmospheric pressure is "the principal (if not the only) cause of this phaenomenon."

Congruity is "a property of a fluid body, whereby any part of it is readily united or intermingled with any other homogeneal or similar, fluid or firm and solid body." By contrast, incongruity is "a property of a fluid, by which it is kept off and hindred from uniting or mingling with any heterogeneous or dissimiliar fluid or solid body." Through this new principle many puzzling phenomena could be accounted for, such as the bubbles of air in water and the form of rain drops.[25] This latter, for instance, was discussed by Galilei, Descartes, and Gassendi, among others. All refused the magical idea of sympathy and antipathy, and looked for mechanical explanations.[26] A "great antipathy for antipathy" supported Galileo's refusal of any internal force of cohesion within the liquid. "I confess," Galileo's spokesman Salviati stated, "that I don't know how that business of sustaining large and elevated globules of water is accomplished; yet I am certain that it does not derive from any internal tenacity existing among their parts, so the cause of this effect should be situated outside."[27] Descartes resorted to the external pressure of celestial globules.[28] Mechanical explanations, such as the "thinness and thickness of mediums" and the different shapes of atoms, according to Gassendi, must replace occult powers, sympathies and antipathies.[29]

Unlike Gassendi, Hooke chose congruity to replace these magical principles. The new principle was consistent with a mechanical view of nature. But in 1661 Hooke did not commit to any specific hypothesis on the nature of congruity, perhaps because of the insufficiency of the traditional explanations provided by Descartes and Gassendi. He chose instead to introduce a new concept, the roots of which are probably in the Baconian "congruity of bodies." Although Bacon provided an explanation of the forms of bubbles and drops of water similar to those rejected by Galileo, the "congruity of bodies," like many other principles of Bacon's natural philosophy, could still be interpreted in corpuscularian terms.[30] Following Galileo, Hooke expressed the belief in the principles of the new philosophy but chose not to limit the theoretical options he could explore within its wider spectrum:

Now from what cause this congruity or incongruity of bodies one to another does proceed, whether from the figure of their constituent particles, or interspersed pores, or from the

---

[25] Id. (1661), 7–8, 25–6.

[26] cf. Fracastoro (1584), 58–65.

[27] Galilei (1890–1909), vol. VII, 115–6; Id. (1974), 74–5.

[28] Descartes (1964–74), vol. VIII.1, 211–1; Id. (1983), 189.

[29] Gassendi (1658), vol. I, 450.

[30] Bacon (1857–74), vol. II, 346–7, 439; see also Manzo (2006), 164, 168–73.

differing motions of the parts of the one and the other, as whether circular, undulating, progressive, etc. Whether I say from one, or more, or none of these enumerated causes, I shall not here determine, it being an enquiry more proper to be followed and explained among the general principles of philosophy.

How these latter could be declined in order to account for the new property was not yet clear. The fundamental question on the cause of congruity remained unanswered. "It [is] likewise sufficient," Hooke concluded in 1661, "for this inquiry to shew that there is such a property, from what cause what soever it proceeds."[31]

A few years later, in 1665, Hooke clarified that to "better finde what the cause of congruity and incongruity is, it will be requisite to consider, first, what is the cause of fluidness." And this latter is the effect of "a vehement agitation" of the constituent parts of a body. To explain his ideas, Hooke gave the example of what happens to sand on a moving surface. The movement transforms the solid grains into a fluid aggregate. Like grains of sand, the particles of bodies tend to cohere when their dimensions, structures, and motions correspond. "Particles that are similar," Hooke added, "will, like many equal musical strings, equally stretcht, vibrate together in a kind of harmony or unison." According to their different degrees of motion, particles form fluid or solid bodies. Heat is nothing else than a "quick and violent motion" that loosens the links among similar particles. The cohesion that produces solid bodies is not a static bond, but a dynamic effect of the correspondence of the motions and forms of the constituent particles. No portion of matter in the universe is at rest. Absolute rest was banned from Hooke's image of nature.

Even particles with different physical properties could produce an aggregate, according to Hooke. Carrying further the musical analogy, Hooke noted that unison is not produced only by "two strings of the same bigness, length, and tension." Inasmuch as several combinations of different "such varieties" make harmonious sound, several combinations of "matter or substance," "figure or shape," and "body or bulk" of particulate constituents could lead to aggregates. A "disproportion" in one respect may be "counter-ballanc'd by a contrary disproportion" in another respect. Only motion cannot vary. Different combinations of the physical features of elementary particles could produce different bodies, only if they are "all agitated by the same pulse of vibrative motion."[32] This is the only permanent physical mark that in Hooke's view differentiates bodies and their qualities. But in *Micrographia* Hooke did not explain where these different motions originate from; nor did he clarify whether the constituent particles he discussed were the elementary component of bodies, or how they were formed. It seems, therefore, that he envisaged the existence of at least two levels of invisible material configurations. The first could be considered as the immediate structure of a body, wherein different particles are bonded by the same "vibrative motion." The second consists in the process of formation of these particles from an originally shapeless matter. Even if in the 1660s Hooke did not engage with the question that had arisen from the formation of the

---

[31] Hooke (1661), 9–10.

[32] Hooke (1665), 12–3, 15–6.

constituents of bodies, the presence of a more fundamental layer in Hooke's matter theory is suggested by his emphasis on motion and the refusal of atomism.

4.3 Atoms or Structures?

The discovery of the spring of air contributed to a long debate on the nature of rarefaction and condensation of bodies. This debate soon turned on the definition of body and the constituents of matter, with regard to which two main mechanical hypotheses competed. As Boyle noted in 1662, "there are three (and for ought we know but three) ways of explicating it." "Aristotle and most of his followers" maintain that "the self-same body does not onely obtain a greater space in rarefaction, and a lesser in condensation, but adequately and exactly fill it." In Boyle's account, the peripatetic hypothesis clearly rejects both mechanical alternatives, i.e., the void of atomists and Descartes' subtle matter. A rarefied body "acquires larger dimension without either leaving any vacuities betwixt its component corpuscles, or admitting between them any new or extraneous substance."[33]

Although Boyle endorsed neither the Cartesian nor the atomist hypothesis because of their theological implications, he clearly preferred a mechanical explanation over the Aristotelian one.[34] For this reason Boyle's *Defence* includes a section on rarefaction that aims to rebut the "endeavours to invalidate the hypothesis of the weight and spring of air" by the Jesuit Francis Line (Linus). Both the mechanical hypotheses, Cartesian and atomist, provide intelligible explanations of the spring as opposed to "the unintelligible hypothesis" of the "strange and imaginary funinculus" introduced by Linus. It suffices to suppose that particles of air are "very long, slender, thin and flexible laminae" with the shape of a spring and "an innate circular motion" to explain all the phenomena of pneumatics that Linus could not explain. The Aristotelian argument against void is fallacious. On the contrary, atomistic principles of void and innate motion of atoms provide a more intelligible account of pneumatics that is consistent with the new property of air discovered by Boyle. "By granting Epicurus his principles, and our supposition of the determinate motion and figure of the aerial particles, all the phaenomena of rarefaction and condensation, of light, sound, heath, etc. will naturally and necessarily follow." If one believes, like Aristotelians, that the atomist void is impossible, then the principles of the "most acute modern philosopher, Monsieur Des Cartes" provide another mechanical and more intelligible explanation than Linus's *funinculus*.[35]

Mislead by the reference to Boyle's hypothesis on the shape of aerial particles, Christiaan Huygens assumed that the atomist explanation was the mechanical

---

[33] Boyle (1999), vol. III, 42.
[34] Clericuzio (1998), 70, 75.
[35] Boyle (1999), vol. III, 83–6.

hypothesis chosen by Boyle.[36] Following Huygens' comments, historians have attributed to the section on rarefaction in the *Defence* an atomist character. Boyle clearly stated in a letter to Huygens that the author of the essay on rarefaction was Hooke, whose name did not appear because an oversight of the printer. Although Huygens, Boyle added, "mistakes the proposer of it, yet rightly apprehends both that the hypothesis is plausible enough, and that tis propos'd (as his letter speaks) as a project or a possible way of solving the phaenomena of rarefaction without having recourse to the unintelligible way of Aristotle."[37] Even in the book itself, Boyle had already distanced himself from the opinion of the "ingenious man" who contributed a defence of the mechanical explanations against Linus, reserving the publication of his "own thoughts concerning the manner of rarefaction and condensation for another treatise."[38] Neither the innate motion postulated by the atomists, nor the vortexes of the Cartesian celestial globules were acceptable hypotheses for Boyle. The choice to distance himself from the "ingenious" author of the essay on rarefaction was not made on philosophical grounds. Hooke's contribution to the *Defense* endorsed neither of the two mechanical hypotheses opposed to Linus'. In a reply to Huygens's criticism of the atomist explanation of the spring of air, Hooke added that he had not undertaken a vindication of atomist principles, "but only those being granted" he aimed to show that "all phaenomena of rarefaction may be at least as well, if not more intelligibly explicated than by those of Aristotle."[39] Hooke did not show any hesitation in recognizing that one of the hypotheses advanced in his contribution to Boyle's *Defence* was based on the Epicurean doctrine of the innate motion of atoms. Regardless of its intelligibility, the atomist hypothesis was not Hooke's point of view.[40]

None of the tenets of the Epicurean hypothesis were consistent with Hooke's opinions on matter, motion, and the nature of scientific explanations expressed since the early 1660s.[41] To respond to all the arguments that Linus provided in support of his Aristotelian views, Hooke undertook an analysis of the wheel paradox of the pseudo-Aristotelian *Quaestiones mechanicae*.[42] This confirms the polemical nature of his contribution to Boyle's work, and offers some insights on his ideas on such a fundamental question as the divisibility of matter. How could two concentric circles complete a revolution along their respective tangents in the same time?[43] Galileo and Gassendi, for instance, proposed two solutions based on two different forms of

---

[36] Huygens (1888–1950), vol. IV, 172.

[37] Boyle (2001), vol. II, 26–7.

[38] Id. (1999), vol. III, 44–5, 61.

[39] Id. (2001), vol. II, 30.

[40] Cf. Clericuzio (1998), 74–5; Gal (2002), 129.

[41] For a different view, see Gemelli (1996), 280–1; Chapman (2005), 44; Ehrlich (1992), 143; Gouk (1980), 585; Wardhaugh (2008), 423; cf. Malet (1996), 143–6.

[42] Boyle (1999), vol. III, 44, 89.

[43] Drabkin (1950), 162–3.

atomism.[44] Interested in neutralising Linus's arguments, Hooke suggested that assuming "an infinite succession of points in the space of an infinite succession of instances" the paradox could be solved "without having recourse to the hypothesis of the determinate number of indivisibles of space and time." In Hooke's account, every point of the diameter of the concentric wheels moves with different "degrees of celerity." Many objections to the "divisibility of quantity in infinitum," of which "the Schollastick Writers are full," can be easily solved, according to Hooke. But we have "very little information of the nature of infinity from our senses."[45]

In Hooke's opinion, the discrete or continuous nature of matter is not a mathematical question. Microscopes provide a decisive instrument of analysis. Significantly, *Micrographia* begins with the observation of a physical point. Even the sharpest of these appears as a confused and complex form to the microscope. "Certainly," Hooke concluded, "the quantity or extension of any body may be divisible in infinitum, though perhaps not the matter."[46] Like Saville, Dee, and Recorde, Hooke maintained that geometrical objects are abstract entities. A point is an indivisible portion of the geometrical space. "This exactnes of definition," Recorde noted, "is more meeter for onlye theorike speculacion, then for practise and outward worke." The definition suggested by Recorde for practical geometry, however, proved inconsistent with the evidence provided by the microscope. For practical purposes, Recorde suggested defining a point as a "small printe of penne, pencyle, or other instumente, whiche is not moved, nor drawen from his fyrts touché, and therefore hath no notable length nor breadthe." In a lecture on geometry read at Gresham College in 1680, Hooke claimed that "since all quantity is divisible in infinitum," there can only be mathematical points, but not physical points because the microscope would "make those points divisible even to sense." Microscopes have proved that "we know not the limits of quantity, matter, and body as to its divisibility and extension, no imagination can comprehend the maximum and minimum naturae." These notions are "compound ideas, and consist of the simple ideas of the sensible maximum and minimum with a proportion annexed." These compound ideas "suppose for a material to be made use of in reasoning," but do not correspond to any physical object.[47] In Hooke's view, the microscope showed that the subtlety of nature exceeds our eyes and imagination. The "vessels of nutrition" of vegetables "are yet so exceedingly small that the atoms which Epicurus fancy'd would go neer to prove too big to enter them, much more to constitute a fluid body in them." Far from being indivisible particles, the atoms of the Epicureans are just notions built to deal with the complexity of nature. For this reason, they cannot be maintained. In Hooke's opinion, the qualities traditionally attributed to them are nothing else than simple notions that account for the physical properties of the

[44] Palmerino (2000), 289–92, 317; Id. (2001), 385, 407–8.

[45] Boyle (1999), vol. III, 92–3.

[46] Hooke (1665), 1–2; see also Power (1664), sig. b2r.

[47] Hooke (1705), 66, 175–6; Savile (1621), 52–3, 61–3; Dee (1570), sig. A1r; Recorde (1551), sig. A1r.

constitutive structures of bodies, intermediate between a shapeless pristine matter and its portions that can be perceived by bare human senses.[48]

The extension of bodies depends on these structures. Hooke was aware that temperature affects the degree of condensation and rarefaction of bodies.[49] Warmed-up bodies usually occupy larger portions of space than cold ones. But a new experiment carried out after 1668 showed that bodies' weight is also affected when the volume of certain bodies reduces or expands, even when their quantity of matter seems to remain the same. Building on an idea of Boyle's, Hooke mixed different substances and observed variations in weight and volume. The mixture of "an equal quantity of oil of vitriol and fair water," for instance, after "a very great heath" forms a composed substance smaller in volume but heavier than the sum of the two separate liquids.[50] Copper and tin produced a very hard alloy whose weight "was found to exceed both copper and tin in their specifick gravity." By melting half an ounce of silver and half an ounce of lead, on the contrary, Hooke produced a substance whose "specifique gravity" was less than "it ought to be according to the bare composition of those two bodys."[51] These experiments, Hooke claimed, can "lead one further into the recesses of nature." The increase in volume and decrease in weight of the mixture of oil of vitriol and water was due, according to Hooke, to the "penetration of texture or dimension."[52] Two substances could mix by "juxtaposition," "penetration," or even resist either of them and "acquire a greater rarefaction of texture than they had before their union."[53] In Hooke's view, therefore, the composition of bodies could not be the effect of a simple mixture of atoms interposed by void spaces. The new evidence provided by the experiments on the "penetration of liquids" contributed to the drafting of a research agenda directed to the study of the microscopic textures of bodies.

4.4   A Map and a Theory

Hooke's hypothesis on capillarity sparked interest in the scientific community. Soon after the *Attempt* was presented to the Royal Society, a "committee for the matters of fact concerning the rising of liquors in small tubes" was formed.[54] In Lancashire, Richard Towneley and Henry Power discussed Hooke's ideas. After finding that "in his Micrographia he hath reprinted the same (soe well he seems to bee pleased with

---

[48] Hooke (1665), 114.

[49] Birch (1756–57), vol. I, 127–30; Hooke (1665), 37–41.

[50] Birch (1756–57), vol. II, 284, 295; see also Ibid., vol. I, 111; vol. II,: 79; Royal Society MS/847, f. 633; Hooke (1935), 37.

[51] Royal Society MS/847, f. 397; British Library Sloane MS 1039, f. 114r; cf. Hooke (1726), 209.

[52] Birch (1756–57), vol. II, 284; vol. III, 509; Royal Society MS/847, f. 667.

[53] Royal Society Classified Papers, vol. XX, f. 169v; British Library Sloane MS 1039, f. 114r; cf. Birch (1756–57), vol. IV, 6; Journal Book of the Royal Society, vol. VIII, 262–68.

[54] Birch (1756–57), vol. I, 18, 25.

it),", Towneley decided to undertake "a serious thorough examination," and in September, 1665 he wrote his "considerations upon Mr Hooke's Attempt."[55] Like Power's, Towneley's predilection for Descartes went hand in hand with a "strongly experimental" approach.[56] By means of "experiments and reason," he claimed to prove that Hooke's hypothesis was mistaken and the true cause of capillarity could be offered by the "Cartesian hypothesis." "The excellencie of that great man's philosophie is such," Tonweley noted, "that it is extendable to all nature's effects." Towneley accepted that there was some "incongruitie" between air and glass on one hand, and water and glass on the other, but could not accept that Hooke's new property provided a satisfactory explanation. In Tonweley's eyes, congruity offered a "refuge" to Hooke that further experiments proved apparent. Despite his intention to support a Cartesian hypothesis by means of experiments, Towneley's "reasons" contain the most effective arguments against Hooke. Townely's criticism eloquently highlights a limitation of Hooke's early matter theory. "That fluiditie consists in motion of its component parts I graunt," he wrote, "but what kind of motion is requiste to make a fluid bodie, with its origine and cause are things yett equally unknown." The comparison between fluid bodies and sand on a moving surface, on the one hand, and the analogy between musical strings and constituent particles of bodies, on the other, were inadequate and obscure for Towneley; they did not "at all illustrate the matter under consideration, which the authore seems tacitly to confesse, adding in the clause of his discourse that conguitie seems nothing but simpatie, and incongruities an antipathie of bodies, i.e. effects whose cause remaine yeet undiscovered." An answer to the questions evaded by Hooke on the formation of globular constituents of bodies and the emergence of their physical properties could only be "assigned out of Descartes his principles."[57] The experiments in pneumatics that Towneley carried out when he composed his *Considerations* were directed to this aim.

In *Micrographia*, Hooke rather focused on the role of the new principle of congruity in the formation of complex bodies in a mechanical *scala naturae*.[58] All constitutive particles of bodies have, according to Hooke, "globular form." By congruity, any "fluid body encompast with a heterogeneous fluid must be protruded into a spherule or globe." All the "regular figures" of bodies "arise from three or four several positions or postures of globular particles." Seal salt, for instance, "is composed of a texture of globules placed in a cubical form." Like salt, all substances consist of different textures of globular.particles tied by a specific "vibrative motion." Hooke was aware that an adequate understanding of the process of formation of the elementary globules and their combination to form different substances was missing in his discussion. Rather than focusing on the emergence of forms, he engaged in a

---

[55] Cushing/Whitney Medical Library, Yale University Towneley MS, ff. 1–2; cf. Webster (1976), 445.

[56] Webster (1966), 75–6; Henry (1982), 211–2.

[57] Cushing/Whitney Medical Library, Yale University Towneley MS, ff. 1, 2, 4, 5, 13.

[58] Boas Hall (1991), 51; Aït-Touati (2011), 137, 142; Id. (2012), 82.

description of the combinations of the structures into bodies of increasing complexity. Following Boyle and Gassendi, Hooke maintained that the physico-chemical properties of all material substances were rooted in the features of these structures.[59]

This research "into the labirinth of nature" takes place by means of comparisons of a "multitude of instances." By means of early inductions, "the principle which nature has made use of in all inanimate bodies" could be reached, "so that knowing what is the form of inanimate or mineral bodies, we shall be the better able to proceed in our next enquiry after the forms of vegetative bodies, and last of all, of animate ones." The steps in Hooke's continuous and mechanical scale of nature are numerous. The "plastic virtue" of nature in the formation of forms and compounds follows geometrico-mechanical principles. Vital "seminal principles" are redundant. The presence and function of any substance in this scale, either inanimate or living, is compared by Hooke to the different components of a clock or a "compounded automaton."[60] "If to a wild Indian that had never heard of or seen [a] clock watch or wheel one should describe or show one inclosed in leather case of a bigness and shape of a small turnep," according to Hooke the Indian "would certainly conclude that there was some very cunning creatures included in that box." And if one showed to him the dial plate and dial ring, "still he would be apt to think some living thing was kept within the yet unopen part." But if "the watch be further opened" and the Indian saw the "concealed parts, he would more truely be informed of its excellent contrivances," and understand that there are not "spirits" hidden in the wheels or the barrel of the spring.[61]

Like the unaccustomed observer from the New World, the readers of *Micrographia* are shown the hidden structures of inanimate and animate bodies. These structures are "the latent scheme[s], (as the Noble Verulam calls it) or the hidden, unknown texture[s] of bodies." Following Bacon, Hooke maintained that the real natural philosophy was the effective knowledge of the forms and structures of bodies. In *Micrographia* he drew a provisional map of them, but did not provide an adequate "theory" explaining their origin and formation from a shapeless pristine state.[62]

4.5 Congruous Subtle Matters

In Hooke's non-atomist theory of matter, thin elastic fluids play a significant role. From the early 1660s he introduced some of them in hypotheses and conjectures to explain a wide range of phaenomena. Hooke's ethereal fluids differ from the subtle

---

[59] Cf. Clericuzio (1997), 233–4; Id. (2010), 341–7; Lolordo (2007), 139, 153, 157; Newmann (2006), 188–9.

[60] Hooke (1665), 86–8, 90–1, 124–5.

[61] Royal Society Classified Papers, vol. XX, f. 183v.

[62] Hooke (1665), 93, 204.

matter that Cartesian philosophers employed. Some of the most relevant differences emerged in the discussion of the so-called anomalous suspension. In an attempt to reproduce some of Boyle's pneumatic experiments, Huygens observed that water and quicksilver, once purged, did not descend in the barometer located within an evacuated receiver of the air pump until some air was introduced. "This strange phaenomenon" seemed to challenge the newly discovered property of the spring of air.[63] The "most likely" explanation consisted for Huygens in the existence "of a substance thinner than air, which easily penetrates glass, water, mercury and all the other bodies that appear impermeable to air." When air was removed from the pump, it was only the pressure of this substance that maintains purged water and mercury at their previous levels in the barometer.[64] Once the experiment was finally reproduced at the Royal Society by Hooke and Power, Boyle concluded that anomalous suspension could be explained without questioning the spring of air. This latter was a "theory which these new discoverys shew'd to be not false but insufficient." "Something else," Boyle concluded, "must be taken in to explain this odd phaenomenon."[65] But Boyle did not suggest any additional hypothesis.[66]

   Hooke was soon interested in the nature of the new fluid introduced by Huygens.[67] "Since I first made the experiment," he wrote, "I saw an absolute necessity of a pressing fluid very much more subtle than the Air, and yet consisting of a determinate bulk, which would easily strain through and pervade the pores of glass."[68] After the publication of Huygens' hypothesis in 1672, the anomalous suspension was again debated at the Royal Society.[69] John Wallis sent a paper in which he rejected Huygens' new fluid because it appeared to him nothing more than the "materia subtilis of Des Cartes," or "Mr. Hobbes's Air." For Wallis, the existence of such a fluid had still to be proved. It would be nonetheless insufficient to provide an account of the anomalous suspension, because it was supposed to be in the receiver of the pump even when there was still air. "Surely," Wallis claimed, "there must be somewhat more in it than this subtle matter to solve the phanemenon."[70] Unlike Wallis, Hooke considered Huygens' experiments as a proof of the existence of the "second element of Descartes." But the experiments also provided "many reasons" to believe that the thin fluid involved in the anomalous suspension had some properties that Descartes' second element lacked. It was, for instance, a substance less congruous to the glass than quicksilver.[71] In what he later described as a "cursory meditation for the solving of the phaenomenon" included in *Micrographia*, Hooke

---

[63] Huygens (1888–1950), vol. IV, 171–2, 432.

[64] Ibid., vol. VII, 204.

[65] Birch (1756–57), vol. I, 310.

[66] Shapin and Schaffer (1985), 250–5.

[67] Birch (1756–57), vol. I, 268; Boyle (2001), vol. II, 100.

[68] Hooke (1705), 365.

[69] Birch (1756–7), vol. III, 58–61.

[70] Wallis (1673), 5161–2.

[71] Hooke (1705), 366–7.

already made clear that a consistent explanation of the anomalous suspension was not possible without a careful consideration of the relative properties of the substances involved.[72] Unlike "several other fluid matters, some more subtle than others," the fluid postulated by Hooke did "not at all penetrate the body of quicksilver," did not have "springy nature like that of air," but "hath a pressure every way analogous to the pressure of air." Such a "pressing fluid or aether," in Hooke's opinion, kept the quicksilver above its usual standard by freely pressing on the surface of the quicksilver's vessel when the air was removed by the pump.

The existence of many thin fluids with distinct properties appeared to Hooke indispensable if one wished to account for the numerous phenomena "of which there hath as yet been no intelligible reason given of their power and original."[73] The transmission of light, magnetic and gravitational attractions, and many other phaenomena could be satisfactorily explained only by assuming that each of them was due to the action of a specific ether with specific relative properties with respect to the bodies on which each operates.[74] In Hooke's opinion, the existence of these fluids was supported by a "multitude of experiments." The air pump provided Hooke with the greatest part of the evidence he used to claim the existence of fluid ethereal substances. While sounds were not transmitted in the *vacuum boyleanum*, light was. The absence of air altered the oscillation of pendulums, but did not stop them.[75] Along with the microscope, the air pump in Hooke's hand was instrumental to the emergence of a certain view of matter, for it was intended as a source of evidence against the existence of an absolute void and of atoms.

4.6 Springs

Hooke's theory of matter emerged and evolved along with a series of experiments and observations that seemed to defy the solutions offered by continental mechanical philosophies. The so-called Prince Rupert's drops were one of these phenomena. These glasses had a peculiar drop shape, and broke into many small pieces when pressed on their thinnest part. Hooke's explanation echoed what Hobbes had suggested in *Problemata Physica*, published in 1662. According to Hobbes, when these glasses are formed, the internal motions of their constituents are directed from the "main part" towards the "small end." All particles of matter are in motion. In hard bodies, this motion is "very swift and in very small circles." The existence of these invisible motions is proved by the springs, which have an "endeavour of restitution" (*conatum restituendi*) to their former postures.[76] In *Micrographia*, Hooke claimed

---

[72] Id. (1665), 16, 31–2.

[73] Id. (1705), 366–7.

[74] Id. (1661), 26–7; Id. (1665), 21, 31, 96–7.

[75] Birch (1756–57), vol. III, 371; vol. IV, 83.

[76] Hobbes (1682), 35–6, 40–1; cf. Bibliotheca Hookiana, 48 n.191.

that "the outward parts of the glass have a great conatus to fly asunder," but are "held together by the tenacity of the part of the inward surface." Like Hobbes, Hooke maintained that no material particle in the universe was at absolute rest. Although he criticized Hobbes' *conatus* in his optics lectures, Hooke clearly employed it in the study of elasticity since the early observations of Prince Rupert's drops.

The hypothesis of an internal tension of bodies was already present in the *Attempt* of 1661. "It is evident, Hooke wrote, that there is an extraordinary and adventitious force" in all the bodies "of any other figure than globular." Hooke compared this force to that required to reduce a round spring into an oval form. The experimental determination of this force and the consequent definition of the nature of the spring were left in 1661 for "another discourse."[77] "I believe,"Towneley wrote in 1667, "Mr Hooke will find it a taske considerably difficult experimentally to calcute what degrees of force are requisite to reduce round springs into longer and longer ovalls, and perhaps impossible untill ye true cause of restitution or Springinesse shall be found out."[78] A few years later, when Hooke "shewed an experiment on the springiness of glass" at the Royal Society, Petty noted "that it was a desirable thing to have a good theory of elasticity," and in November, 1674 presented to the Society a "discourse concerning the importance and usefulness to human life of the consideration of duplicate and subduplicate proportion."[79] In the discourse, Petty presented a "new theory" of springs that was, in his opinion, "mechanically explicated" by the "subduplicate and duplicate proportions."[80] Petty's new theory was built on an atomist view of matter.[81] Like planets, atoms move around their axes and around a common centre. Because of these motions, atoms form particles and visible aggregates. The springiness of bodies is due to the innate motions of atomic constituents of bodies, operating according to the "subduplicate and duplicate proportions."[82] Petty provided a list of natural, mechanical, and social phenomena to which this rule applies as well.[83]

Published in the same year as Petty's *Discourse*, Mayow's *Tractatus quinque* included an alternative explanation of the physical cause of elasticity. Following the "very ingenious Descartes," Mayow maintained that the elastic force could not be explained by an innate motion; rather it was the effect of some invisible substance. "The illustrious Descates," Mayow added, "has already remarked that there is such a substance always in motion, whose existence cannot be doubted." The motions and pressure of this *materia subtilis* within the small pores of solid bodies are the source of their different degrees of elasticity.[84]

---

[77] Hooke (1661), 20–1.

[78] Cushing/Whitney Medical Library, Yale University Towneley MS, f. 5.

[79] Birch (1756–57), vol. III, 109, 156.

[80] Register Book of the Royal Society, vol. IV, f. 246.

[81] Kargon (1965), 64–5.

[82] Register Book of the Royal Society, vol. IV, ff. 247–67.

[83] McCormick (2009), 217.

[84] Mayow (1674), 87–9.

Hooke could hardly accept Petty's conclusions and Mayow's Cartesian hypothesis. In the early 1675, he read to the Society a discourse on what he called the "helioscope," a telescope "for observing the sun without offending the tenderest eye."[85] When he printed the discourse in 1676, Hooke added "the true nature and principle of springs" in the "universal and real character" of Wilkins. His theory, Hooke claimed, had provided "the physical and geometrical ground" for the application of springs to the balance of watches "in the year 1664."[86] The "true theory of elasticity or springiness," however, was disclosed only in a discourse read at the Royal Society in 1678 and soon after published[87]: *Ut tensio sic vis*, "the power of any spring is in the same proportion to the tension thereof." According to what is today known as Hooke's law, the power of "every springing body" to restore its "natural position" is in direct proportion to the "distance or space it is removed therefrom." In a springing body, "whether fluid or solid," whose form is altered, the frequency of the vibration of the constituent particles increases inasmuch as their space reduces and they become closer. Alteration of shape changes the structure of elastic bodies. *Ut tensio sic pondus*, Hooke added: increases in elastic power are mainly due to the compression of bodies' structure, i.e. condensation. When bodies are rarefied, on the contrary, their textures are larger than before, and the vibrations of their constituent particles are slower.[88]

In Hooke's natural philosophy, *ut tensio sic vis* meant something different than what in classical mechanics is known as Hooke's law. The concept of elastic limit, for instance, is missing in Hooke's study of elastic bodies, and the mechanical model employed by Hooke is quantitatively incompatible with the law that still bears his name.[89] Like Petty, Hooke thought that the new law was just a form of a more general principle of nature. "This is," Hooke noted, "the rule or law of nature, upon which all manner of restituent or springing motions doth proceed, whether it be of rarefaction or extension, or condensation and compression." The variation of volume of air, for instance, could be explained by the same law. "If therefore a quantity of this body be inclosed by a solid body, and that be so contrived as to compress it into less room, the motion thereof (supposing the heat the same) will continue the same, and consequently the vibrations and occursions will be increased in reciprocal proportion." In Hooke's mechanical model, the pressure of a gas consists in the frequency of the vibrations of its particles. Like the compression of the length of a common spring, the diminution of the volume of a gas produces an increase in the vibrations of the particles. The elastic force, in other words, is in inverse proportion to the extension of the spring, and, vice versa, the force employed to reduce the volume of a gas is in direct proportion to its pressure. "This explanation," Hooke

---

[85] Birch (1756–57), vol. III, 179.

[86] Hooke (1676), 29–31.

[87] Birch (1756–57), vol. III, 429–30.

[88] Hooke (1678), 1, 4–5, 13–4.

[89] Williams (1956), 77–9; Truesdell (1960), 53–8; Moyer (1977), 271; Gal (2002), 92; Gal and Chen Morris (2012), 459.

added, "will serve mutatis mutandis for explaining the spring of any other body whatsoever."[90] The spring of air was one of the many phaenomena to which his new law applied, because the "ela[s]ter" of air is in "reciprocal proportion" to its "extension," as Boyle's "hypothesis" established.[91]

The degrees of tension of a spring are infinite, since they correspond to the infinite portions of space it can occupy. To every degree of "flexure" corresponds a degree of elastic force. Since the former are infinite, the latter are also infinite. The aggregate of these infinite degrees of elastic force is, according to Hooke, "in duplicate proportion to the space bended or degree of flexure." Thus "a spring bent two spaces in its return receiveth four degrees of that tension from which it returned."[92] It is significant that this "arithmetical proportion" coincides with Hooke's "general rule of mechanics," according to which "the proportion of the strength or power of moving any body is always in duplicate proportion of the velocity it received from it." This principle explains the motion of "springs and other vibrating bodies" and "of all other mechanical and local motion." Hooke, in fact, was ready to abstract from the "impediment of air of other fluid mediums" through which bodies move, but not from their elastic nature.[93] Hooke's law applies to all "impressed and received motions" of bodies, because these latter are not the inelastic aggregates of matter of geometrical physics, but real physical bodies whose texture is composed by congruous vibrating particles.[94] With no knowledge of the "quantity and quality of matter," along with the rule of mechanics, there cannot be understanding of the motion of bodies.[95] The "reflection of motion" in a collision depends for Hooke "upon the springiness of bodies, so that where there is no spring, there can be no reflection."[96] Of these concrete elastic bodies and their motion we can have direct experience. This makes the rule of mechanics and the principles of mechanical philosophy "easily to be understood" and "most obvious and clear to sense."[97]

4.7 Matter and Motion

In the *Attempt* of 1661, Hooke introduced two new principles "called by the names of congruity and incongruity" and "promised a further explanation" of them.[98] As we have seen in this chapter, Hooke's ideas on this topic were shaped by some

---

[90] Hooke (1678), 334, 347–8; cf. Id. (1726), 11.

[91] Id. (1665), 227; cf. Birch (1756–57), vol. III, 384–5; See also Sacco (2019).

[92] Hooke (1678), 349.

[93] Hooke (1677), 31–2. cf. Bertoloni Meli (2006), 234; Westfall (1971), 206–9.

[94] Hooke (1678), 355.

[95] Id. (1677), 34.

[96] Birch (1756–57), vol. II, 328, 333.

[97] Hooke (1677), 32–3.

[98] Id. (1678), 338.

puzzling experimental questions. In 1678 this intellectual process led to a consistent matter theory that Alexandre Koyré described as kinetic.[99] "By congruity and incongruity," Hooke wrote, "I understand nothing else but an agreement or disagreement of bodys as to their magnitude and motions." This definition does not differ from the explanation offered in *Micrographia*. In the *Lectures de Potentia Restitutiva*, however, Hooke undertook a discussion of the role of congruity in the formation of bodies and a definition of body, matter, and motion. Hooke seemed to distinguish between a macroscopic congruity operating among perceivable bodies, on the one hand, and a microscopic congruity among invisible particles, on the other. This latter was decisive in the formation of invisible textures and schematisms. All the intermediate aggregates of matter between perceptible bodies and shapeless pristine matter were due to the congruity between particles or groups of them. Particles "of a like nature, when not separate by others of a differing nature will remain together, and strengthen the common vibration of them all against the differing vibrations of the ambient bodies." But at each level the physical property of these different portions of matter change. "Two or more of these particles," Hooke explained, "joined immediately together and coalescing into one become of another nature, and receptive of another degree of motion and vibration." Bodies are formed of aggregates "kept together by the differing or dissonant vibrations of the ambient bodies or fluid." Inasmuch as it depends on the congruity of similar particles or aggregates, cohesion of bodies is also due to the incongruity between these aggregates and a "subtle matter that incompasseth and pervades all other bodies." This "common menstruum" is interspersed among the particles of fluid bodies, whose "vibrative motions" it transmits. On the contrary, the texture of solid bodies is such that it does not allow any significant quantity of this subtle fluid to enter. The internal motions that form the textures of solid bodies are transmitted by direct contact among the constituent particles. These motions are usually incongruous to that of the "heterogeneous fluid menstruum." Variations in temperature can alter the texture of solid bodies by increasing the quantity of ether among their particles, whose "vibrative motions" are confused and weakened. This menstruum, therefore, plays a decisive role in the formation of bodies and in the definition and permanence of their physical state. "All bodies whatsoever," Hooke noted, "would be fluid were it not for the external heterogeneous motion of the ambient." It is because of "some prevailing heterogenous motions" of this menstruum that fluid bodies are not "unbounded and have their parts fly from each other."[100]

In the *Lectures de Potentia Restitutiva* Hooke expressed also some innovative ideas on the definition of extension and the nature of matter. Since Gassendi distinguished two opposite positions within the mechanical worldview, the topic became more impelling because of some questions raised by advances in the physical sciences.[101] Descartes identified extension with the three-dimensional space occupied

[99] Koyré (1968), 223 n.2.

[100] Hooke (1678), 341–4.

[101] Gassendi (1658), vol. I, 258.

by particles at relative rest. To many natural philosophers this solution did not seem adequate to explain the communication of motion. To provide a philosophical foundation for mechanics, the otherwise Cartesian Huygens, for instance, in 1690 embraced some atomistic principles.[102] "I do not agree with Mr. Des Cartes," Huygens wrote in the *Discour de la cause de la pesanteur*, "who claims that only extension defines the nature of bodies, and I add also the perfect hardness that makes them impenetrable."[103] By contrast, a different question than the communication of motion stopped Boyle from questing for an alternative definition of extension to the one offered by Descartes.[104] Boyle maintained that bodies are formed by a "catholick or universal matter" which is a "substance extended, divisibile and impenetrable." The "bare rest" and "intricate textures" of particles and aggregates of this "catholick" matter produce solid and impenetrable bodies. Rather than a Cartesian point of view, Boyle was defending a corpuscular philosophy consistent with the principles of Christian religion. The most relevant mechanical alternative to this view, in his eyes, was atomism. But the "origine of concretions assign'd by Epicurus" stood in contradiction to the Mosaic narration, and as such was unacceptable to Boyle.[105]

Hooke's anti-atomism, on the contrary, was rooted in his image of nature. Since matter appears infinitely divisible, there cannot be impenetrable atoms with immutable forms, dimensions, and motions. He maintained that "the whole of reality that any ways affect our senses" is body and motion. Body is "nothing else but reality that has extension every way, positive, and immutable, not as figure, but as to quantity." Influenced by Bacon, Hooke believed that "there is no necessity to suppose atoms" perfectly solid with immutable dimensions and forms. "The essence of body," Hooke stated, "is only determinate extension or a power of being unalterably of such a quantity, and not a power of being unalterably of such a quantity and a determinate figure, which the atomists suppose."[106] Like Huygens, he believed that the communication of motion could not be explained by Descartes' principles. For Hooke, the cause of the emergence of particles, aggregates, and bodies was to be found in "vibrative motion" rather than relative rest. And this motion was the physical cause of extension. Bodies preserve their shapes according to the power of the "vibrative motion" that form their textures. Extension is the "sphere of activity" of this motion. Not an innate motion, though. "This vibrative motion," Hooke noted, "I do not suppose inherent or inseparable from the particles of body, but communicated by impulses given from other bodies in the universe." Matter is not imbued with motion, but with the property to receive it according to its magnitude. "Every particle of matter according to its determinate or present magnitude is receptive of this or that peculiar motion and no other, so magnitude and receptivity of motion

---

[102] Clarke (1989), 88; Mormino (1993), 33, 70; Id. (1994), 828–9, 831–2, 847–8.

[103] Huygens (1888–1950), vol. XXI, 473.

[104] Clericuzio (1997), 228.

[105] Boyle (1999), vol. II, 151,165, 230; vol. V, 305, 316.

[106] Hooke (1705), 171–2; Bacon (2004), 212–5.

seems the same thing."[107] Receptivity of motion seems a sort of microscopic congruity. This property depends on the quantity of matter. But since the same quantity of matter may receive "any assignable quantity" of motion, the selective reception of "vibrative" motions emerges only when any portion of matter coalesces into particles and assumes determinate shapes. Nevertheless, only "abstractely considered" matter could be separated from motion and thought of as inert and shapeless.[108] In reality, matter and motion are intermingled and indistinguishable. "These two," Hooke noted, "do always counterbalance each other in all the effects, appearances, and operations of nature, and therefore it is not impossible that they may be one and the same." The motions of bodies are in "reciprocal proportion" to their dimensions, so "a little body with great motion is equivalent to a great body with little motions as to all its sensible effects in nature."[109]

Such a matter-motion continuum had a beginning in the divine creation. "It may be possible still demanded," Hooke wrote in 1682, "what is matter and what is motion? To which I can only answer that they are what they are, powers created by the omnipotent to be what they are, and to operate as they do." As "immediate products of the omnipotent creator," matter and motion are "immutable" and cannot be altered by any physical process following the Creation. They rather "act in a regular and uniform geometrical or mechanical method" which may be discovered and "reduced under certain rules and geometrically demonstrated." These rules are the universal and immutable "laws of nature" established by God.[110] A new experimental and mechanical philosophy is the only means to gain access to these principles and understand the effects they have produced since their creation. The matter-motion continuum drawn by Hooke was, therefore, linked to his dynamic cosmology. These two mutually stand and provide, as will be apparent in the following chapters, a foundation for the entire construction of Hooke's natural philosophy.

References

Aït-Touati, Frédérique. 2011. *Fictions of the cosmos: Science and literature in the seventeenth century*. Chicago/London: Chicago University Press.

———. 2012. "Give me a telescope and I shall move the earth": Hooke's attempt to prove the motion of the earth from observations. *History of Science* 50: 75–91.

Bacon, Francis. 1857–74. *Works*, 7 vols., ed. Robert L. Ellis, James Spedding, Douglas D. Heath. London: Longman.

———. 2004. The Instauratio magna. In *Part II: Novum organum*, ed. Graham Rees with Maria Wakely. Oxford: Oxford University Press.

Bertoloni Meli, Domenico. 2006. *Thinking with objects: The transformation of mechanics in the seventeenth century*. Baltimore: Johns Hopkins University Press.

---

[107] Id. (1678), 341–4.

[108] Id. (1705), 172–3.

[109] Id. (1678), 339–42.

[110] Id. (1705), 172–3.

Birch, Thomas. 1756–57. *The history of the Royal Society of London*, 4 vols. London.

Boas, Marie. 1952. The establishment of the mechanical philosophy. *Osiris* 10: 412–541.

Boas Hall, Marie. 1991. *Promoting experimental learning: Experiment and the Royal Society 1660–1727*. Cambridge: Cambridge University Press.

Boyle, Robert. 1999. *The works of Robert Boyle*, 14 vols., ed. Michael Hunter and Edward Davies. London: Pickering and Chatto.

———. 2001. *The correspondence of Robert Boyle*, 6 vols., ed. Michael Hunter, Antonio Clericuzio and Laurence Principe. London: Pickering and Chatto.

Centore, Floyd. 1970. *Robert Hooke's contributions to mechanics*. The Hague: Martinuus Nijhoff.

Chapman, Alan. 2005. *England's Leonardo: Robert Hooke and the seventeenth-century scientific revolution*. Bristol/Philadelphia: Institute of Physics Publishing.

Clarke, Desmond. 1989. *Occult powers and hypotheses: Cartesian natural philosophy under Louis XIV*. Oxford: Clarendon.

Clericuzio, Antonio. 1993. From Van Helmont to Boyle. *The British Journal for the History of Science* 26: 303–334.

———. 1997. L'atomisme de Gassendi et la philosophie corpuscolaire de Boyle. In *Gassendi et l'Europe*, ed. Sylvia Murr, 226–235. Paris: Vrin.

———. 1998. The mechanical philosophy and the spring of air: New light on Robert Boyle and Robert Hooke. *Nuncius* 13: 69–74.

———. 2000. *Elements, principles and corpuscles: A study of atomism and chemistry in the seventeenth century*. Dordrecht: Kluwer.

———. 2010. "Sooty empiricks" and natural philosophers: The status of chemistry in the seventeenth century. *Science in Context* 23: 329–350.

Clucas, Stephen. 1997. The infinite variety of forms and magnitudes: 16th and 17th-century English corpuscular philosophy and Aristotelian theories of matter and form. *Early Science and Medicine* 11: 251–271.

Dee, John. 1570. Mathematicall preface. In *The elements of geometrie of the most ancient philosopher Euclide of Megara*, Trans. Henry Billingsley. London.

Descartes, René. 1964–74. *Oeuvres*, 12 vol., ed. Charles Adam and Paul Tannery. Paris: Vrin.

———. 1983. *Principles of philosophy*, Trans. Valentine Rodger Miller and Rees P. Miller. Dordrecht: Reidel.

Drabkin, Israel. 1950. Aristotle's wheel: Notes on the history of a paradox. *Osiris* 9: 162–198.

Ehrlich, Mark. 1992. *Interpreting the scientific revolution: Robert Hooke on mechanism and activity*. PhD dissertation ., University of Wisconsin. University Microfilms Internationals, Ann Arbor, MI.

———. 1995. Mechanism and activity in the scientific revolution: The case of Robert Hooke. *Annals of Science* 52: 127–151.

Emerton, Norma. 1984. *The scientific reinterpretation of form*. Ithaca/London: Cornell University Press.

Fracastoro, Girolamo. 1584. *Opera omnia*. Venice.

Gabbey, Alan. 1993. Between ars and philosophia naturalis: Reflections on the historiography of early modern mechanics. In *Renaissance and revolution: Humanists, scholars, craftsmen and natural philosophers in early modern Europe*, ed. J.V. Field and Frank A. James, 133–145. Cambridge: Cambridge University Press.

Gal, Ofer. 2002. *Meanest foundations and nobler superstructures: Hooke, Newton and the "compounding of the celestial motions of the planetts"*. Dordrecht: Kluwer.

Gal, Ofer, and Raz Chen-Morris. 2012. Nature's drawing: Problems and resolutions in the mathematization of motion. *Synthese* 185: 429–466.

Galilei, Galileo. 1890–1909. *Opere*, 20 vols., ed. Antonio Favaro. Florence: Barbera.

———. 1974. *Two new sciences, including Centres of gravity and Force of percussion*, Trans. Stillman Drake. Madison/London: University of Wisconsin Press.

Gassendi, Pierre. 1658. *Opera omnia*, 6 vols. Leiden.

Gemelli, Benedino. 1996. *Aspetti dell'atomismo classico nella filosofia di Francis Bacon e nel Seicento*. Florence: Olschki.

Giglioni, Guido. 2011. *Francesco Bacone*. Rome: Carocci.

———. 2013a. Francis Bacon. In *The Oxford handbook of British philosophy in the seventeenth century*, ed. Peter Anstey, 41–72. Oxford: Oxford University Press.

———. 2013b. How Bacon became Baconian. In *The mechanization of natural philosophy*, ed. Daniel Garber and Sophie Roux, 27–54. Dordrecht: Springer.

Giudice, Franco. 2006. Newton e la tradizione dei principi attivi nella filosofia naturale inglese del XVII secolo. In *Scienza e teologia fra Seicento e Ottocento*, ed. Chiara Giuntini and Brunello Lotti, 39–55. Florence: Olschki.

Gouk, Penelope. 1980. The role of acoustic and music theory in the scientific work of Robert Hooke. *Annals of Science* 37: 573–605.

———. 1999. *Music, science and natural magic in seventeenth-century England*. New Haven/London: Yale University Press.

Henry, John. 1982. Atomism and eschatology: Catholicism and natural philosophy in the interregnum. *British Journal for the History of Science* 15: 211–239.

———. 1986. Occult qualities and the experimental philosophy: Active principles in pre-Newtonian matter theory. *History of Science* 24: 335–381.

———. 1989. Robert Hooke, the incongruous mechanist. In *Robert Hooke: New studies*, ed. Michael Hunter and Simon Schaffer, 149–180. Woodbridge: Boydell Press.

———. 2013. The reception of Cartesianism. In *The Oxford handbook of British philosophy in the seventeenth century*, ed. Peter Anstey, 116–143. Oxford: Oxford University Press.

Hesse, Mary. 1966. Hooke's vibration theory and the isochrony of springs. *Isis* 57: 433–441.

Hobbes, Thomas. 1682. *Seven philosophical problems and two propositions of geometry*. London.

Hooke, Robert. 1661. *An attempt for the explication of the phaenomena*. London.

———. 1665. *Micrographia*. London.

———. 1676. *A description of helioscopes*. London.

———. 1677. *Lampas*. London.

———. 1678. *Lectures de potentia restitutiva*. London.

———. 1705. *Posthumous works*, ed. Richard Waller. London.

———. 1726. Philosophical experiments and *observations*. ed. William Derham. London.

———. 1935. In *The diary of Robert Hooke 1672–1680*, ed. Henry W. Robinson and Walter Adams. London: Taylor and Francis.

Huygens, Christiaan. 1888–1950. *Oeuvres completes*, 22 vols., ed. Société hollandaise des sciences, The Hague: Martinus Nijhoff.

Kargon, Robert. 1965. William Petty's mechanical philosophy. *Isis* 56: 63–66.

Kassler, Jamie, and David Oldroyd. 1983. Robert Hooke's Trinity College 'Music Script', his music theory and the role of music in his cosmology. *Annals of Science* 40: 559–595.

Koyré, Alexandre. 1968. *Newtonian studies*. Chicago: University of Chicago Press.

Lolordo, Antonia. 2007. *Pierre Gassendi and the birth of early modern philosophy*. Cambridge: Cambridge University Press.

Lynch, Michael. 2005. A society of Baconians? The collective development of Bacon's method in the Royal Society of London. In *Francis Bacon and the refiguring of early modern thought: Essays to commemorate the advancement of learning (1605–2005)*, ed. Julie Robin Salomon and Chaterine Gimelli Martin, 173–202. Aldershot: Ashgate.

Machamer, Peter, J.E. McGuire, and Hylarie Kochiras. 2012. Newton and the mechanical philosophy: Gravitation as the balance of the heavens. *The Southern Journal of Philosophy* 50: 370–388.

Malet, Antoni. 1996. *From indivisibles to infinitesimals: Studies on seventeenth-century mathematization of infinitely small quantities*. Bellaterra: Universitat Autònoma de Barcelona Servei de Publicacions.

Manzo, Silvia. 2006. *Entre el atomismo y la alquimia: la teoria de la materia de Francis Bacon*. Buenos Aires: Editorial Biblos.

Massignat, Corinne. 2000. Gassendi et l'élasticité de l'air: une etape entre Pascal et la loi de Boyle-Mariotte. *Revue d'histoire des sciences* 53: 179–204.

Mayow, John. 1674. *Tractatus quinque medico-physici*. Oxford.

McCormick, Ted. 2009. *William Petty and the ambitions of political arithmetic*. Oxford: Oxford University Press.

McLauglhlin, Peter. 2006. Mechanical philosophy and artefact explanation. *Studies in History and Philosophy of Science* 37: 97–101.

Mormino, Gianfranco. 1993. *Penetralia motus: la fondazione relativistica della meccanica in Christiaan Huygens*. Florence: La Nuova Italia.

———. 1994. Atomismo e meccanicismo nel pensiero di Christiaan Huygens. *Rivista di Storia della Filosofia* 4: 829–863.

Moyer, Albert. 1977. Robert Hooke's ambiguous presentation of Hooke's Law. *Isis* 68: 266–275.

Newman, William. 2006. *Atoms and alchemy: Chemistry and the experimental origins of the scientific revolution*. Chicago/ London: University of Chicago Press.

Newman, William, and Lawrence Principe. 1998. Alchemy vs chemistry: The ethymological origins of a historiographical mistake. *Early Science and Medicine* 3: 32–65.

Palmerino, Carla Rita. 2000. Una nuova scienza della materia per la Scienza Nova del moto. La discussione dei paradossi dell'infinito nella prima giornata dei Discorsi galileiani. In *Atomismo e continuo nel XVII secolo*, ed. Egidio Festa and Romano Gatto, 275–319. Naples: Vivarium.

———. 2001. Galileo's and Gassendi's solutions to the Rota Aristotelis paradox: A bridge between matter and motion theories. In *Late medieval and early modern corpuscular theories of matter*, ed. Christoph Lüthy, John Murdoch, and William Newman, 381–422. Leiden: Brill.

Power, Henry. 1664. *Experimental philosophy*. London.

Pumfrey, Stephen. 1987. Mechanizing magnetism in restoration England: The decline of magnetic philosophy. *Annals of Science* 44: 1–22.

Pyle, Andrew. 1995. *Atomism and its critics: Problem areas associated with the development of atomic theory of matter from Democritus to Newton*. Bristol: Thoemmes.

Recorde, Robert. 1551. *The pathway to knowledge*. London.

Rees, Graham. 1977. The fate of Bacon's cosmology in the seventeenth century. *Ambix* 24: 27–38.

———. 1980. Atomism and 'subtelty' in Francis Bacon's natural philosophy. *Annals of Science* 37: 549–571.

Rogers, Graham Allen. 1997. Charleton, Gassendi et la réception de l'atomisme épicurien en Angleterre. In *Gassendi et l'Europe*, ed. Sylvia Murr, 213–225. Paris: Vrin.

Sacco, Francesco G. 2019. Bacon and the virtuosi: Experimental contingency and mechanical laws in the early Royal Society. In *Contingency and natural order in early modern science*, ed. Pietro Daniel Omodeo and Rodolfo Garau, 219–237 Chem: Springer.

Savile, Henry. 1621. *Praelectiones tresdecim in principium elementorum Euclidis*. Oxford.

Schaffer, Simon. 1987. Godly men and mechanical philosophers: Souls and spirits in restoration natural philosophy. *Science in Context* 1: 55–85.

Shapin, Steven, and Simon Schaffer. 1985. *Leviathan and the air pump: Hobbes, Boyle, and the experimental life*. Princeton: Princeton University Press.

Truesdell, Clifford. 1960. *The rational mechanics of flexible or elastic bodies 1638–1788*. Zurich.

Wallis, John. 1673. An extract of letters … concerning the suspension of quicksilver well purged of air. *Philosophical Transactions* 7: 5160–5170.

Ward, Seth. 1654. *Vindiciae academiarum*. Oxford.

Wardhaugh, Benjamin. 2008. Formal causes and mechanical causes: The analogy of the musical instrument in late seventeenth-century natural philosophy. In *Philosophies of technology: Francis Bacon and his contemporaries*, ed. Claus Zittel, Gisela Engel, Romano Nanni, and Nicole Karafyllis, vol. II, 411–428. Leiden: Brill.

Webster, Charles. 1976. Richard Towneley (1629–1707). In *Dictionary of scientific biography*, ed. Charles Gillispie, vol. XIII, 444–445. New York: Scribner.

Westfall, Richard. 1971. *Force in Newton's physics: The science of dynamics in the seventeenth century*. New York: American Elsevier.

Williams, E. 1956. Hooke's law and the concept of elastic limit. *Annals of Science* 12: 74–83.

Willis, Thomas. 1659. *Diatribae duae medico-philosophicae*. The Hague.

Wren, Stephen. 1750. *Parentalia or memories of the family of the Wrens*. London.

Chapter 5
Vibrant Matters

5.1 From Oxygen to Aerial Nitre

What may not be expected, Hooke asks in the preface to *Micrographia*, from human faculties assisted by a new set of ministrations? The mechanical inventions and philosophical discoveries listed in Hooke's answer include, among others, an instrument to find longitude at sea and a "way of flying in the air." "'Tis not unlikely, also," Hooke adds, "but that chymists, if they followed this method, they might found out their so much sought for Alkahest."[1] Van Helmont first described this *liquor alkhaest* as a universal solvent of metallic nature. But generations of chemical adepts never succeeded in finding it. This originated scepticism towards Helmont's largest ideas on the nature of matter.[2] Hooke's criticism was not limited to Helmont's matter theory. "Van Helmont and the rest of the chymists" employed a sterile method of inquiry.[3] An experimental scrutiny of the properties of bodies could lead, according to Hooke, to a different notion of "alkahest." The "universal menstruum" is a substance "which dissolves all sorts of sulphureous bodies." The alkahest "has not been before taken as such." Like Descartes in the *Essays*, in *Micrographia* Hooke presents his discovery as a proof of the efficacy of his newly reformed methodology.[4] A coherent narrative of the discovery, along with its numerous epistemological consequences, emerges *prima facie* from Hooke's published papers.

Observed through a microscope, the sparks produced by the striking of flint against steel appear as "long thin chip[s] of iron." Depending on the different degrees of heat produced during the striking, some of the sparks are vitrified, others

---

[1] Hooke (1665), sig. d1r-v.

[2] Debus (1977), 326–7; Ducheyne (2008), 127, 132–5.

[3] Hooke (1705), 9.

[4] Descartes (2001), 5; Descartes (1964–74),vol. VI, 4.

© Springer Nature Switzerland AG 2020
F. G. Sacco, *Real, Mechanical, Experimental*, International Archives
of the History of Ideas Archives internationales d'histoire des idées 231,
https://doi.org/10.1007/978-3-030-44451-8_5

just melted into small balls. This seems consistent with the experience of metallurgists and chemical practitioners. Iron is turned into steel after being heated, vitrified, and suddenly cooled. Apart from gold and silver, all the metals vitrify when subjected to fire, "that is," Hooke adds, they "are corroded by a saline substance which elsewhere I shew to be the true cause of fire." These are also "thereby, as by several other menstruums, converted into scoria." This is what "chimists" call calcination. "Most kind of vitrifications and calcinations" are due to the action of certain salts on the particles of metals. All the sparks produced by the striking of flint and steel "are very apt, as it were, to take fire, and are presently red hot." In Hooke's view, this offers a remarkable instance in favour of his hypothesis of combustion, and against the existence of igneous atoms. For Hooke those metals contain a "combustible sulphuresous body." When this is freed by the striking, it is dissolved by the "aereal incompassing menstruum." This simple chemical interaction ends the interminable quest for the atoms of fire supposedly hidden in the pores of bodies. "Nor need we trouble ourselves to examine," Hooke concludes, "by what Prometheus the element of fire comes to be fetched down from above the regions of air, in what cells or boxes it is kept, and what Epimetheus lets it go."[5]

Combustion and calcination involve a sort of saline dissolvent of sulphuric particles, but what is this menstruum? As Boyle and Hooke observed in some experiments with the air pump, both combustion and flame suddenly stop in the *vacuum boyleanum*. But gunpowder and other combustible substances burn in it.[6] Saltpetre was traditionally employed in the production of gunpowder.[7] Boyle distinguished two main components of saltpetre, one volatile and another fixed. In contact with sulphur, this latter produces combustion and calcination.[8] This happens also within the void of the air pump, as Boyle and Hooke observed in March 1665.[9]

Building on Boyle's experiments, Hooke concluded that combustion consists in the dissolution of burning bodies by a saline substance "inherent and mixed with the air, that is like, if not the very same, with that which is fixt in saltpeter." The combustible bodies consist of a component dissolved by the saline menstruum and another "which remains behind in a white body call'd ashes, which contains a substance, or salt, which the chymists call alkali." These chemical properties of bodies are not primary characters of different kinds of matter, but depend on the structure of their elementary components. The difference between volatility and fixedness, for instance, "consists only in this, that one is of a texture, or has component parts that will be easily rarify'd into the form of air, and the other, that it has such as will not, without much ado, be brought to such a constitution."

Like combustion, respiration is impossible in the absence of air. The experiments with the air pump showed that the increase in rarefaction within the pump makes

---

[5] Hooke (1665), 45–6, 51–2.

[6] Boyle (1999), vol. I, 184–92.

[7] Biringuccio (1540), 35v, 149v; Ercker (1951), 292.

[8] Boyle (1999), vol. II, 94–5, 102.

[9] Birch (1756–57), vol. II, 15–9.

respiration more and more difficult. In the *New Experiments*, Boyle described air as a fluid body made of small corpuscles "of a springy nature." Since the lungs do not have muscles and nerves, Boyle noted, it is the "motion of the thorax," or rather the diaphragm, that is mainly responsible for respiration. The springiness of air and the difference of pressure between the cavities of the lungs and the external atmosphere do the rest. In Boyle's view, this mechanism is consistent with the description of respiration as "ventilation (not of the heart, but) of the blood." It contrasts with the hypotheses of the "Hermetick Philosopher" Paracelsus, who maintained that air contains a "quintessence" nourishing through the blood the vital flame burning in the heart. "But though this opinion is not (as some of the same author) absurd," Boyle notes, "yet besides that, it should not be barely asserted, but explicated and prove'd." Support to the "resemblance betwixt flame and life" could be found in some pneumatic experiments, showing the flame of a lamp lasting "as little after the execution of the air, as the life of an animall." However, the cogency of the analogy between the pump and the thorax, along with the numerous anatomical objections raised against the Paracelsian hypothesis, led Boyle toward a mechanical rather than chymical explanation of respiration.[10]

When in *Micrographia* Hooke referred to the possibility to regenerate air, he probably had in mind the work of "that famous mechanician and chymist Cornelis Drebbell," credited, as Boyle noted, with the invention of a mechanism to purify air based on the Paracelsian hypothesis.[11] The analogies described by Boyle between flame and life in the *vacuum boyleanum* led Hooke to suggest that the chymical hypothesis of combustion could satisfactorily explain the nature of respiration. "In this place," Hooke writes in *Micrographia*, "I have onely time to hint an hypothesis, which, if God permit me life and opportunity, I may elsewhere prosecute, improve and publish."[12] In November, 1664, the book was already in printing. Early that month, at the Royal Society, Hooke "proposed an experiment to be made upon a dog by displaying his whole thorax, to see how long, by blowing his lungs, life might be preserved." The experiment, carried out after a few days, aimed to clarify the role of air and blood in the lungs. Hooke could observe that, despite the dog's diaphragm being pierced, the motion of the heart continued. This offered an argument against the mechanical hypothesis. Hooke, though, admitted that he "could not perceive any thing distinctly." Does the air or part of it mix with blood in the lungs or in the heart? Does the motion of the lungs depend on that of the heart, or vice versa? To answer these questions Hooke and Richard Lower carried out a new experiment of insufflation in October, 1667. But, as George Ent observed, even the new experiment "shewed what was not the use of respiration, but not what it was."[13] Thus, Hooke's chymical hypothesis of respiration was prosecuted but not improved and published, as announced in the *Micrographia*.

[10] Boyle (1999), vol. I, 170, 277–8, 282–3, 287–9.

[11] Hooke (1665), 140; Boyle (1999), vol. I, 287–8; cf. Keller (2011), 130; Harris (1961), 174–5.

[12] Hooke (1665), 105.

[13] Birch (1756–57), vol. I, 482, 485–6; vol. II, 198; cf. Boyle (2001), vol. II, 399.

   Hooke's narrative has generally passed through historical scrutiny unquestioned. Many scholars have been deceived by the similarities between Hooke's saline quintessence of the air and oxygen. Others have emphasized the originality of Hooke's chemical approach as opposed to Boyle's attachment to traditional mechanical models.[14] However consistent it seems with the methodology drawn in the preface of *Micrographia*, Hooke's narrative contrasts with the evolution of his thought in the first half of the 1660s. The opposition between Boyle's mechanic approach to respiration and Hooke's chymical hypothesis is also misleading. Boyle was sceptical towards Hooke's and Mayow's aerial nitre. Nonetheless, he changed his mind about the role of air in respiration and combustion after the *New Experiments*. Furthermore, the general association between the two should be questioned here. It is true that Hooke was associated with Boyle since 1657, when as a student at Christ Church College he became Boyle's paid assistant. But, at Oxford, Boyle was not the only figure who helped to introduce the young Hooke to the new philosophy. Since Hooke's pivotal role in building Boyle's first efficient air pump, scholars have focused almost exclusively on their relationship, neglecting for instance the influence of Wilkins, Ward and Willis.[15]

   In 1653, after attending Westminster School in London, Hooke enrolled at Christ Church College as servitor, i.e., a poor student whose duties included attending to richer students. He also served as a chorister. As a brilliant but poor student, Hooke soon entered into contact with Wilkins' scientific group based at Wadham College. In 1655 he was employed as paid assistant by Thomas Willis.[16] At this time Hooke begun to read Descartes.[17] Willis and the other virtuosi, therefore, influenced Hooke's reception of the new philosophy. A Fellow of the Christ Church and Sedleian Professor of natural philosophy, Willis was then one of the leading figures of the so-called Oxford physiologists. In *Diatribae duae* (1659), he drew an eclectic system linking helmontian chymistry and corpuscularianism.[18] In Willis' corpuscular chymistry, elementary constituents of matter have different chymical qualities, according to their degrees of motion. While mercury or "spirit" is a very subtle and ethereal substance, sulphur is rather heavy. Mercurial particles are mainly aerial and active, whereas the sulphuric ones are mainly fixed and less active. For this reason the interaction between these substances produces combustion, which Willis describes as the dissolution of the links among sulphuric constituents by a tiny, aerial and active spirit. Unlike other salts, saltpetre is a dissolvent of sulphuric bodies. According to Willis a process of fermentation takes place in the heart. Fermentation is "the internal motion of the constituent particles of a body." Building on Nathaniel Highmore's corpuscular definition of blood, Willis maintained that

---

[14]Lysaght (1937), 107–8; McKie (1953), 474; Purrington (2009), 103; Frank (1980), 127–8, 137–9, 144, cf. Chapman (2005), 29.

[15]Inwood (2002), 18; Jardine (2003), 65–73.

[16]Gunther (1968), VI, 69, Frank (1980), 129.

[17]Hooke (1665), 44.

[18]Frank (1980), 113–4, 160, cf. Emerton (1984), 110–1; Clericuzio (1993), 318, 326–7; Id. (2000), 100–1.

within the left ventricle a ferment constantly operates on the particles of the blood and thereby causes its circulation. Through his theory of fermentation, Willis therefore offered a possible unique chymical explanation for both respiration and combustion.[19] Because it appeared too similar to that of Paracelsus, Willis' theory was rejected by Boyle in the *New Experiments*. But in 1661 the anatomical observations of Marcello Malpighi offered significant support to chymical approaches to respiration. In *De pulmonibus*, the Italian anatomist announced the microscopic discovery of the lung air cells. Within these "circular vesicles" the blood's fermentation takes place, because the air, according to Malpighi, mixes the lighter and darker parts of the blood.[20] The main function of respiration, Malpighi concluded, takes place in the lungs rather than in the heart as Willis and other chymists maintained.

The chymical idea of a nitrous component of air necessary for combustion was shared also by Bacon.[21] The founding members of the Royal Society considered Bacon the *artium instaurator*, as he was presented in the frontispiece of Sprat's *History* published in 1667.[22] It is not unlikely that his position on combustion supported the chymical approach of some of the younger fellows of the Society, such as Hooke, Lower and Mayow.[23] Even Kenelm Digby expressed similar ideas in a lecture delivered at the Royal Society in January, 1660. The nitrous component of some kind of airs, Digby wrote, "is the food of the lungs and the nourishment of the spirits."[24]

Willis' corpuscular reinterpretation of the chymical philosophy and Malpighi's discovery of the lung air cells were decisive for the emergence of a new chymical approach to combustion and respiration by the younger generation of Oxford physiologists.

5.2   Cosmic Nitre

When Malpighi's decisive discovery was published, Hooke was still dealing with some questions discussed in Boyle's *New Experiments*, i.e., the nature of condensation and rarefaction, and the raising of water in tin pipes. Published in 1661, the *Attempt for the Explication of the Phaenomena Observable in an Experiment published by the Honourable Robert Boyle* contains more than the title suggests. In one of the "queries," Hooke implies that the principle of congruity can account for the dissolutions of bodies by saline substances.[25] Although he admits that "there are in

---

[19] Willis (1659), 5–8, 11, 17, 25–7; cf. Frank (1980), 98, 166–7.

[20] Malpighi (1686), vol. II, 133, 137–8, cf. Belloni (1975), 101.

[21] Bacon (1857–74), vol. II, 351.

[22] Hunter (2017), 183.

[23] Debus (1977), 494.

[24] Digby (1661), 65; Birch (1756–57), vol. I, 13, 28.

[25] Hooke (1661), 40.

some dissolutions some other causes concurrent," congruity clearly represents for Hooke the answer to the question of chemical interactions. It is against this theoretical background that we should read the chymical hypothesis of the *Micrographia*. The pneumatic experiments carried out by Hooke and Boyle in the late 1650s showed clear analogies between respiration and combustion. The "principle of congruity" offered a theoretical explanation for chymical qualities, whereas Malpighi's discovery provided the anatomical foundation for a new approach to the physiology of respiration. After 1662 Hooke disposed of all the theoretical and experimental elements necessary to formulate a corpuscular and chymical theory of air. The hypothesis on combustion published in 1664 was only a part of a more general conjecture on fluids that could explain at once many physical, physiological and cosmological questions. This general theory is entailed by Hooke's use of the proprieties of vibrant fluids to explain different phenomena related to light and air. On February 23, 1663, Hooke submitted to the Royal Society a "scheme of inquiries concerning air." The role of air in the circulation of blood and the formation of vital spirits is listed among the hypotheses worth inquiring. Does "the air," Hooke asks, "circumvest or inclose any of the planets, or other great bodies of the world? as the moon, which many suppose; or the sun, as Kepler thinks"?[26] The presence of the aerial nitre beyond the terrestrial atmosphere was decisive for Hooke's hypothesis. The light of the sun and comets was for Hooke the result of their combustion. For this reason, in 1663 he drew a general plan of inquiry largely dependent on the existence of a nitrous component in atmospheric air. Such a plan overcomes disciplinary boundaries as they are conceived nowadays. Astronomical, optical and physiological questions are all dependent on the presence of a substance whose actions are consonant with the principle of congruity introduced in 1661.

The reason why in *Micrographia* Hooke published only the hypothesis of combustion is perhaps that he could provide a coherent hypothesis only for some of the phenomena that he was examining. The analogy between respiration and combustion, in fact, was questioned by the results of some experiments carried out in 1662. In the early pneumatic experiments, the effects of the boylean void on flame and life were observed separately. On January 28, 1662, at the Royal Society, "Mr. Hooke made the experiment of shutting up in an oblong glass a burning lamp and a chick; and the lamp went out within two minutes, the chick remaining alive and lively enough." A repetition of the experiment confirmed that the effects of the lack of fresh air on flame and life were different at least in terms of time duration, since a new "experiment of respiring the same air" realized on 4 February showed that fresh air was as much needed for respiration as for combustion. The experiments seemed to confirm the insufficiency of non-mechanical explanations in Boyle's eyes, and led Hooke to limit his chymical hypothesis to combustion only.[27]

This should not suggest that Hooke's conjecture was entirely dependent on experimental confirmation, or that it arose only from experiments, as his

---

[26] Birch (1756–57), vol. I, 202–5.

[27] Ibid., 180, 192; cf. Frank (1980), 152.

methodological writing set out. His hypothesis on respiration could never explain
the different length of respiration and combustion in the Boylean void. Nonetheless
he never abandoned it. In a later meeting at the Royal Society in 1690, Hooke
affirmed that like "a candle covered with an extinguisher," the "vital flame" is annihilated by "want of a supply of fresh air."[28] Although the experiments carried out in
January, 1662, were not favourable to non-mechanical explanations of respiration,
in *Micrographia* Hooke secured part of his chymical hypothesis and hinted at its
general significance for physiology, optics and cosmology. In the *General Scheme*,
probably written soon after the *Micrographia*, Hooke asks "whether the fire in the
sun is not maintain'd by the air that incompasses it." In this view, the sun's spots can
be considered as "clouds of smoke or vapours rais'd up in that atmosphere," and the
light of comets may be ascribed to the dissolution of celestial bodies "by the encompassing air, which is somehow more condensed near them." Isn't air, Hooke asks,
composed by "a kind of volatile or small ramify'd bodies swimming in the aether,
like a tincture in water or in some such fluid bodies"? Since air is such a compound
of volatile particles, it can be produced anew and regenerate when some of its components are exhausted. Does not respiration consist in the fermentation of the blood
produced by one of these components of air, i.e. the aerial nitre, in the lungs? For
Hooke this was "somewhat more than probable."[29] However, out of his general conjecture based on aerial nitre after the *Micrographia*, he published only those parts
focusing on the terrestrial and celestial combustions, but not those dealing with
animal physiology.

In the second half of the 1660s, new support to non-mechanical hypotheses of
respiration was provided by Malpighi. In *De polypo cordis*, published in 1666,
Malipighi abandoned the explication of respiration contained in *De pulmonibus*.
The air contains a very active "balsam of life." In the air cells of the lungs this "salt
of life" originates the fermentation of the blood.[30] Malpighi's new hypothesis was
due to the influence of the telesian physician Tommaso Cornelio. In *Progymnasmata
physica*, published in 1663, Cornelio argued that a salt with "vital force" is responsible for the respiration of the blood and the production of vital spirits.[31] It is significant that Hooke owned a copy of the first edition of Cornelio's book in his library.[32]
Malpighi also provided, although indirectly, an experiment confirming the existence
of this "balsam of life" and its effects on blood. In the late 1650s he observed the
alteration of blood's colour in open air. For the mechanist Malpighi the alteration of
a secondary quality such as colour did not entail any structural variation in the substance of the blood.[33] Nor did the observation influence the new theory of respiration maintained in *De polypo cordis*, since it mainly depended on the influence of

---

[28] Journal Book of the Royal Society, vol. IX, 13.

[29] Hooke (1705), 30–3, 50.

[30] Malpighi (1686), vol. II, 130.

[31] Cornelio (1663), 102; cf. Adelmann (1966), vol. I, 196–7, 312–3.

[32] Bibliotheca Hookiana, 10 n.126.

[33] Bertoloni Meli (2011), 125–6.

Cornelio. Thus the observation was published for the first time only in 1665, when Carlo Fracassati in his *De cerebro* employed it to reject the existence of a "melancholic humor" in the blood. An extract of Fracassati's book was published in the *Philosophical Transaction* of 1666.[34]

At the Royal Society, Malpighi's new hypothesis and experiment reinforced the programme of chymical physiology embraced by Hooke and Lower. In Hooke's opinion the insufflation experiments, though not decisive, supported the hypothesis of the aerial nitre.[35] But many questions were still unsolved. Until the publication of Lower's *De corde* in 1669, the passage of the aerial nitre into blood, and the difference between venous and arterial blood were still puzzling for Hooke. According to Oldenburg, in late 1664 Hooke shared the opinion that the inhaled air mixes with the venous blood in the lungs and "passes along with it into the left ventricle of the heart, and then into the arteries."[36] But on 30 April 1667 at the Royal Society, Hooke proposed to observe the colour of the blood passing from the right ventricle to the left, "before it enters into the great artery." The presence of a colour as florid as that observed by Malpighi, "would be an argument that some mixture of air with blood in the lungs might give that floridness." In October he suggested to pass the blood of an animal "from one side to the other out of the vena arteriosa into the aorta, without passing into the lungs." But no accounts of this experiment followed. A new experiment by Hooke and Walter Pope in December, 1667, did not prove decisive in understanding respiration and circulation. The two virtuosi dissected a pregnant "mastiff-bitch with puppies to see whether foetuses live in the womb by their own or their mother's respiration." In order to avoid access to air, Hooke and Pope cut or tied the trachea of the small whelps, which soon died. According to Hooke, the results suggest "that is not the want of motion of the blood through the lungs, or the imaginary stopping of it there, that kills the foetus." The small whelps died because of "the want of ventilation of the blood, or whatever other operation respiration may be proved to work on the blood."[37] Despite the fact that he firmly believed in the existence and decisive role of the aerial nitre, Hooke had not advocate his chymical hypothesis for the physiology of respiration and circulation. He maintained that ventilation occurs, since it was not inconsistent with a non-mechanical hypothesis of respiration. Air is a medium that brings into the lungs "something essential to life" and carries out "something that was noisome to it." In the lungs the "pabulum of animal spirits" passes into the venous blood, which releases a "steam and fuliginous matter" and assumes "an inlivening florid arterial colour."[38] Many of the experiments that he realized did not allow going further than a rejection of a purely mechanical explanation. None of these experiments provided a decisive confirmation of the alternative view based on the chymical properties of the aerial nitre.

---

[34] Fracassati (1666), 492.

[35] Birch (1756-57), vol. II, 200.

[36] Boyle (2001), vol. II, 401.

[37] Birch (1756–57), vol. II, 232–3.

[38] Ibid., vol. III, 55, 405.

The cause of circulation continued to puzzle Hooke and other fellows of the Royal Society. Hooke and Lower maintained that the heart is a muscle pumping the blood, but neither of them explained the cause of that motion. Hooke suggested "that there must be a very subtle volatile spirit, which enters into the muscles, and the same must very quickly be discharged again to cause the contraction and expansion of the muscles." Using the effects of heath and cold on air, he built a mechanical model for muscular motion very similar to Papin's steam pump.[39] Building on Hooke's experiments, in *De corde* Lower described respiration as the mixing of the *spiritus aereus nitrosus* with the venous blood in the lungs. Once the inhaled air is freed of the aerial nitre, it can take out the fumes produced by this process of fermentation.[40]

Lower's book provided an anatomical foundation for Hooke's chymical hypothesis, but it did not take into account the different duration of respiration and combustion in the Boylean void. This was the subject of a new series of experiments that Hooke made in July, 1668. In February, 1671, he built a machine to test the effect of rarefaction of air on humans. He himself went into a "vessel for rarefying the air," carrying a burning candle. As rarefaction increased, Hooke felt pain in the ears till losing the hearing. But the fire of the candle "went out long before he felt any of that inconvenience in his ears."[41] The most important obstacle to his chymical hypothesis remained. Even John Mayow, the brilliant young Oxford physiologist, proved unable to overcome it. In *Tractatus duo* (1668), Mayow outlined a chymical physiology based on aerial nitre, whereas in *Tractactus quinque*, the aerial nitre became the key for almost all natural phenomena. Adopting mechanical models, Mayow linked the elasticity of air to the structural properties of the nitric particles included in it.[42] But he could not satisfactorily explain why life lasted longer than fire in the Boylean void.

5.3   The Survival of the Aerial Nitre

In Hooke's eyes, Mayow's works confirmed his "hypothesis of fire and flame."[43] Despite the similitudes between the ideas of Hooke and Mayow on physiology, the two young Oxford physiologists had different opinions on chymical qualities. Hooke's chymical physiology was based on the principle of congruity or selective interactions among the constituents of bodies. Mayow, on the contrary, postulated the high speed of aerial nitre particles to explain their dissolutive power. Far from

---

[39] Birch (1756–57), vol. II, 412, 436–7; cf. Lower (1669), 90, Birch (1756–57), vol. III, 179–80, 402–3.

[40] Lower (1669), 177–80.

[41] Birch (1756–57), vol. II, 304, 469–70, 472–3.

[42] Mayow (1674), 12–3, 21, 96–7; Cf. Frank (1980), 261; Clericuzio (2000), 150–1; Guerlac (1954), 246.

[43] Hooke (1677), 1; Royal Society Classified Papers, vol. XX, f. 169v.

being the expression of the Society's general programme to restore spirits and immaterial beings in natural philosophy, as Simon Schaffer has argued, the aerial nitre adopted by Hooke, Lower and Mayow was criticized by many fellows.[44] In the *Philosophical Transactions* of 1667, Hooke published the results of the second experiment of insufflation. In the first experiment of 1664, Hooke opened the thorax of a dog, cut the trachea and diaphragm, and used a bellow to pump air in the lungs. The animal survived for more than an hour. Since "eminent physicians had affirmed that the motion of the lungs was necessary to life upon the account of promoting the circulation of the blood," it could have been still argued that the cause of death was the cease of the motion of the lungs. Therefore Hooke carried out a new version of the experiment. Through a second bellow the lungs were constantly kept full of air, and not moving. The dog survived almost as long as in the previous experiment. Despite the absence of motion, the blood was still circulating through the lungs. Hooke concluded that the "immediate cause" of death of the dogs in both experiments was not the absence of motion in the lungs, but "the want of a sufficient supply of fresh air."[45]

In a letter to Oldenburg following the publication of the *Transactions* of 1667, the anatomist Walter Needham criticized Hooke's conclusions. According to Needham, Hooke's experiment was decisive neither against the mechanical hypothesis, nor in favour of the chymical one. He proposed to alter the experiment by avoiding any access of fresh air in the lungs. Using a bladder instead of the bellow, the lungs could be always inflated by the same air. "You shall see," Needham predicted, "ye circulation of the blood as well continu'd as if there were a constant repetition of blasts." In Needham's opinion, the chymical properties of air or of any of its components were not relevant for respiration. Air is just a mechanical fluid. Any other fluid, such as "water or the like" would produce the same mechanical effect.[46]

Needham's letter was read at the Royal Society, and Hooke, Lower, and Edmund King carried out the experiment proposed by Needham. Unlike the previous insufflation, in almost eight minutes the dog seemed "very near to death or just suffocated." The animal recovered only when the bladder was removed and fresh air was pumped through the trachea. "We all concluded," the account reads, "that if we had stayed one minute more, before we let in fresh air, in all probability the dog's life would have been quite lost."

For Hooke, Lower, and King, the result clearly justified rejecting the mechanical hypothesis in favour of the chymical one. Still some fellows raised objections. King and Hooke were ordered "to repeat this experiment and then let the dog lay two or three minutes longer, when they should judge him as deader as they did at this time,

---

[44] Schaffer (1987), 63.

[45] Hooke (1705), 539–40; cf. Birch (1756–57), vol. II, 200.

[46] Oldenburg (1965–86), vol. IV, 237–43.

that so the trial may be beyond exception." The experiment was repeated on 26 May: the dog died after only 10 minutes.[47]

For Needham and other fellows, the defence of the mechanical hypothesis of respiration was one and the same with the refusal of Hooke's aerial nitre. Boyle, on the contrary, in the early 1670s abandoned the mechanical explanation of respiration proposed in the *New Experiments*. But the experiments carried out by Hooke, Lower and King in the late 1660s were not sufficient to persuade him that the aerial nitre was the cause of the fermentation of the blood. The *New Experiments touching the relation betwixt flame and air* confirmed both the need of air for respiration and combustion, and the different length in time of these phenomena in the pneumatic void.[48] Air is not a primary element, but "a confus'd aggregate of effluviums." In 1674 Boyle admitted "that there may be dispers'd through the rest of the Atmosphere some odd substance, either of a Solar, or Astral, or some other exotic, nature, on whose account the Air is so necessary to the subsistence of Flame." The "vital substance" necessary for life and fire can "be a volatile Nitre, or (rather) some yet anonymous substance, Sydereal or Subterraneal."[49] In the *General History of Air* Boyle admitted that air contains many "saline corpuscles." In many places it "is impregnated with corpuscles of a nitrous nature." But the vital nitre, of which "divers learned men, some physicians, some chymists, and some also philosophers speak," is just an "ingenious supposition" lacking "any positive proof."[50]

Boyle's ideas were consistent with the scepticism of many fellows about the chymical hypotheses of Hooke and Mayow. After a long discussion on respiration in May 1678, the fellows of the Royal Society seemed to embrace Hooke's opinions about the passive role of the lungs. But new objections were raised.[51] Hooke himself seemed to believe that it still lacked a decisive experimental confirmation. As in *Micrographia*, in a lecture read in 1689 he described his hypotheses as "hints or heads of inquiry" for an experimental research on physiology that he never completed.[52] After the experiments with the "vessel for rarfying the air" made in 1671, Hooke privileged the astronomical aspects of his general conjecture on combustion and fermentation. In 1678 he published *Cometa*, a small book consisting mainly of some lectures delivered in 1665.[53] Building on the analogies between earth, moon and comets, Hooke concluded that the nucleus of these celestial bodies consists of the same sulphuric substance that causes terrestrial earthquakes. Referring to *Micrographia*, Hooke affirmed that an ether carrying earthly particles, such as aerial nitre, fills the celestial space. The light of comets originates from the dissolution of

---

[47] Birch (1756–57), vol. II, 274, 282–3, 292.

[48] Boyle (1999), vol. VII, 83, 103, 109–14, 117–20.

[49] Ibid., vol. VIII, 121, 129–30.

[50] Ibid., vol. XII, 32.

[51] Birch (1756–57), vol. III, 407; vol. IV, 172–3.

[52] Royal Society Classified Papers, vol. XX, f. 173v; cf. Journal Book of the Royal Society, vol. IX, f. 13.

[53] Birch (1756–57), vol. II, 19; Hooke (1678), 223, 260.

their sulphuric nucleus by the aerial nitre of their encompassing atmosphere. Hooke's inferences about comets were based on analogies and similitudes, as the structure of comets could not be directly observed. The belief that comets have a sulphuric nucleus rested also on some cosmological arguments. The common material constitution of all celestial bodies could be considered as the effects of the same formative process.[54] But Hooke was aware that the other principle on which he built his cometary theory was questionable. If the "nitrous air" does not expand beyond the terrestrial atmosphere, he asked in 1682, what does originate the dissolution of comets' sulphuric nucleus?[55] Like the experimental obstacles to his chymical physiology, this objection did not stop Hooke from maintaining his original conjecture on combustion along with its numerous applications. He clearly broke the methodological rules that he himself proposed for a reform of natural philosophy.[56] Hooke's chymical conjecture showed more heuristic power than his epistemology. It could grasp what this latter could not. It was able to reach the uppermost celestial spaces, and to fill the void of the air pump. Such a vigorous heuristic power relied on its links with the rest of Hooke's natural philosophy, especially the hypothesis on the nature of light and colours, thought of as ethereal vibrations produced by chymical processes of combustion and dissolution.

5.4   "A Body or a Motion of a Body It Seems to Be"

As Robert Boyle noted in 1664, the debate about the nature of light in the mid-seventeenth century mainly focused on the alternative between body and motion.[57] Gassendi, for instance, explained combustion as the dissolution of the structure of bodies by the "atoms of heat." Due to the high motion of these atoms, the constituents of burning bodies are reduced into ashes or mixed into soot. Light is also a product of combustion and consists of types of atoms very similar to the "atoms of heat."[58] On the contrary, Descartes, building on his second law of nature, described light as the "effort" (*conatus*) to move of the particles of the second element. This effort is a "certain pressure which occurs along straight lines" originating from the source of light and reaching the eyes in "an instant." Since its speed is indefinite or rather infinite, light cannot be a concrete motion nor "something material," but just a pressure.[59]

---

[54] Hooke (1678), 225, 227–8, 231, 248–250, 260, 263–4, 267.

[55] Hooke (1705), 164; cf. Partington (1961), vol. II, 560–1.

[56] Royal Society Classified Papers, vol. XX, f. 179r.

[57] Boyle (1999), vol. IV, 59.

[58] Gassendi (1658), vol. I, 277–8, 395, 422.

[59] Descartes (1964–74), vol. VI, 84–5, 88; vol. VIII.1, 108, 115, 217; Id. (2001), 67–8, 70; Id. (1983), 112, 117, 194.

Both atomists and Cartesians rejected the Aristotelian distinction between real and apparent colours.[60] Even though different theories of light entailed different definitions of colours, in the new corpuscular philosophy, colours are all apparent, and consist, as Boyle noted, of "modifi'd light."[61] In Hobbes' opinion colours are "troubled light" whose "perturbed motion" originated from refraction.[62] According to Descartes, the particles of the *matière subtile* that transmit white light have a tendency to move in straight lines. Refraction transforms these motions in a variety of tendencies to rotate around their axis; colours correspond to the different degrees of the new impressed circular tendencies. In Descartes' eyes, this explanation "perfectly" linked reason and experience, because the perception of colours consisted of the ethereal pressure on eyes. Thus, different kinds of pressure produced different perceptions. Applying this consistent theory to the productions of colours in the rainbow or by prisms, Descartes noted that in both phenomena light is twice refracted. The decisive refraction, though, takes place when the ray exits from these bodies, because it finally transforms the straight tendencies of the ethereal globules into circular ones. If these persist when the ray meets the eyes, they cause the sensation of colours.[63]

For the atomist Charleton, "the aspiring wit of Descartes" has brought optics "to the familiar view of our reason." But his "grand hypothesis" on the nature of light was "merely præcarious, and never to be conceded by any, who fears to ensnare himself in many inextricable difficulties, incongruities, and contradictions." Like Descartes, Charleton admits that as "sensible qualities" colours are not in the "first matter." But he affirms that the sensation of colours originates from the modification of the "atoms of light, or rayes consisting of them."[64] Even in Hooke's eyes, the concept of *conatus* was confusing, and Descartes' account needed experimental verification.[65] The colours of thin plates provided a new phenomenon to challenge Descartes' theory. In the *Experimental History of Colours*, Boyle noticed that the so-called Muscovy glass was a "remarkable instance" for the study of light. The mineral muscovite is generally opaque, but the thin plates in which it usually breaks are "the most transparent sort of consistent bodies."[66] Although Boyle described the colours produced by thin water bubbles, he did not refer to those produced by the thin plates of Muscovite. Commenting on Boyle's book, Huygens referred to the colours produced by two thin plates of glass pressed together that Boyle affirmed to have observed "several times" before the publication of his book.[67] Thanks to the microscope, in early 1664, Hooke clearly observed the alternating series of irregular

---

[60] Nakajima (1984), 261, 277–8.

[61] Boyle (1999), vol. IV, 28; cf. Baker (2015), 550–51.

[62] Hobbes (1839–45a), vol. I, 374; Id. (1839–45b), vol. I, 459.

[63] Descartes (1964–74), vol. VI, 325–30, 333; Id. (2001), 332–5, 337; cf. Buchwald (2008), 19, 42.

[64] Charleton (1654), 190–1, 197.

[65] Hooke (1665), 44.

[66] Boyle (1999), vol. IV, 51–2.

[67] Huygens (1888–1950), vol. V, 107, 558.

coloured and black rings formed when light passes through the thin films of muscovite. In Hooke's opinion, this was an "experimentum crucis" against the Cartesian
theory of light and colours, because it showed the presence of colours where,
according to his reading of Descartes, there should not be any. Through the thin
plates of muscovite the rays of light are refracted twice in two equal though opposite
angles, so that any alteration produced by the first refraction is annulled by
the second.

The phenomena of muscovite and other thin films confirmed Hooke's scepticism
towards the whole Cartesian theory of light. Descartes described the coloured ray as
the line that conveys a certain "vorticity of the globules" of the ether. According to
Hooke, Descartes' account neglected the particles contiguous to those of the line of
light. These particles can have different motions or tendencies, which are likely to
influence those of the ethereal globules forming the line of light. "Whence according to the Cartesian hypothesis, there must not be distinct colour generated, but a
confusion." In Hooke's view, the limits of Descartes' account of the formation of
colours originate from the underlying definition of light as a tendency or pressure
instantaneously transmitted. Despite the fact that light is propagated "to the greatest
imaginable distance in the least imaginable time," there is "no reason to affirm that
it must be in an instant." Hooke lamented the lack of experiments to measure the
speed of light, and suggested that the observation of eclipses "may seem very much
to favour this supposition of the slower progression of light then most imagine."[68] In
Micrographia the criticism of Cartesian pressure is linked to the assumption of the
finite speed of light. Hooke seems to believe that, if transmission is not instantaneous, light cannot be a tendency to motion, but it should rather be a concrete motion.

One would expect, therefore, that Hooke welcomed Ole Rømer's calculations of
the finite speed of light based on the eclipses of the satellite Io by Jupiter.[69] But in a
series of lectures at the Royal Society between 1680 and 1682, Hooke questioned
Rømer's estimate. According to Hooke, Rømer's representation of the motion of the
satellite could be inaccurate, because it was influenced not only by the sun and
Jupiter, but also by the other three satellites. "And therefore," Hooke concluded,
"unless we are assured of the true intermediate times between the eclipses of it, we
cannot make a certain conclusion." The propagation of light, Hooke now states, is
"infinitly swift" and "is made in a point or instant of time." But these new ideas
about the speed of light did not influence Hooke's approach to the Cartesian theory.
On the contrary, new arguments against Descartes' "propension to motion" were
provided in the lectures. The "impression" of light on the eye is "momentary," hence
"temporary," because we can distinguish a "terminus a quo" and a "terminus ad
quem." It can originate only from the contact with a "body moved in local motion."
Although its speed is instantaneous, the motion of light can be understood only by
means of "similitudes" with the "local motions" that fall under human senses, "they
having both the same proprieties and differing only in space and time."

---

[68] Hooke (1665), 48, 54, 57, 59–61.

[69] Rømer (1676), 233–6.

Hooke's anti-Cartesian arguments echo his ideas on the nature of knowledge and the use of analogies and similitudes in natural philosophy. These ideas, however, do not support the refusal of Rømer's estimate, or the uncritical assumption of the principle of the instantaneous transmission questioned in 1664. Rather than epistemological reasons, Hooke's changing attitudes on this question seem due to the defence of his theory of light after the polemics with Newton in the 1670s. In 1680 Hooke judged Rømer's estimate of light's speed as "beyond imagination. "If so," he notes, "why it may not be as well instantaneous I know no reason, unless it may be said 'tis inconceivable any body can be infinitely fluid, which yet how it can be denied, I know not, unless we will allow a vacuity." *Micrographia*'s crucial experiment led Hooke to reject the two main principles of Cartesian theory, viz. the principle of instantaneous transmission and the *conatus*. This opened the way to the definition of light as a concrete motion of the ether. The emergence of the Newtonian theory of light in 1672, which Hooke opposed as atomistic, pressed Hooke's defence of the ether and its role in the transmission of light. It was perhaps also for this reason that he never accepted Rømer's proof. "This being what is light," Hooke wrote in 1680, "sure if anything, it may be call'd the Anima Mundi." Since it moves in an instant, "it may be said tota in toto and tota in qualibet parte, possibly with some kind of plausibleness." However, the instantaneous transmission did not exclude light from the realm of mechanical intelligibility, because the imperceptible ether conveying light was supposed to be subject to the universal laws of motion. Indeed, it was motion, and its study could be reduced to "calculation and mathematical exactness."[70]

Hooke's *experimentum crucis* consisted in an indubitable "negative instance" about what light was not, viz. a pressure or tendency of the Cartesian subtle matter. To "see likewise what affirmative and positive instructions" it provided, Hooke formulated an alternative explanation of the colours of muscovite. Since he maintained the principle of homogeneity of white light, his explanation was based on the idea that colours were caused by the refraction of light. Hooke stressed the irregular thinness of muscovite. This caused the combined effects of refractions and internal reflections that eventually produced the alternate black and coloured rings that he first observed. This explanation of the colours of muscovite was based on a hypothesis that could be extended to all the phenomena of thin films and colours in general. It was the assumption that light is a "very short vibration," as it is able to pass also through hard bodies like diamonds. This motion is propagated from the source of light in all directions through a homogeneous medium by direct or straight lines like the rays from the centre of a sphere.[71] Hooke compared the propagation of light *in orbem* to the waves of water caused by the impact of a stone or to the flowing of new water forced out of a pipe.[72] The rays of light are physical parallelograms whose longest sides are the "lines of radiations," and whose shortest sides are, respectively,

---

[70] Hooke (1705), 76–9, 81, 116, 130–1.

[71] Id. (1665), 54–7, 63, 65, 68; cf. Boyer (1987), 234.

[72] Hooke (1705), 117.

the source and the "pulse or wave." A ray is "not a mathematicall line, but a physical one of some latitude." The motion that produces the effect of light is propagated within this space, and corresponds to a line which joins the two sides of the phys-ical ray.

Hooke's optical definitions are not consistent throughout his writings. The "lines of radiations" are referred to even as "mathematical rays," or just "pulses." The front of the parallelogram is described as the "pulse," the "wave," the "orbicular impulse," or even the "undulation."[73] The need of clarity and the suggestive character of Hooke's vibratory theory have led historians to apply to it a uniform terminology proper to a clearly structured wave theory. Thus, for instance, Hooke's "pulse" or "orbicular impulse" has been considered as the normal or the front of the wave, because in white light it is perpendicular to the "mathematical rays" or the "lines of radiations."[74] Yet, in spite of the highly suggestive similitudes, the Hookian theory of light was a vibratory rather than a wave theory. Hooke conceived it as such with respect to Huygens' wave theory.[75] Refraction alters the relationship between the "orbicular impulse" and the "lines of radiations" of a white ray of light. This altera-tion takes place when the ray passes through mediums of different optical density. When the ray is not perpendicular to the ideal plane of the new medium, one of its sides enters it earlier than the other. According to the variation of optical density, the orbicular impulse becomes oblique to the lines of radiations. Colours are the effect of such "oblique and confus'd" pulses of light on the retina. Red and blue have a special role in Hooke's theory, because they seem to correspond to the two extreme degrees of obliquity of the pulse to the lines of radiation.[76] However, Hooke never undertook a quantitative description; despite the apparent use of geometrical notions, he never really abandoned the qualitative level of sense-perception.[77] Blue, therefore, is described in terms of "an impression on the retina of an oblique and confused pulse whose weakest part precedes, and whose strongest follows." Red, on the other hand, is produced when the "strongest part precedes" and the "weakest follows." All other colours are intermediaries between these two, or rather are formed by a mixture of red, blue and, as in the peripatetic theory, black and white. Furthermore, according to Hooke the external parts of the bodies include "tinging substances," particles that have a specific degree of refrangibility of light. But he made no effort to determine these degrees with respect to colours, or just red and blue. If the external texture does not contain such particles, the air enclosed in it just reflects white light.[78]

Hooke presented his hypothesis as a radical alternative to Cartesian and Hobbesian theories. But some of Hooke's contemporaries, and optics historians,

---

[73] Id. (1665), 57–9, 64–5; Hooke (1677), 39; Cf. Blay (1981), 101–3.

[74] For instance Boyer (1987), 235; Sabra (1981), 192, 254.

[75] Hall (1951), 221–2.

[76] Hooke (1665), 57–9.

[77] Blay (1981), 110–1; Sabra (1981), 258.

[78] Hooke (1665), 64, 68, 74, 84.

have underlined the debt that the vibratory hypothesis owes to Descartes and Hobbes. All three of these natural philosophers shared the idea that white light is originally homogeneous, and colours originate from refraction. But they had different opinions on the nature of light's rays. Hobbes, in particular, conceived light as the instantaneous motion originating from the diastole of the sun and propagated by the ether. To explain how such a motion could produce straight rays, Hobbes introduced the concept of "line of light," describing the part of the ray that conveys light and is always perpendicular to the rest of it. Although colours are "troubled light," the Hobbesian "line of light" remains perpendicular to the sides of the rays after diffraction.[79] Unlike Hobbes, Hooke believed that there could not be colours without at least a small amount of physical alteration of the white rays. For this reason he did not maintain the Hobbesian "front of light," and refused Huygens' "normal" to the waves, namely the most important conceptual tool of early kinematic optics. This would have been enough to mark a significant difference between Hobbes' optics and Hooke's vibration hypothesis. But Hooke probably felt the need to increase this divide, and in the early 1680s he attributed to Hobbes a Cartesian definition of light as "conatus ad motum."[80]

Hooke's chymical hypothesis entailed a clear refusal of atomistic theories of fire and light. The dissolutive nature of combustion was the basis of Hooke's ideas on the nature of light and colours. Instead of the diastolic motion of the sun, postulated by Hobbes, Hooke focused on the dissolutive motions of all burning bodies. Although produced under different conditions, light is always due to the "internal motion of the parts which shine."[81] The constituents of burning, putrefying, or fermenting bodies are "exceeding nimbly and violently moved." Such a very "short" and exceedingly "quick" motion is communicated to the surrounding ether.[82] "And this may be further argued from this," Hooke observes, "that faster and quicker the dissolution is made, the stronger and vivid is the light."[83] Both this vibrating motion and its means of transmission have specific features distinguishing them from other motions of the remaining fluid bodies filling the space. Sound, for instance, is conveyed from a different medium, as some experiments with the air pump had shown. The body conveying light is a specific, homogeneous fluid and thin substance susceptible only of this specific vibrating motion. Even though Hooke did not call it into account, this fundamental property of the luminiferous ether seems based on the principle of congruity. Only implicitly assuming this principle, he could postulate the existence of a substance that is selectively susceptible of motion. It goes without saying that a quantitative description of this motion is lacking in Hooke's work. Because the cause of light is the dissolution of bodies, other products of this process, such as heat, is also conveyed by the same medium. To account for some

---

[79] Hobbes (1839–45a), vol. V, 221–6; cf. Shapiro (1973), 144–51.
[80] Hooke (1705), 130–1, 136; cf. Giudice (1999), 6, 60; Darrigol (2012), 53–4, 58.
[81] Hooke (1678), 46.
[82] Id. (1665), 55.
[83] Id. (1705), 111.

problematic phenomena, like the interposition of light rays, Hooke simply attributed to the luminiferous ether the property of conveying the motion in different directions at the same time, and defined optical density in terms of the quantity of this ether present in a unit of space.[84]

5.5 Mathematicians and Naturalists

"Though Descartes may be mistaken so is Mr Hooke in confuting his 10 Sec. 38 Cap. Meteorum."[85] This criticism opens Newton's notes on the theory of light and colours of the *Micrographia*. Probably composed in early 1665, Newton's notes draw on his early optical studies carried out between 1661 and 1664. The entries *Of Light* and *Of Colours* of the *Quaestiones quaedam philosophicae* were the main results of these studies. According to the young Newton, the different sensations of colours depend on the different speeds of the rays. Slow rays produce light colours, while fast rays generate the dark ones. Perception of black, grey and white light is due to a mix of slow and fast rays, and refraction alters white light only inasmuch as it separates rays of different speeds. Despite questioning the principle of original homogeneity of white light, Newton stressed the function of speed, an alterable physical property. It could be modified, for instance, by the interaction of light with coloured bodies. However, even at this early stage he was able to scrutinize the theories of his contemporaries from a privileged point of view. "Light," he writes, "cannot be by pressure."[86] According to the young Newton, Hooke was right when he affirmed that Descartes' hypothesis cannot explain the colours of muscovite, "or that ye turbinated motion of ye Globuli signifys nothing unless they did not only endeavour but also move to ye eye." But a radical objection could be moved also to Hooke's hypothesis. If it is a concrete vibratory motion, "why yn may not light deflect from streight lines as well as sounds &c"?[87] In Newton's opinion, the hypotheses of Descartes and Hooke share the assumption that white light is originally homogeneous, and colours originate from refraction. These mechanical hypotheses aimed to account for the physiological perception in a simple way, and just by employing the principles of matter and motion, regardless of the physical differences among coloured and white rays. On the contrary, Newton saw himself as a philosopher whose ideas on the property of bodies were constantly under the scrutiny of experiments.[88] In the mid-1660s Newton undertook a series of experiments with two prisms. The original aim of these experiments was perhaps to verify that

---

[84] Id. (1665), 55–7, 96–7, 99–100; Id. (1705), 113, 116, 130, 184.

[85] Newton (1962), 403.

[86] Id. (1983), 380–3, 430–43.

[87] Id. (1962), 403.

[88] Shapiro (2004), 188, 215; Janiak (2008), 24–5.

white light consisted of coloured rays of different speeds.[89] The manuscript sent to the Royal Society in 1672, *The New Theory of Light and Colours*, was based on a "specimen" of these experiments. In the *Experimental history of colours*, as the young Newton knew well, Boyle observed that colours formed by refraction of white light through a prism where not altered by following reflections or refractions. A similar phenomenon had already been described by Johannes Marcus Marci.[90] Employing two prisms, Newton let the coloured rays obtained by the first refraction be refracted again. The colours were not altered because light "is a heterogeneous mixture of differently refrangible rays."[91] The experiment provided the immutable physical property of coloured rays that was missing from Newton's earlier notes.

Newton presented this *specimen* as the new and real *experimentum crucis* of optics. He employed a term that Hooke first used in optics. As one moves from Bacon's *instantia crucis* to the optical *experimenta crucis* of Hooke and Newton, the epistemological role of the instance changed according to the different images of science within which it was employed. As we have seen, for Bacon it provided a negative instance, a step in a longer process towards the discovery of the schematisms and textures of bodies. Hooke, on the other hand, claimed that it was a necessary examination of hypotheses. Although the *experimentum crucis* of 1665 concluded with a new hypothesis of light, it retained the primary feature of negative instance. Hooke's alternative to the Cartesian theory was indeed a hypothesis. The experiment did not demonstrate it, but did suggest it as an explanation worthy of further inquiries.[92] What can be described as the contingent character of Bacon's *instantia crucis*, definitely disappeared in Newton's double prism experiment. This latter was crucial not just because of the refusal of mechanical hypotheses (a label including also Hooke's hypothesis), but even more because it concluded with a theory rather than a new hypothesis. The new "doctrine" of light and colours, Newton claimed, is "the most rigid consequence" of the experiment, "concluding directly and without any suspicion of doubt." It provided, in other words, an unquestionable description of the true nature of light, since it proceeded as a mathematical demonstration. The study of colours had traditionally pertained to naturalists, like Hooke. With the new *experimentum crucis*, it became part of mathematical optics.[93] The necessary inferential link between experimental data and theoretical principles in Newton's crucial experiment relied to some extent on the literary construction of his specimen. Like Descartes' account of the experiment of refraction by a glass bowl,[94] Newton's historical narration is an idealized version of a more complex process, both at the experimental and theoretical levels. From the early 1660s, Newton was engaged in gathering observations, elaborating different hypotheses,

---

[89] Giudice (2009), 92; Darrigol (2012), 80–3.

[90] Boyle (1999), vol. IV, 103; Boyer (1987), 220–2; Solc and Smolka (2004), 300–1.

[91] Newton (1959–77), vol. I, 95.

[92] Hooke (1665), b1r, 54, 67, 84; cf. Jalobeanu (2015), 311–4.

[93] Newton (1959–77), vol. I, 96–7.

[94] Buchwald (2008), 7–9, 13–4.

and conceiving new experiments. In the *New Theory* this long process was hidden by an apparently linear sequence of matters of facts and necessary inferences.[95]

On February 9, 1672, the *New Theory* was read at the Royal Society. Along with Boyle and Ward, Hooke was asked to report his opinion to the Society. Thanks to "many hundreds of tryalls" Hooke confirmed that coloured rays produced by the refraction of white light retain different degrees of refrangibility through following refractions. But they also proved neither "that all colours were actually in every ray of light before it has suffered a refraction," nor "that those proprietys of coloured rayes, which we find after their first refraction, were not generated by the said refraction."[96] For Hooke, the double prism experiment could not be considered as an *experimentum crucis*, and Newton's explanation was not a theory "as certain as mathematicall demonstrations," but just a "hypothesis of saving the phænomena of colours," like many other similar hypotheses. To provide an equally adequate explanation of the double prism experiment, Hooke seemed to undertake a suggestive remodelling of his vibratory hypothesis of light. In *Micrographia* he described the physical quality of a coloured ray, produced by refraction of white light, as "an inseparable concomitant" only if "not streightened by a contrary refraction." In 1672 he suggested instead that it could be considered innate. Thus, the coloured rays would be vibrations with different refractive properties, while white light could be conceived as the uniform motion resulting from the composition of different vibrations. "But," Hooke notes, "why there is a necessity, that all these motions, or whatever els it be that makes colours, should be originally in the simple rays of light, I doe not yet understand." Just like the sounds produced by the air passing through the pipes of an organ are not in the bellows or in the fingers of the musician, the vibrations producing the different perceptions of colours are not in the white light before refraction. Atomism, in Hooke's opinion, was the fundamental hypothesis underlying Newton's conclusions on the heterogeneous nature of white light. It made it possible to ground upon different kinds of corpuscles the physico-mathematical qualities of light revealed by the experiment. What else are the "original and connate" properties that Newton conferred to the different coloured rays? Thus, in Hooke's words, the Newtonian definition of light as "an heterogeneous mixture of differently refrangible rays" became instead "an heterogeneous aggregate of bodys."[97]

According to Hooke, if one put aside Newton's atomist assumptions, the double prism experiment did not question his optical hypothesis. Nevertheless, Newton's fundamental objection remained. How could the vibrations of any fluid like the rays of light be propagated in a straight line "without a continuall & very extravagant spreading & bending every way into ye quiescent medium"? For this reason, coloured rays could not be different vibrations of the luminiferous ether as Hooke

---

[95] Shapiro (1980), 214; Bechler (1974), 116; Sabra (1981), 246–8; Blay (1985), 372; Dear (1985), 155; Jalobeanu (2015), 238.

[96] Newton (1959–77), vol. I, 110, 202; Birch (1756–57), vol. III, 9–15, 193–4, 313–4; cf. Schaffer (1989), 85–6.

[97] Newton (1959–77), vol. I, 95, 97, 110, 113, 195, 111, 199–200, 202.

suggested, because they could not "coalesce into one uniform motion" of the white light.[98] To resist Newton's objections, Hooke employed a newly discovered optical phaenomenon, viz. diffraction. "Besides the progresse of light in a straight line," Hooke observes, "there is also a propagation of light by straight lines which deflect into the shadow dark or quiescent medium from the superficies or term of the opacus body."[99]

Hooke's criticism highlighted some real problems of the *New Theory*.[100] Newton's rejection of the homogeneity of white light preceded the *experimentum crucis*.[101] This latter provided experimental support for the idea that light was a mixture of differently refrangible rays. Newton's argument was based on the fact that the degrees of refrangibility of coloured rays could not be altered after the initial break-up of white light. On this evidence he concluded that these constant properties were innate, and that white light was heterogeneous. In the *New Theory*, he presented this brilliant inference as a matter of fact. In 1675 he was still sure to have "shewn that *de facto* the rays of light are indued with those proprieties," leaving to the hypotheses-makers the easy task to explain it through mechanical principles.[102]

The Newtonian inference from the observable phenomenon of coloured rays produced by refraction to the unobservable nature of light before refraction was supported by his mathematical atomism.[103] According to this view, the properties of matter and its elementary components have mathematical characters that experiments can show. In Newton's early notes, the rejection of the abstract optical mechanism of the hypotheses-makers was accompanied by the assumption of atomism and the rejection of the principles of Cartesian philosophy.[104] Some scholars have acknowledged that atomism played an important role in the formation of the Newtonian views of light and colours.[105] As Hooke noted, in the *New Theory* Newton affirmed that, given his hypothesis, "it can be noe longer disputed whether there be colours in the dark nor whether they be the qualitys of the objects we see nor perhaps whether light be a body." He was aware that the determination of the nature of light was "not so easie," whereby he aimed not to "mingle conjectures with certainties."[106] But after Hooke's criticism, between 1672 and 1675 Newton publicly accepted to debate hypotheses.[107] Building on the great "affinity" between the

---

[98] Id. (1959–77), vol. I, 175–7, cf. Id. (1779–85), vol. I, 445; Id. (1999), 776.

[99] Id. (1959–77), vol. I, 200–1; Hooke (1705), 186–90; Birch (1756–57), vol. III, 52–4, 63, 194–5; cf. Dilworth and Sciuto (1984), 32–5; Mamiani (2000), 148, Boyer (1987), 239; Shapiro (2008), 431–5.

[100] Sabra (1981), 295; Guicciardini (2011), 52–3, 78–9.

[101] Shapiro (1980), 215.

[102] Newton (1959–77), vol. I, 386, 264; cf. Id. (1779–85), vol. IV, 231–8.

[103] Hall and Boas Hall (1970), 60.

[104] Newton (1983), 336–45, 420–5.

[105] Westfall (1970), 90; Mamiani (2000), 146; Shapiro (1980), 230; Id. (2002), 231, 245, 249–50; Sabra (1981), 242, 296; Giudice (2009), 149.

[106] Newton (1959–77), vol. I, 100, 199–200.

[107] Casini (1983), 23; Mamiani (2000), 148; Guicciardini (2018), 91–2.

vibratory and atomistic hypotheses, Newton assumed that, if light consists of particles emitted from shining substances, like stones thrown into water, these should produce vibrations in the encompassing ether. Newton's new *Hypothesis explaining the Properties of Light discussed of in my severall Papers* maintained that light is a heterogeneous mixture of different bodies, and used ethereal vibrations to explain almost all optical phenomena then known, including the colours of thin films and diffraction.[108] On the one hand, Newton admitted that in the *New Theory* he argued for the "corporeity of light," and that the proprieties of light that emerged from the crucial experiment could be explained by many "mechanicall hypotheses." On the other, he introduced a significant distinction between the vibratory and the atomistic hypothesis. Only the latter could be considered a "very plausible consequence" of the crucial experiment. "The best and safest way of philosophizing" consists first in inquiring and establishing the proprieties of phenomena by means of experiments, and then in explaining these by means of hypotheses. Those that disagree with the results of the experiments should be rejected, the others further developed by means of new experiments. The vibration hypothesis contrasts with the results of the crucial experiment, while on the contrary, the idea that light could be a body is consistent with them.[109] The first is a mechanical hypothesis, the second an experimental query. In the dispute with Hooke, therefore, the expression "mechanical hypothesis" seems already to be for Newton something like heresy, a label that never applies to ourselves, but only to others.[110]

Bibliography

Adelmann, Howard. 1966. *Marcello Malpighi and the evolution of embryology*, 5 vols. Ithaca/London: Cornell University Press.

Bacon, Francis. 1857–74. *Works*, 7 vols., ed. Robert L. Ellis, James Spedding, and Douglas D. Heath. London.

Baker, Tawrin. 2015. Color and contingency in Robert Boyle's works. *Early Science and Medicine* 20: 536–561.

Bechler, Zev. 1974. Newton's 1672 optical controversies: A study in the grammar of scientific dissent. In *The interaction between science and philosophy*, ed. Yehuda Elkana, 114–142. Atlantic Highlands: Humanities Press.

Belloni, Luigi. 1975. Marcello Malpighi and the founding of anatomical microscopy. In *Reason, experiment and mysticism in the scientific revolution*, ed. Maria Luisa Righini Bonelli and William Shea, 95–110. Canton: Science History Publications.

Birch, Thomas. 1756–57. *The history of the Royal Society of London*, 4 vols. London.

Biringuccio, Vannoccio. 1540. *De la pirotechnia*. Venice.

Blay, Michel. 1981. Un exemple d'explication mécaniste au XVIIe siècle: l'unité des théories hookiennes de la couleur. *Revue d'histoire des sciences* 34: 97–121.

---

[108] Newton (1959–77), vol. I, 174–5, 362–3; cf. Id. (1779–85), vol. I, 259, Id. (1999), 625–6.

[109] Id. (1959–77), vol. I, 164, 167, 173, 210, 266,

[110] Koyré (1968), 52; cf, Shapiro (2002), 227–8, 231–2; Cohen (1956), 575–8; Domski (2010), 526, 532.

————. 1985. Remarques sur l'influence de la pensé à la Royal Society: pratique et discours scientifiques dans l'étude des phénomènes de la couleur. *Les Études philosophiques*: 359–373.

Boas, Marie. 1952. The establishment of the mechanical philosophy. *Osiris* 10: 412–541.

Boyer, Carl. 1987. *The rainbow from myth to mathematics*. Princeton/London: Princeton University Press.

Boyle, Robert. 1999. *The works of Robert Boyle*, 14 vols., ed. Michael Hunter and Edward Davies. London: Pickering and Chatto.

————. 2001. *The correspondence of Robert Boyle*, 6 vols., ed. Michael Hunter, Antonio Clericuzio and Laurence Principe. London: Pickering and Chatto.

Buchwald, Jed. 2008. Descartes' experimental journey past the prism and through the invisible world to the rainbow. *Annals of Science* 65: 1–46.

Casini, Paolo. 1983. *Newton e la coscienza europea*. Bononia: Il Mulino.

Chapman, Alan. 2005. *England's Leonardo: Robert Hooke and the seventeenth-century scientific revolution*. Bristol/Philadelphia: Institute of Physics Publishing.

Charleton, Walter. 1654. *Physiologia epicuro-gassendo-charletoniana*. London.

Clericuzio, Antonio. 1993. From Van Helmont to Boyle. *The British Journal for the History of Science* 26: 303–334.

————. 2000. *Elements, principles and corpuscles: A study of atomism chemistry in the seventeenth century*. Dordrecht: Kluwer.

Cohen, I. Bernard. 1956. *Franklin and Newton*. Philadelphia: American Philosophical Society.

Cornelio, Tommaso. 1663. *Progymnasmata physica*. Venice.

Darrigol, Olivier. 2012. *A history of optics: From Greek antiquity to the nineteenth century*. Oxford/New York: Oxford University Press.

Dear, Peter. 1985. Totius in verba: Rhetoric and authority in the early Royal Society. *Isis 76*: 145–161.

Debus, Allen George. 1977. *The chemical philosophy*. New York: Science History Publishing.

Descartes, René. 1964–74. *Oeuvres*, 12 vols., ed. Charles Adam and Paul Tannery. Paris: Vrin.

————. 1983. *Principles of philosophy*. Trans. Valentine Rodger Miller and Rees P. Miller. Dordrecht: Reidel.

————. 2001. *Discourse on method, optics, geometry, and meteorology*, ed. Paul Olscamp. Indianapolis/Cambridge: Hackett Publishing.

Digby, Kenelm. 1661. *A discourse concerning vegetation of plants*. London.

Dilworth, Craig, and Maurizio Sciuto. 1984. Hooke, Grimaldi and the diffraction of light. In *Atti del IV congresso nazionale di storia della fisica*, ed. Pasquale Tucci, 31–37. Milan: Clued.

Domenico, Bertoloni Meli. 2011. The color of blood: Between sensory experience and epistemic significance. In *Histories of scientific observation*, ed. Loraine Daston and Elisabeth Lunbeck, 117–134. Chicago/London: University of Chicago Press.

Domski, Mary. 2010. Newton as historically-minded philosopher. In *Discourse on a new method: Reinvigorating the marriage of history and philosophy of science*, ed. Mary Domski and Michael Dickson, 65–89. La Salle: Open Court.

Ducheyne, Steffen. 2008. A preliminary study of the appropriation of Van Helmont's oeuvre in Britain in chemistry, medicine and natural philosophy. *Ambix* 55: 122–135.

Emerton, Norma. 1984. *The scientific reinterpretation of form*. Ithaca/London: Cornell University Press.

Ercker, Lazarus. 1951. *Treatise on ores and assaying*, ed. Annaliese Grünhaldt Sisco and Cyril Stanley Smith. Chicago: University of Chicago Press.

Fracassati, Carlo. 1666. An experiment of signior Fracassati upon bloud grown cold. *Philosophical Transactions* 2: 492.

Frank, Robert Greg. 1980. *Harvey and the Oxford physiologists: A study of scientific ideas and social interaction*. Berkeley/Los Angeles/London: University of California Press.

Gassendi, Pierre. 1658. *Opera omnia*, 6 vols. Leiden.

Giudice, Franco. 1999. *Luce e visione: Thomas Hobbes e la scienza dell'ottica*. Florence: Olschki.

————. 2009. *Lo spettro di Newton: la rivelazione della luce e i colori*. Rome: Donzelli.

Guerlac, Henry. 1954. The poet's nitre. *Isis* 45: 243–255.

Guicciardini, Niccoló. 2011. *Newton*. Rome: Carocci.

———. 2018. *Isaac Newton and natural philosophy*. London: Reaktion Books.

Gunther, Robert T. 1968. *Early science in Oxford*, 15 vols. London: Dowsons of Pall Mall.

Hall, Alfred Rupert. 1951. Two unpublished lectures of Robert Hooke. *Isis* 42: 219–230.

Hall, Alfred Rupert, Boas Hall, and Marie. 1970. Newton and the theory of matter. In *The annus mirabilis of Isaac Newton 1666–1966*, ed. Robert Palter, 54–68. Cambridge, MA: The MIT Press.

Harris, L.E. 1961. *The two Netherlanders. Humphrey Bradley and Cornelis Drebbel*. Leiden: Brill.

Hobbes, Thomas. 1839–45a. *Opera philosophica*, 5 vols., ed. William Molesworth. London.

———. 1839–45b. *The English works*, 11 vols., ed. William Molesworth. London.

Hooke, Robert. 1661. *An attempt for the explication of the phaenomena*. London.

———. 1665. *Micrographia*. London.

———. 1677. *Lampas*. London.

———. 1678. *Lectures and collections*. London.

———. 1705. *Posthumous works*, ed. Richard Waller. London.

Hunter, Michael. 2017. *The image of Restoration science: The frontispiece of Thomas Sprat's History of the Royal Society 1667*. London: Routledge.

Huygens, Christiaan. 1888–1950. *Oeuvres completes*, 22 vols., ed. Société hollandaise des sciences. The Hague: Martinus Nijhoff.

Inwood, Stephen. 2002. *The man who knew too much: The strange and inventive life of Robert Hooke 1653–1703*. London: Macmillan.

Jalobeanu, Dana. 2015. *The art of experimental natural history: Francis Bacon in context*. Bucharest: Zeta Books.

Janiak, Andrew. 2008. *Newton as philosopher*. Cambridge: Cambridge University Press.

Jardine, Lisa. 2003. *The curious life of Robert Hooke: The man who measured London*. London: Harper Collins.

Keller, Vera. 2011. How to become a seventeenth-century natural philosopher: the case of Cornelis Drebbel (1572–1633). In *Silent messengers: The circulation of material object of knowledge in the early modern Low Countries*, ed. Sven Dupré and Christoph Lüthy, 125–151. Berlin: Lit Verlag.

Koyré, Alexandre. 1968. *Newtonian studies*. Chicago: University of Chicago Press.

———. 1978. *Galileo studies*. Atlantic Highlands: Humanities Press.

Lower, Richard. 1669. *Tractatus de corde*. Amsterdam.

Lysaght, D.J. 1937. Hooke's theory of combustion. *Ambix* 1: 93–108.

Malpighi, Marcello. 1686. *Opera omnia*. Leyden.

Mamiani, Maurizio. 2000. The structure of a scientific controversy: Hooke versus Newton about colors. In *Scientific controversies: Philosophical and historical perspectives*, ed. Marcello Pera, Peter Machamer, and Aristides Baltas, 143–152. Oxford: Oxford University Press.

Mayow, John. 1674. *Tractatus quinque medico-physici*. Oxford.

Mckie, Douglas. 1953. Fire and the flamma vitalis: Boyle, Hooke and Mayow. In *Science, medicine and history: Essays in the evolution of scientific thought and medical practice*, ed. E. Ashworth Underwood, 469–488. Oxford: Oxford University Press.

Nakajima, Hideto. 1984. Two kinds of modification theory of light: Some new observations on the Newton-Hooke controversy of 1672 concerning the nature of light. *Annals of Science* 41: 261–278.

Newman, William, and Lawrence Principe. 1998. Alchemy vs. chemistry: The etymological origins of a historiographic mistake. *Early Science and Medicine* 3: 32–65.

Newton, Isaac. 1962. *Unpublished scientific papers*, ed. Alfred Rupert Hall and Marie Boas Hall. Cambridge: Cambridge University Press.

———. 1983. *Certain philosophical questions: Newton's Trinity notebook*, ed. J.E. McGuire and Martin Tamny. Cambridge: Cambridge University Press.

———. 1999. *The Principia: Mathematical principles of natural philosophy*, ed. I. Bernard Cohen and Anne Whitman. Berkeley/Los Angeles/London: University of California Press.

Oldenburg, Henry. 1965–86. *The correspondence of Henry Oldenburg*, 13 vols., ed. Alfred Rupert Hall and Marie Boas Hall. Madison: University of Wisconsin Press.

Partington, James. 1961. *A history of chemistry*, 4 vols. London: Macmillan.

Purrington, Robert. 2009. *The first professional scientist: Robert and the Royal Society of London.* Berlin/Basel: Birkhäuser.

Rømer, Ole. 1676. Démonstration touchant le mouvement de la lumière trouvé par M. Römer de l'Académie Royale des Sciences. *Journal des Sçavans*: 233–236.

Sabra, Abdelhamid Ibrahim. 1981. *Theories of light from Descartes to Newton*. Cambridge: Cambridge University Press.

Schaffer, Simon. 1987. Godly men and mechanical philosophers: Souls and spirits in restoration natural philosophy. *Science in Context* 1: 55–85.

———. 1989. Glass works: Newton's prism and the uses of experiment. In *The uses of experiment: Studies in natural sciences*, ed. David Gooding, Trevor Pinch, and Simon Schaffer, 67–104. Cambridge: Cambridge University Press.

Shapiro, Alan. 1973. Kinematics optics: A study of wave theory of light in the seventeenth century. *Archive for the History of Exact Sciences* 11: 134–266.

———. 1980. The evolving structure of Newton's theory of white light and color. *Isis* 71: 211–235.

———. 2002. Newton's optics and atomism. In *The Cambridge companion to Newton*, ed. I. Bernard Cohen and George E. Smith, 227–255. Cambridge: Cambridge University Press.

———. 2004. Newton's "Experimental philosophy". *Early Science and Medicine* 9: 185–217.

———. 2008. Twenty-nine years in the making: Newton's Opticks. *Perspectives on Science* 16: 417–438.

Solc, Martin, and Joseph Smolka. 2004. Optical and mechanical experiments in the books "Thaumantias" (1648) and "De proportione motus" (1639, 1648) by Johannes Marcus Marci. *Acta historiae rerum naturalium nec non technicarum* 8: 295–306.

Westfall, Richard. 1970. Uneasy fitful reflections on fits of easy transmission. In *The annus mirabilis of Isaac Newton 1666–1966*, ed. Robert Palter, 88–104. Cambridge, MA: The MIT Press.

Willis, Thomas. 1659. *Diatribae duae medico-philosophicae*. The Hague.

Chapter 6
A New System of the World

6.1 From Congruity to Magnetic Philosophy

In his inaugural speech for the chair of astronomy of Gresham College in 1657, the young Christopher Wren expressed confidence that, thanks to "the industry of some writers of our age," the foundation of Copernican astronomy, laid by Galileo and Kepler, would soon be completed and perfected.[1] A few years later, in 1670, another Gresham professor delivered a lecture which attempted to provide an unquestionable proof of Copernican astronomy. Although unsuccessful in this respect, Hooke's *Attempt to Prove the Motion of the Earth by Observations*, published in 1674, did present three new principles of the Copernican "system of the world." The first principle introduced the idea "that all celestial bodies whatsoever, have an attraction or gravitating power towards their own centres." This attraction is not limited to the constituents of each body, rather it extends "to all the other celestial bodies that are within the sphere of their activity." Terrestrial gravity is part of a wider physical system, since the sun, the moon and all the other planets "have an influence" over the earth, as this latter does on them.

The second principle described the motions of celestial bodies as curved trajectories around the centre of the system. Through this principle Hooke extended terrestrial mechanics to celestial bodies. "All bodies whatsoever," Hooke wrote, "that are put into direct and simple motion will continue to move forward and in a straight line, till they are by some other effectual powers deflected and bent" into a curved path. The motion resulting from such a composition might be "a circle, an ellipsis or some other more compounded curved line." Two implicit assumptions link the first two principles. The "effectual power" mentioned in the second is nothing else than the "gravitating power" introduced in the first. The former operates on celestial

---

[1] Ward (1740), 35.

© Springer Nature Switzerland AG 2020

F. G. Sacco, *Real, Mechanical, Experimental*, International Archives of the History of Ideas Archives internationales d'histoire des idées 231, https://doi.org/10.1007/978-3-030-44451-8_6

bodies originally set in straight motion. Hooke's system of the world, therefore, was built on the physical principles of inertia and gravitational attraction.

The effects of this interaction depend on the extension of the "sphere of activity." The third principle provides an incomplete quantitative description of these effects: the closer the bodies, the stronger the attraction between them. There are reasons to believe that in 1674 Hooke thought of gravity in terms of the so-called inverse square law, despite his claims that it was not "yet experimentally verified."[2]

These three principles describe a "system of the world differing in many particulars from any yet known, answering in all things to the common rules of mechanical motions." The second principle builds on the "nature of circular motion." Once experimentally determined, the third principle "will mightly assist the astronomer to reduce all the celestial motions to a certain rule."[3] In a lecture published in 1677, Hooke described the attracting power introduced in the first principle as a "mechanical motion" consisting of "continual repetitions indefinitely swift" of "the same force, pressure, indeavour, impetus, strength, gravity, power, motion, or whatever you will call it."[4] Stressing the novelty of his system of the world, Hooke clearly distanced himself from orthodox mechanical cosmologies, such as Descartes'. The idea of "sphere activity" employed by Hooke traces back to Wilkins' "sphere of vigor" and Gilbert's *orbis virtutis*.[5] In the mid-seventeenth century, as Boyle acknowledged, magnetical philosophy offered the main alternative to the Cartesian vortexes.[6] Hooke's reference to circular motion confirms an essentially centripetal approach to attraction in the new system of the world. Nevertheless, such an active power operates through mechanical means. The coexistence of mechanical and non-mechanical elements in Hooke's cosmology has puzzled scholars. To solve this apparent inconsistency some scholars have ignored or undermined the role of the mechanical mediums in the new system.[7] Thus freed from its supposedly cumbersome and inconsistent mechanical components, Hooke's cosmology has been considered as merely an incomplete astronomical outcome of the magnetical philosophy of seventeenth-century England.[8]

A better understanding of all the components of Hooke's cosmology cannot be attained without an historical reconstruction of its genesis and development. In the *Attempt* published in 1661, Hooke discussed congruity and its relationship with gravity. He clearly thought of them as two distinct powers operating on matter. Congruity is the cause of the spherical form of bodies, either terrestrial or celestial. Gravity might interfere with the action of congruity. But gravity itself might be explained by the effect of an incongruous ether. In Hooke's hypothesis, a very thin

---

[2] Hooke (1674a), sig. A4v, 27–8. Cf. Gal and Chen-Morris (2005), 391–3.

[3] Hooke (1674a), 27–8.

[4] Id. (1677), 30–1.

[5] Bennett (1981), 174.

[6] Boyle (1999), vol. III, 249.

[7] Gal (2002), 40; Bennett (1989), 227–8; Purrington (2009), 176; cf. Patterson (1949), 331.

[8] Ducheyne (2012), 147; Gouk (1980), 575–7; Wang (2016), 717.

fluid, all-pervading, and "heterogeneous" to all terrestrial bodies could detrude them "as far from it as it can, and partly thereby, and partly by other of its properties (…) move them towards the center of the earth." Hooke was aware that the existence of such a heterogeneous fluid needed to be proved, although he firmly believed in it. Gravity, though, could be explained by "a more likely hypothesis" which, Hooke added, "I can make out by experiment."[9]

When the whole section of the *Attempt* of 1661 on gravity and congruity was reprinted in *Micrographia*, the reference to the new hypothesis was dropped.[10] In November, 1664, when *Micrographia* was completed and published, Hooke's early speculations and experiments on gravitation assumed a new and more consistent form.[11] In a paper read at the Royal Society in December, Hooke noted that the variations of gravity make the isochronal pendulum a useless instrument for "universal measure." These variations suggest an analogy with magnetism. "If the gravitation of the earth," Hooke concludes, "be magnetical, that may also alter."[12] Was the magnetic nature of gravity suggested in 1664 the "more likely hypothesis" referred to a few years earlier? Magnetical philosophy was also at the core of Hooke's experiments on gravity that had begun in late 1662 at the Royal Society. Discussing the experiments on the weight of bodies at the base and on the top of Westminster Abbey, Hooke called upon the "magnetical attraction of the parts of the earth." As Hooke declared in the report to the Society, the programme was intended as a "prosecution of my lord Verulam's experiment concerning the decrease of gravity, the farther the body is removed below the surface of the earth."[13] A comparable idea, advanced by Jean de Beaugrand, had been debated in the Mersenne circle in 1636. Descartes suggested testing the hypothesis by comparing the weight of bodies at the top of a tower and the bottom of a well. But he did not mention Bacon, who a few years earlier had proposed a similiar experiment.[14] In *Sylva*, Bacon considered as "very probable that the motion of gravity worketh weakly both far from the earth and also within the earth."[15] The virtuosi of the Royal Society decided to test Bacon's hypothesis. The fellows intended this research as a contribution to a wider Baconian institutional revival, and the experiments were assigned, as usual, to the curator. Hooke linked Bacon's ideas on gravity to magnetic cosmology:

> Gilbert began to imagine it a magnetical attractive power, inherent in the parts of the terrestrial globe. The noble Verulam also, in part, embraced this opinion; Kepler (not without good reason) makes it a property inherent in all celestial bodies, sun, stars, planets.[16]

---

[9] Hooke (1661), 13, 28.

[10] Id. (1665), 22.

[11] Bennett (1975), 44.

[12] Birch (1756–57), vol. I, 506–7.

[13] Birch (1756–57), vol. I, 164; cf. Hooke (1726), 232.

[14] Roux (2004), 36–41.

[15] Bacon (1857–74), vol. II, 354.

[16] Birch (1756–57), vol. II, 70.

Bacon, however, was not a magnetical philosopher. In his view, the Aristotelian definition of gravity as a property of space was false. "For this reason," Bacon noted, "not unskilfully Gilbert introduced magnetic forces." But magnetism in Gilbert's hands almost became the universal principle of nature. By describing the earth as a big magnet, Gilbert had transformed a thole in a big ship.[17] On this ground, Bacon refused the magnetic rotation of the earth, and "revised and limited" Gilbert's notion of magnetic variation.[18] Bacon maintained magnetism and gravity as distinct forces, but acknowledged the existence of significant analogies.[19] In *Sylva,* gravity and magnetism are listed among the phenomena "that work at distance, not at touch." Both operate "by the agreement with the globe of the Earth," and are due to the "universal passions of matter."[20] Like magnetism, gravitational attraction depends on distance. In Bacon's opinion this variation consisted in a diminution of gravity, both towards and away from the centre of the Earth. On the contrary, in the magnetical cosmology the attraction is seen as stronger when bodies are closer.

Hooke was aware of the differences between Bacon's and Gilbert's ideas. As he noted, Bacon's view of gravity was only "in part" consistent with magnetic philosophy. His experimental programme on gravitation was rather based on a heterogeneous mixture of different ideas. Hooke accepted Gilbert's idea of magnetism as a force operating through a sphere of action. Following Kepler, he extended this force to all celestial bodies. Bacon's natural philosophy reinforced his belief that gravity is an attractive cosmic power, although for Hooke it does not operate as action-at-distance.[21]

For a long time historians have considered the reception of Bacon at the Royal Society as limited to his methodological claims. Ernst Mach, for instance, ignored the Baconian roots of Hooke's experimental programme on gravitation.[22] According to Graham Rees, Bacon's semi-Paracelsian cosmology was neglected by the Royal Society, and Hooke's cosmological work was incompatible with the core principles of Bacon's natural philosophy.[23] Seventeenth-century Baconianism emphasized a limited part of Bacon's oeuvre.[24] The reading of the virtuosi was indeed selective. Hooke's observations on magnetism show that they did not focus only on method, rather they appropriated also significant elements of Bacon's natural philosophy.

Hooke read and interpreted Bacon's works according to the changing demands of mid-seventeenth-century natural philosophy. The same can be said of his reading of Gilbert's work. In *De Magnete*, gravity was not described as a magnetic force, but as an electric one because "a loadstone appeals to magneticks only; towards

---

[17] Bacon (1857–74), vol. II, 80

[18] Bacon (1857–74), vol. III, 52, 58.

[19] Manzo (2006), 204–9.

[20] Bacon (1857–74), vol. II, 453–5; vol. III, 735.

[21] Cf. Wang (2016), 713, 717.

[22] Mach (1919), 532.

[23] Rees (1975), 81, 90–2; Rees (1977), 27–30.

[24] Giglioni (2013), 51; cf. Vickers (2007), 17–25.

electricks all things move."[25] Gilbert did not consider electric repulsion.[26] Gilbertian electricity was an attractive force consisting of humid effluvia directed towards the centre of the Earth. It is ironic that the founding principle of magnetical philosophy – the magnetic nature of gravitational attraction – was one that Gilbert specifically excluded.[27] Rather than Gilbert's work, therefore, Hooke focused on the use of some of Gilbert's ideas by Bacon, Kepler and the English Keplerians.[28]

At the end of his warfare on Mars, Kepler's new elliptical astronomy assumed a manifest anti-hypothetical character. Kepler's estimate of the motions of celestial bodies was not independent of the physical nature of the forces operating among them.[29] In *Astronomia nova*, all planets of the solar system have a magnetic nature. The "motive force" of the sun is described as a constant magnetic attraction operating through incorporeal emanations spread during the rotation around its axis. This constant force proved inconsistent with the elliptical orbits of planets. "It is therefore necessary," Kepler concluded, "that the planets themselves, rather like the skiffs, have their own motive powers, as if they had riders or ferrymen." Thanks to these powers the distances of the planets from the sun vary and describe elliptic orbits.[30] Although a decisive component of Kepler's celestial dynamics was an Aristotelian notion of inertia, in Kepler's magnetical cosmology gravity is no longer a natural tendency of bodies towards the centre; rather it is the effect of earth's magnetic attraction within its sphere of virtue.[31] The extension of a revised form of magnetical philosophy to the Copernican cosmos was completed in the *Epitome*. Here Kepler explained the varying distance of the planets from the centre as the effect of the magnetic virtue emitted by the sun. This virtue is "corporeal, nor animal, neither mental." It is attractive and repulsive. "The solar body," Kepler notes, "has the force to attract the planet with respect to its friendly part and to repulse it with respect to its unfriendly part." The distance from the sun and the (Aristotelian) inertia of each planet contribute as well to this new magnetical celestial dynamic.[32]

Among the early readers and followers of Kepler, John Wilkins most influenced Hooke's early approach to magnetical philosophy. Wilkins was the first to link Bacon's ideas on the decrease of gravity with Gilbert's notion of a magnetic orb of virtue. In Wilkins' view, Gilbert greatly contributed to Copernican astronomy because he introduced magnetic forces into the new cosmology. Gilbert, Kepler and Galileo, Wilkins wrote, "have much beautified and confirmed this hypothesis, with

---

[25] Gilbert (1600), 30; Gilbert (1900), 53.

[26] Heilbron (1979), 175–7.

[27] Bennett (1989), 222; Pumfrey (1989), 53.

[28] Bennett (1981), 166–7; Pumfrey (1994), 247–50.

[29] Applebaum (1996), 473–5; Lombardi (2008), 37, 52, 59–61.

[30] Kepler (1937–98), vol. III, 255; Id. (1992), 405; cf. Palmerino (2007), 149.

[31] Granada (2010), 124.

[32] Kepler (1937–98), vol. VII, 299–300, 337–8; Id. (1995), 57–8, 99–100.

their new inventions." Although he embraced Kepler's magnetic cosmology, Wilkins maintained that gravity and magnetism are distinct, albeit similar forces.[33]

In mid seventeenth-century England, magnetic cosmology was a heterogeneous and eclectic system of ideas. Gilbert was credited as its founding father, and Bacon as an important contributor. The relationship between magnetism and gravity was not clear or univocal. These two forces were often considered similar, rather than identical. In spite of a lack of experimental evidence, this similitude led to the introduction of an approach focusing on attractive physical forces alternative to the Cartesian vortices. Magnetical philosophy and mechanism, however, were not mutually exclusive. Magnetical philosophers often accepted the corporeal nature of magnetic forces, while mechanical philosophers employed magnetic analogies. Walter Charleton, for instance, maintained that like a big loadstone "doth the Earth uncessantly emit certain invisible streams" that attract terrestrial bodies within a certain distance. These streams are corporeal effluvia, like those emitted by loadstones. But gravity is not a magnetic force. Charleton believed that the primary defenders of the opposite view were Kepler and Gassendi.[34] Even Charleton's reading of Gassendi was selective. He focused on some aspects of Gassendi's work and ignored other equally relevant ones. The *Syntagma* indeed includes magnetic analogies. Gassendi even described the rays emitted from the sun as "almost magnetic bodies." Nonetheless, he maintained that magnetism and gravity are different forces operated by effluvia emitted by the attractive bodies.[35]

The wide circulation of analogies between gravity and magnetism in the mid-seventeenth century suggests that Hooke's use of these arguments in the early 1660s should not be read as a straightforward refusal of mechanical ideas or corpuscular explanations. Around the same time, Hooke was engaged in his project to lay the ground for a new natural philosophy at once mechanical and experimental. Influenced by Bacon, Kepler, and Wilkins, Hooke thought of the physical nature of gravity as a force originating from the attractive body. Was it a magnetic force? The experimental programme carried out by Hooke at the Royal Society aimed to answer this question. Since 1662, the results had proved disappointing.[36] The greatest number of experiments was performed from August to October in 1664. The absence of significant alterations in the weight of bodies at different distances from the earth's surface, and the existence of perturbing factors, such as weather conditions, decisively contributed to leaving the question undecided.[37] But the programme did not end in 1664. The Society often asked for reports or ordered new trials. In Hooke's experimental agenda, these experiments were just part of a wider programme examining the speed of falling bodies through different mediums and the transmission of

---

[33] Wilkins (1802), vol. I, 115–6, 144.

[34] Charleton (1654), 269, 283.

[35] Gassendi (1658), vol. I, 345–8, 639.

[36] Birch (1756–57), vol. I, 164.

[37] Boyle (2001), vol. II, 342, 537.

magnetic and electric forces.[38] From the autumn of 1664 to the spring of 1666, only one experiment was performed.[39] At that time, Hooke was mainly engaged in the study of comets. After the last series of disappointing results, in March, 1666 Hooke concluded that "this opinion, how probably soever it might seem to Gilbert, Verulam, and divers other learned men, is not at all favoured by the experiments."[40] In the spring of 1666 Hooke's larger research programme continued, but was focused mainly on the transmission of magnetic forces. Thanks to his studies on cometary motions, in 1665 Hooke's cosmological speculations entered a new phase.

6.2 Comets

A very bright comet was first observed in Spain on November 17, 1664. Information about the comet reached the Royal Society in December. Hooke was able to observe it on December 23. The comet was visible until early March. Soon after, a new comet was first sighted in the south of France, disappearing from the skies of Europe at the end of April. Comets soon became a much debated topic among astronomers and natural philosophers. The Royal Society received discordant reports and contrasting works on the comets from all over Europe, and it was called on to settle the controversy between the astronomers Adrien Azout and Johannes Hevelius on the positions of the first comet.[41] At the Society, Hooke could access these works, easily compare different observations, and discuss the opinions of astronomers from different parts of Europe. In the early months of 1665, he mostly wrote on the nature and motion of comets. The occasion was offered by a new series of public lectures at Gresham College, financed by the merchant John Cutler.[42] Despite the early requests made by some fellows, the papers were published, as *Cometa,* only in 1678, after a new comet appeared in 1677. The study of these comets represents a decisive phase in the development of Hooke's early cosmological thinking. His writings on comets are chronologically and theoretically preliminary to the notes on the compound nature of circular motion read at the Royal Society in the spring of 1666. The delay in publication, therefore, renders problematic the reconstruction of the origins and development of Hooke's new system of the world. It is not unlikely, for instance, that he later inserted new parts among the "several scattered papers and lectures" written in 1664 and 1665. He probably excluded from *Cometa* those sections whose contents did "agree with others since published."[43] Drawing on internal inconsistencies, Jim Bennett has distinguished and dated different sections of

[38] Birch (1756–57), vol. I, 167, 172, 174–6, 218, 455–6, 460–8, 473.

[39] Boyle (2001), vol. I, 362–3, 492–4; Royal Society MS/847, f. 35.

[40] Birch (1756–57), vol. II, 71.

[41] Shapin (1994), 266–90; Milani (2013), 198–201.

[42] Birch (1756–57), vol. I, 453.

[43] Hooke (1678), sig. A3r, 223.

Cometa. In Bennet's reconstruction, only the central section of the book reproduces Hooke's early papers. The initial and final sections, on the contrary, seem to have been revised after the summer of 1666.[44] References to the subject in other published and unpublished works help to refine such a reconstruction. In a hitherto unpublished *Text of a lecture concerning natural history*, for instance, Hooke used "the comet or blazing star which has soe lately abrightned the greatest part of the world" as an illustration of his new "philosophicall history."[45] In the posthumously published *General Scheme*, comets are listed as the first "head of inquires" in natural history. The corresponding queries to this head were to be found in "a following discourse to the late comets, and of the nature of comets in general."[46] These papers were probably composed and "read in the beginning of 1665," along with the central section of *Cometa*.[47] The Society requested Hooke to publish them when the first comet was being observed. On February 15, "Sir Robert Moray moved that Mr. Hooke's lecture might be perfected and printed; which was assented to."[48] Moray seems to be referring to the first of Hooke's lectures on comets, since a "second very curious lecture on the late comett" was read, according to Samuel Pepys, on March 1, 1665.[49] The same day, at the Society's weekly meeting, the fellows agreed that "Mr. Hooke should extract out of his lecture a discourse upon the late comet, and fit to the press with the necessary schemes."[50] Hooke drew the first comet's early path at the beginning of 1665 using different sets of observations. The scheme was published in *Cometa* along with Hooke's "thoughts about those comets which appeared in 1664 and 1665." "Soon after" the first observation on December 23, 1664, Hooke wrote some queries.[51] A few days later, Hooke reported "in writing" on the observation to the Society. The queries are published in the opening section of *Cometa*. At the Society meeting on 28 December 1664, Hooke "exhibited also a scheme of the hypothesis, whereby he conceived, that all the irregular motions of the star towards the west (…) might be explained by the motion of the earth, without ascribing any or but a very little motion to the comet." Although this account suggests a strong similarity with Kepler's hypothesis, Hooke did not embrace Kepler's cometary view. A perfectly straight motion proved insufficient to explain the positions of the comet that he had observed for the first time on December 23. Thus, Hooke introduced "a little of comets motion westward."[52]

New observations reinforced Hooke's opinion. In early January, 1665, Auzout sent to the Royal Society his *Ephéméride du comète*, which provided new

---

[44] Bennett (1975), 56–7.

[45] Guildhall Library, London MS 1757.11, f. 106r.

[46] Hooke (1705), 22, 29.

[47] Id. (1678), 44.

[48] Birch (1756–57), vol. II, 16.

[49] Pepys (1995), vol. VI, 48.

[50] Birch (1756–57), vol. I, 19; Royal Society MS/847, f. 41.

[51] Hooke (1678), 7, 44.

[52] Birch (1756–57), vol. I, 511; Royal Society MS/847, f. 14.

observations and a description of cometary motion as a "great circle."[53] In early March, Auzout himself also sent a letter of the astronomer Gian Domenico Cassini, along with Cassini's *Hypothesis motus cometae novissimi*. In Cassini's view the circular path of comets was so big that any visible portion of it could hardly be distinguished from a straight line.[54] In the early lectures of 1665, Hooke described the path of the comet of 1664 as slightly curved, and attributed a uniform, circular and cyclic motion to comets. This hypothesis seemed to explain "very regularly and very naturally" the "appearances" of comets. "Yet," Hooke concluded, "– 'tis not the only hypothesis for that design; nor do I believe it so evident a demonstration for that end." Influenced by Auzout, Hooke excluded that comets could move "in an uncertain curve, with unequal degrees of velocity."[55] At the same time, he was aware that in a Copernican cosmos, comets, like planets, could move with unequal motions. In Kepler's cosmology, in fact, the sun was the motive power of all celestial bodies. For Kepler this entailed the rejection of the dogma of circularity and the attribution of straight paths to comets.[56] Hooke thought that the path of a comet could be an "elleipsis" like the planetary orbits, and built this analogy on the fact that both comets and planets are subjected to the sun's central attraction:

> The physical reason indeed seems pretty difficult by what means it should be confin'd or bound as to move in a circle; but this is no more than is usually supposed in all the planets, and without supposing a kind of gravitation throughout the whole vortice or cœlum of the sun, by which the planets are attracted, or have a tendency towards the sun, as terrestrial bodies have towards the centre of the earth. I cannot imagine how their various motions can with any satisfaction be imagine, but that being granted (…) not only the reason of all the irregular motion of the planets may be easily found, but the reason also of the strange and various motions of the comets.[57]

Solar attraction was not the only physical reason employed in Hooke's cometary theory. Since 1664, Hooke had shown significant interest in the physical nature of comets. Building on the chymical hypothesis of combustion published in *Micrographia*, to the Society Hooke presented the comet of 1664 as a solid "body that was dissolved in the æther."[58] In Hooke's approach, chymical and dynamical aspects of cometary phenomena were neither separate nor independent of each other. The nature of comets was strictly linked to their motion, since the direction of the tail was part of the data that competing hypotheses had to account for. From this point of view, for instance, Kepler's cometary theory proved inconsistent with the direction of the "beard" of comets, whereas Hooke's hypothesis could explain "why the beard grows broader and broader, and fainter and fainter towards the top; why

---

[53] Auzout (1665), 1–3, 7–8.

[54] Oldenburg (1965–86), vol. II, 359–61.

[55] Hooke (1678), 25–7, 30; Auzout (1665), 6.

[56] Ruffner (1971), 178; Boner (2006), 34; Id. (2013), 126–7.

[57] Hooke (1678), 31.

[58] Birch (1756–57), vol. I, 511.

there is a halo about the body, (...) why the beard becomes a little deflected from the body of the sun."[59]

Hooke's criticism of Kepler's views was influenced by the works of Seth Ward. It was probably Ward himself who introduced the young Hooke to Kepler's work in Oxford in the 1650s. Following Kepler, Ward maintained that the sun is the source of planetary and cometary motions. In *De Cometis* he rejected the view of comets as ephemeral bodies. In Ward's opinion, comets are sulphureous and nitrous bodies. Their combustion produces light and the tail. Since these bodies move in circular paths, their motion is cyclic and their nature as long-lasting as the planets.[60] Even Gassendi described comets as part of the "eternal works of nature."[61] And so did Hooke. "The body of the comet," Hooke wrote in an early lecture of 1665, "may be both as ancient and as lasting as the world."[62] For Hooke and Ward the long-lasting nature of comets depended on their circular and cyclic motion. Gassendi, on the contrary, rejected the idea that comets move along curves and are bodies in combustion. Indeed, Gassendi maintained Kepler's view that comets move in straight lines.[63]

In the initial and final sections of *Cometa*, likely revised after spring 1666, the fundamental elements of Hooke's early approach to comets – solar attraction and slightly curved path – led to a clear description of the compound nature of cometary motion. "I suppose," Hooke wrote, that "the attractive power of the sun, or other central body may draw the body towards it, and so bend the motion of the comet from the straight line, in which it tends, into a kind of a curve." The curvature of the paths of comets is so minimal that they seem "almost direct motion[s]," though they are "not exactly straight" because of the attractive central force which all celestial bodies are subject to. Such a description of cometary motions is also "most agreeable" with Hooke's "physical notion of comets."[64] The parts of the comet that are dissolved by the ether recede from the centre of the curve, whereas those undissolved are more affected by the attractive force and occupy the concave part of the curve.

After 1665, in Hooke's hands Kepler's cometary straight path became a rectilinear tendency rather than a concrete motion. "I rather incline," Hooke stated, "to the incomparable Kepler's opinion that its motion tends towards a straight line."[65] The dynamic consequences of Hooke's new approach to cometary motion are significant for the emergence of his new system of the world. Like comets, all planets tend to move in straight lines and are forced in curved orbits by the attractive central power of the sun. The basic elements of this new celestial dynamic were present in Hooke's notes on the first observation of the comet of 1664. As a result of many new

---

[59] Hooke (1678), 32.

[60] Ward (1653), 17–8, 30–1, 35–6.

[61] Gassendi (1658), vol. II, 710–11.

[62] Hooke (1678), 34.

[63] Gassendi (1658), vol. II, 706, 710.

[64] Hooke (1678), 13–4, 37.

[65] Ibid., 43.

observations, Hooke in 1665 could refine his hypothesis and finally introduce the idea of inertia into the study of cometary motion.

How this happened is difficult to reconstruct. It is well known, for instance, that in early 1665, Hooke and Wren exchanged ideas and data on the two new comets. Both virtuosi were influenced by Kepler's work. But Wren was seemingly ahead of Hooke in the understanding of celestial dynamics. In an annotation to his copy of Galileo's *Discorsi*, Wren had already suggested that planetary orbits might result from the combination of their rectilinear motion and the attraction of the sun.[66] Wren's note followed a Galileian cosmologic speculation in the corresponding section of *Discorsi*. In order to establish a link between his new terrestrial mechanics and Copernican astronomy, Galileo introduced the image of God as "planets-thrower."[67] To Galileo's premise that the planets so "thrown" during the creation initially move along straight lines, Wren added the magnetic attraction of the sun. In his opinion the works of Galileo and Gilbert were complementary.[68] But Wren's Galileian speculation did not turn into a coherent celestial dynamics.[69] Following Kepler, Wren claimed that the motion of comets was rectilinear. On February 1, 1665, Wren "produced some observations of the comets, with a theory" to the Royal Society.[70] The "theory" consisted of a diagram offering a hypothetical rendering of the comet's positions in space, based on the assumption that it moves in a straight line and with uniform speed.[71] Wren's diagram was published in *Cometa*. From early February, Hooke used the scheme to check new observations against the successive positions expected according to Wren's Keplerian hypothesis. "By tracing the way of this comet of 1664," Hooke wrote in a lecture read in early 1665, "it is evident that either the observations are false, or its appearances cannot be solved by that supposition."[72] Wren's scheme, therefore, did not play any significant role in directing Hooke towards a clearer understanding of the dynamic components of cometary motion. In Hooke's eyes it was not different from Kepler's cometary theory. And since December, 1664, Hooke's observations had not supported Kepler's view. Did Wren share his Galileian speculations with Hooke in the spring of 1665? Although very suggestive, this hypothesis seems implausible. No documentary evidence shows that Wren pursued this speculation and shared it with Hooke.[73] Since April, 1664, Wren also had access to a letter in which Jeremy Horrocks depicted the creation of comets in terms similar to Wren's Galileian speculation.[74] According to Horrocks comets are produced and then reabsorbed by the body of the sun. Once

---

[66] Bennett (1975), 38–9.

[67] Bucciantini (2003), 115–6.

[68] Wren (1750), 204.

[69] Bennett (1982), 62.

[70] Birch (1756–57), vol. II, 12.

[71] Johnston (2010), 384, 387.

[72] Hooke (1678), sig. A4r, 44.

[73] Bennett (1989), 229.

[74] Birch (1756–57), vol. I, 455–6, 470–1; Oldenburg (1965–86), vol. II, 163, 231.

expelled by the sun, these celestial bodies have the tendency to move in straight lines. But the solar "magnetic virtue" transforms their paths into curves ending in the sun itself. Horrocks extended this explanation even to the elliptical orbits of planets.[75] But when in December the new comet appeared, Wren turned to Kepler's cometary theory.

Unlike Wren, who had examined Horrocks' papers with Wallis in Oxford since April 1664, Hooke could access them only after his early observations of the first new comet were completed. The original of Horrocks' letter on comets was sent by Wallis to the Royal Society on January 21, 1665. Wallis also sent an illustration of the path of the comet of 1577 that Horrocks probably drew "according to Tycho's observations."[76] Horrocks' diagram has not been found, but in *Cometa* Hooke published an illustration of *Horoxij Hypothesis* based on the comet of 1577. In a likely revised version of the early lectures, Hooke criticized Horrocks' hypothesis. The motion of a body "shot out of the sun" would not be that described by Horrocks. In Hooke's opinion the body's initial speed would be as great as supposed by Horrocks. "But then," Hooke adds, "it ought likewise to have accelerated its motion in the same manner in its return back to it again, which it does not in his hypothesis." Only the resistance of the fluid medium in which the body moves could alter its initial speed. To support his opinion, Hooke referred to the motion of a pendulum moving upwards. The pendulum, Hooke concludes, "will never rise again on the opposite side to an equal height, with that it descended from, on that side towards which it was thrown."[77] The reference to the pendulum, albeit tangential to the argument, is significant. In the letter to Crabtree, Horrocks for the first time employed a circular pendulum to explain the motion of celestial bodies. Comparing the action of the sun to that of a hand holding the thread, Horrocks attributed to the pendulum's bob a "rectilinear tendency" that the action of the hand transforms into a curved motion.[78] In 1666, Hooke explained the motions of planets and comets by means of pendulums.[79]

6.3   From Comets to Planets

As curator of experiments, from November, 1664 Hooke was involved in the tests on Huygens' "universal measure" through pendulums. The experiments did not succeed in confirming Huygens' findings, and Hooke for the first time expressed the belief that pendulums cannot provide any universal measure.[80] In 1665, Hooke's

[75] Horrocks (1673), 26, 183, 310–1.

[76] Oldenburg (1965–86), vol. II, 353; Birch (1756–57), vol. II, 11.

[77] Hooke (1678), 35.

[78] Horrocks (1673), 312–3.

[79] Birch (1756–57), vol. II, 91.

[80] Ibid., vol. I, 495–6, 499–500, 508–9; see also Hooke (1726), 4–5.

interest in pendulums increased. During the dispute with Huygens for the claim to have first invented spring watches, Hooke dated his studies on pendulums to 1665.[81] The results of these researches appeared in 1666. In May, at the Royal Society, he explained the "inflection of a direct motion into a curve by a supervening attractive principle" by means of "some experiments with a pendulous body," and in November he read "an account of inclining pendulums." "Circular motion," Hooke noted, "is compounded of an endeavour by a direct motion by the tangent, and of another endeavour tending to the centre." The first "endeavour" consists of the inertial tendency, while the second is a central attractive force. Despite some relevant differences, this principle applies to both pendulums and celestial bodies. Once moved, a pendulum's bob has a "conatus to return to the centre" directly proportional to its distance from the centre itself. On the contrary, the attraction of the sun over celestial bodies decreases with distance. Another significant difference between the motion of simple pendulums and that of celestial bodies is the fact that the former are circular, the latter elliptical. To account for this difference Hooke did not provide a mathematical explanation. He rather provided a practical demonstration by means of a conic pendulum:

> For which purpose there was a pendulum fastened to the roof of the room with a large wooden ball of lignum vitae on the end of it. And it was found, that if the impetus of the endeavour by the tangent at the first setting out was stronger than the endeavour to the centre, there was then generated an elliptical motion, whose longest diameter was parallel to the direct endeavour of the body in the first point of impulse. But if that impetus was weaker than the endeavour to the centre, there was generated such an elliptical motion, whose shorter diameter was parallel to the direct endeavour of the body in the first point of impulse. And if they were both equal, there was made a perfect circular motion.[82]

Hooke, in other words, reduced the main problem of celestial physics to a question of applied mechanics.[83] To include satellites in his mechanical demonstration, Hooke added a second bob to the lower part of the wire. The second bob was expected to describe an elliptical motion around the first, but Hooke observed an elliptical motion described by "a certain point, which seemed to be the centre of gravity of these two bodies."[84]

Hooke's conclusion seems to echo an idea of John Wallis. In a paper read at the Royal Society few days before Hooke's, Wallis described the earth and the moon as an "aggregate of bodies which have a common center of gravity." Building on Galileo's work, Wallis explained the tides as the effect of the motions of this "aggregate."[85] Hooke's claim "that the motion of the celestial bodies might be represented by pendulums" was communicated to the Society after Wallis' paper on tides was read.[86] But in the *History of the Royal Society*, Thomas Sprat attributed to

---

[81] Hooke (1674b), 69; Id. (1677), 42–3; cf. Birch (1756–57), vol. II, 4–5, 19, 21, 23, 26.

[82] Birch (1756–57), vol. II, 90–2, 126–7.

[83] Kuhn (1957), 249–52; Bennett (1980), 42.

[84] Birch (1756–57), vol. II, 92; cf. Patterson (1952), 291–2.

[85] Wallis (1666), 272.

[86] Birch (1756–57), vol. II, 90.

Wren the "discovery" that "by a complication of several pendulums depending one upon another, there might be represented motions like the planetary helical motions."[87] The influence of Wallis' hypothesis on Hooke's paper seems limited to the suggestion of an alternative explanation of the results of his double pendulum experiment. Further experiments failed to support this hypothesis.[88] Moreover, Hooke's focus on attractive forces was not consistent with Wallis' approach to cosmology. Although Wallis compared the common centre of gravity of the earth and the moon to the "common center of virtues" of two magnets, he presented his Galileian hypothesis as an alternative to the magnetic hypothesis of Kepler and Horrocks.[89]

In the paper on circular motion, Hooke presented two hypotheses on celestial mechanics, but pursued only the one introducing attractive forces. The alternative hypothesis was based on the "unequal density of the medium thro' which the planetary body is to be moved." Planetary orbs are seen as areas of continuous uniform density due to the varying effect of solar heat on the ether with distance. The internal part of each orbit is less dense than the external part. Thus, the planets are pushed towards the sun rather than attracted by it. In Hooke's view, however, this hypothesis raised too many problems.[90]

In the spring of 1666, when Hooke read his paper to the Royal Society, the *Theoricae mediceorum planetarum* of Giovanni Alfonso Borelli had just been published in Florence. Like Hooke, Borelli introduced the main question of celestial mechanics, asking why the planets did not break their orbits.[91] In Borelli's view the problem arose because planetary orbits were elliptic rather than circular.[92] For Borelli, elliptical orbits were compounded by the planets' "natural desire to unite with the globe about which they revolve in the Universe" and their "impetus to move away from the centre of revolution." It is worth noting that Borelli attributed this inertial "impetus" to the "circular motion" of planets. If the two forces were equals, the planets would follow stable circular orbits around the centre. But planetary orbits are elliptic. Borelli thus introduced solar whirlpools of heterogeneous density. The elliptical orbits are due to the periodical imbalance between the "natural instincts" of the planets "to approach the sun by a direct motion" and their "impetus" to move away from the centre, caused by "the whirlpool of solar rays moving from West to East on the periphery of the circles." This centrifugal "impetus" varies, although the "natural instinct" towards the centre is constant.[93]

---

[87] Sprat (1667), 312–4; Bennett (1975), 45–6.

[88] Birch (1756–57), vol. II, 95, 99, 101, 105–6.

[89] Wallis (1666), 280.

[90] Birch (1756–57), vol. II, 90.

[91] Borelli (1666), 45–6; Birch (1756–57), vol. II, 90; cf. Aiton (1972), 11, 90.

[92] Gal (2002), 29.

[93] Borelli (1666), 47, 60–3, 77–9; cf. Aiton (1972), 92–4; Bertoloni Meli (2006), 196–7.

Hooke did own a copy of the *Theoricae*, but perhaps was not able to read it before May, 1666.[94] Like Wallis, Borelli maintained a Galileian view of celestial mechanics in which there is no place for attractive forces.[95] As he noted in *De vi percussionis*, only heavy bodies (*gravia*) and animals move because of an "inward principle." This latter is not an attraction but a "natural tendency" (*naturalis instinctus*). Unlike animals, planets are not living beings. Nonetheless, they are not attracted by the centre of their motions, rather they move towards it.[96] The physical cause of this natural tendency is not discussed by Borelli. Following Galileo, Borelli considered gravity as a natural propriety of all bodies, whose effects only are worthy of inquiry.[97] Borelli's planets are Archimedean solids, whose tendency to come closer to the sun is a constant force.[98] Hooke, on the contrary, derived the idea of a centripetal force from the magnetic tradition rejected by Wallis and Borelli. For this reason, he could describe the "conatus" to the centre as a force whose effects vary with distance.

According to Alexandre Koyré, the influence of Borelli on Hooke's paper is "unmistakable."[99] The ethereal hypothesis introduced and soon excluded by Hooke, in fact, developed from an earlier "conjecture" discussed in a paper presented to the Royal Society in January, 1663. Hooke speculated on the existence of an ether "whose parts are of different densities, according as they are nearer or farther from the sun." Planets, then, would follow orbits at a distance from the sun "according as they are more or less massy."[100] The existence of such an ether was first introduced by Gilles Roberval in the *Aristarchi Samii de mundi systemate*. In Roberval's opinion, the heavens were filled with a substance of different density due to the action of the sun, described as a body that produces heat and light, and attracts and repels. According to their specific gravity, planets and their satellites occupy different areas of the celestial ether. The earth and the moon, for instance, form a "system" similar to the greater solar system of which they are part. Roberval listed two causes of planetary motions, one "internal" and one "external," both originating from the sun. The first is the "specific force of the sun," that attracts all celestial matter according to distance. The second is the motion of each body solicited by solar heat.[101] In Roberval's view, the sun's rays rarefy the constituent parts of the planets and favour their motion. Given these causes, Roberval attributes the elliptical form of planetary orbits to the inclination of their axes and to the inconsistent rotation of the sun. The anti-mechanical character of Roberval's cosmology emerged also in a discussion on

---

[94] Bibliotheca Hookiana, 17; Armitage (1950), 276.

[95] Westfall (1971), 342; Gal (2002), 30–1.

[96] Borelli (1666),5, 46–7, 76.

[97] Henry (2011), 14–6.

[98] Koyré (1973), 502.

[99] Koyré (1968), 232; cf. Id. (1973), 467.

[100] Birch (1756–57), vol. I, 176.

[101] Roberval (1644), 12–3, 17–18, 21, 35–6.

gravity at the Académie royale des sciences in the early 1660s.[102] In contrast to Huygens, Roberval considered as prejudicial the refusal of attraction, but did not commit to any physical cause of gravity.[103] In 1644, on the contrary, he grounded his celestial mechanics on a living cosmology.[104] Nontheless, he did not change his mind about the existence of different gravities. Each planet had still a limited attractive force, which coexists with the general gravity of the sun.[105]

Hooke owned Roberval's *Aristarchi Samii*.[106] Influenced by the magnetical philosophy, he distinguished and separated the main elements of Roberval's cosmology. In the paper of 1666, an ether of different density and the central attraction became competitive hypotheses. And Hooke clearly privileged the latter. Hooke's congruity echoes Roberval's emphasis on the tendency of matter to aggregate in spherical bodies. According to Roberval, the sun and all the celestial bodies originate as aggregations of celestial matter in solid spheres. The force that causes this process is the same that governs the solar system.[107] In 1661, Hooke suggested that the "globular form" of some bodies might be "caused by the protrusion of the ambient heterogeneous fluid." This could lead to considering gravity as the effect of this subtle fluid's "endeavour to detrude all earthly bodies as far from it as it can and [...] move them towards the center of the earth."[108] In Hooke's hypothesis, the physical cause of gravity is the congruity between the ether and "earthly bodies." In *Micrographia*, congruity provides an explanation even for magnetic and electric phenomena.[109] Like Descartes, Hooke made use of an ethereal fluid that operates on bodies subject to attraction.[110] But Hooke explained this interaction by means of a property that no orthodox mechanical philosopher would accept, namely congruity and incongruity.

The paper read in May, 1666, shows that, rather than Descartes or Borelli, Hooke followed Roberval in focusing on the central attraction of the sun. Like Roberval, Hooke saw gravity as the effect of another fundamental property of matter, the tendency to aggregate in spherical bodies. Magnetic philosophy was not the only source of Hooke's early cosmological ideas. The decisive influence of this tradition was limited to the years 1662–65, while congruity played an enduring role in Hooke's cosmology and natural philosophy.[111] The oval form of the Earth, Hooke noted in 1687, is due to the "indeavour to recede from the Axis" of the rotation of the planet. This force contrasts with the tendency originating from "the congruity of

---

[102] Jullien (2006), 192–7, 206–7; Boantza (2011), 81–2.

[103] Huygens (1888–1950), vol. XIX, 629–30, 640.

[104] Roberval (1644), 80, 140; cf. Westfall (1971), 265–7.

[105] Roberval (1644), 16; Huygens (1888–1950), vol. XIX, 628.

[106] *Bibliotheca Hookiana*, 37; Aiton (1972), 91–2.

[107] Roberval (1644), 3, 9–10.

[108] Hooke (1661), 27–8.

[109] Hooke (1665), 31.

[110] Descartes (1964–74), vol. VIII.1, 112–3.

[111] Westfall (1971), 269.

the matter, which," Hooke notes, "I have many years since in a small treatise, printed in the year 1660, proved, doth shape the glass into a true spherical figure, and so maketh every part to indeavour towards the center of the whole."[112]

In 1665, the works of Kepler, Wilkins and Horrocks played a decisive role in directing Hooke towards the analysis of the compound nature of circular motion. As Hooke noted, his results could explain "the phenomena of the comets as well as of the planets."[113] Horrocks' letter, in particular, contributed most to the evolution of Hooke's approach to cometary motion. The criticism of Horrocks' hypothesis in *Cometa* suggests that it did confirm Hooke's opinion that cometary motions are curved and that the attractive magnetic power of the sun is a component of their motion. The other component envisaged by Hooke derived from mechanical philosophy. Both Hooke and Wren had access to Horrocks' letter while they were studying the new comets. Despite Wren's earlier Galilean speculation, the letter does not seem to have had any apparent effect on Wren's cometary theory. Even his insight on the use of pendulums to represent planetary motions did not translate into concrete results. In their long-lasting collaboration, Hooke and Wren shared many interests. In the mid-1660s they had frequent exchanges of ideas about comets and planetary motion as well. Although Wren's ideas might have influenced Hooke, the latter's hypothesis of 1666 cannot be seen just as an extensive commentary on Wren's meagre ideas.[114] If one shifts the focus from irresolvable questions of priority to the reconstruction of the cultural context, the evolution of Hooke's early cosmology into his "new system of the world" clearly appears as the result of his refinement and clarification of conflicting opinions and data within the Royal Society. He looked at any such new contribution through the prism of his ideas, and developed these latter with his own insight and intuitions.[115]

Hooke's theory developed at the crossroads of two philosophical traditions.[116] In celestial mechanics, Hooke preferred the centripetal force of the magnetical philosophy to the vortices of Descartes. The earlier hypothesis was consonant with a core principle of his view of nature, congruity. In Hooke's early cosmology, however, the contrast between magnetical and mechanical principles is only apparent. Hooke's programme can be described as the endeavour to combine the idea of attraction with the principles of Galileian mechanics.[117] At the Royal Society, this programme was not unconventional. Wilkins, for instance, described Gilbert, Kepler and Galileo as the founders of the elliptical astronomy.[118] And Wren considered

---

[112] Hooke (1705), 351.

[113] Birch (1756–57), vol. II, 91.

[114] Bennett (1982), 63, 68; Id. (1989), 225.

[115] Id. (1975), 60.

[116] Id. (1981), 167, 172.

[117] Koyré (1968), 256.

[118] Wilkins (1802), vol. I, 144

Gilbert's study of earthly magnetism as complementary to Galileo's work on falling bodies.[119]

The principle of inertia is a fundamental component of Hooke's paper on the compound nature of circular motion. The evolution of Hooke's early cometary theory into a new celestial dynamics between 1665 and 1666 was largely due to the application of this principle to the study of comets. As Hooke wrote in 1687, this principle "will not admit of any other demonstration than that of induction from particular observations in natural motions."[120] This process of "induction," as Hooke describes it, required the emancipation from gravity to provide a principle of inertia applicable to the motion of celestial bodies. In the works of Galileo, the inertial motion of bodies was circular, since their weight was not considered the result of any external influence.[121] Descartes maintained that in a universe filled with matter, bodies are always forced to move along curved paths, but they have a constant tendency to recede from their circular motions.[122] Around 1659, Huygens defined this tendency as centrifugal force.[123] In the mechanical cosmology of Descartes and Huygens, this force is responsible of gravity, for it causes vortices of celestial matter that push bodies towards the centre of their circular motions. For this reason, Descartes described comets as "stars which migrate from one vortex to another," and their paths as lines "whose curvature varies according to the diverse movements of the vortices" through which they pass.[124]

Hooke rejected vortices and did not refer to centrifugal force.[125] "Any body moved circularly with any degree of velocity," he stated, "will at the instant that containing power is remov'd, proceed to move directly in the straight line of its tendency." In Hooke's opinion, this centripetal "conteining power" is a component of circular motion, along with the inertial tendency of bodies.[126] It seems, therefore, that in the study of cometary and circular motions in 1665, Hooke used the principle of inertia as formulated by Gassendi. For the first time Gassendi considered gravity as an obstacle, like resistance, to the rectilinear inertial motion of bodies.[127] Emancipated from gravity, the rectilinear tendency of Gassendi was employed by Hooke as the force contrasting the gravitational attraction of comets towards the centre of their motions.

---

[119] Ward (1740), 35.

[120] Hooke (1705), 355.

[121] Galileo (1890–1909), vol. II, 159–60; vol. VII, 118–9; Id. (1974), 77.

[122] Descartes (1964–74), vol. VIII.1, 111; vol. XI, 44; Id. (1983), 114; Id. (1984–91), vol. I, 96.

[123] Huygens (1888–1950), vol. XVI, 258–62; cf. Mormino (1993), 59–61.

[124] Descartes (1964–74), vol. VIII.1, 168, 176; Id. (1983), 150, 156–7.

[125] Hooke (1705), 167, 178–9; Westfall (1972), 486.

[126] Hooke (1705), 355.

[127] Gassendi (1658), vol. II, 354–5.

6.4 Attraction and Ether

The Society's programme to find out "whether gravitation be somewhat magneti-cal" continued in 1666. Hooke made some experiments on a *terrella*, and the fel-lows discussed the magnetic theories of Descartes and Gassendi.[128] In a lecture read in 1668, Hooke described the "circular motion of all planets" as the composition of "a direct motion which makes the indeavour to recede from the sun or the centre, and a magnetick or attractive power that keeps them from receding." But he later stated that the "attractive power of the earth" is "quite differing from that of a loadstone."[129] Although different, magnetic and gravitational attractions remained for Hooke similar forces. Echoing Roberval, in 1678 Hooke noted that the magnetic virtue "might be called an emanation of the Anima mundi, as gravity may be called another," because both seem to be *tota in toto & tota in qualibet parte*.[130] In Hooke's eyes, the two forces operate through subtle and imperceptible ethereal mediums and appear, therefore, spiritual rather than physical. Even "light may be said to be *tota in toto & tota in qualibet parte*." It may be called, therefore, the *anima mundi*. "And yet," Hooke concludes, "after all this we may prove it to be purely corporeal, and subjected to the same laws that bulky, tangible, and gross bodies are subject to." Light and gravity are "active," albeit corporeal, principles. Gravity, in particular, is neither immaterial nor "an innate quality." The analogies between gravity and light are intended by Hooke as decisive epistemological instruments. Like light, gravity is seen as a pressure originating from the internal motion of the emitting body, con-veyed by an ethereal medium. Thanks to this analogy, Hooke reached a quantitative determination of gravity:

> For this power propagated, as I shall then shew, does continually diminish according to the orb of the propagation does continually increase, as we find the propagations of the media of light and sound also to do, as also the propagation of undulation upon the superficies of water. And from hence I conceive the power thereof to be always reciprocal to the area or superficies of the orb of propagation, that is duplicate of the distance, as will plainly follow and appear from the consideration of the nature thereof, and will hereafter be more plainly evidenced by the effect it causes at such several distances.

Hooke was never able to demonstrate the inverse square law, nor did he succeed in "show[ing] experiments at distance sufficient to prove it experimentally and posi-tively." The analogy with light provided "a certain agreement and coherence" with similar "operations in nature."[131] In Hooke's experimental and mechanical philoso-phy, this proved sufficient.[132]

---

[128] Birch (1756–57), vol. II, 74–5, 85–6, 88.

[129] Hooke (1705), 183, 312–3, 322.

[130] Id. (1678), 11; Roberval (1644), 141.

[131] Id. (1705), 166, 176–8, 185.

[132] Nauenberg (1994), 333–4.

Hooke's focus on centripetal attraction and his refusal of Cartesian vortices did not lead to the belief that gravity is a force operating through void spaces.[133] In Hooke's cosmology, in fact, the emphasis on centripetal attraction coexisted with the belief that gravity is conveyed by a subtle ethereal medium. The apparent inconsistency between these two principles is largely due to historians privileging a Newtonian point of view of classical mechanics. *A parte ante*, i.e., from Hooke's pre-Newtonian point of view, there is no inconsistency.

Discussing the nature and properties of the ether, in the mid-1660s, Hooke asked "whether it permeates all bodies, be the medium of light, be the fluid body in which the air is but as a tincture." Significantly, Hooke also asked "whether it causes gravity in the earth or other celestial bodies," and "whether it assists in the action of fire and burning, and in the dissolution of other bodies by menstruums."[134] These queries suggest that ether played a major role in Hooke's natural philosophy. Although a physical explanation of gravity can be found in the *Lectures de potentia restitutiva*, a complete answer to the queries of the 1660s was provided only in the *Discourse of the Nature of Comets* read in 1682.[135] Gravity is not the effect of the pressure of atmospheric air over bodies. Rather it is due to the action of the "aether in which the atmosphere or air is but a kind of dissolution." This "thin and rarify'd, and exceeding fluid medium" does not operates on the external surface of bodies. Since it permeates them, gravitational ether affects only their internal structure. "The bodies most receptive of it," Hooke notes, "are such as have their particles of the greatest bulk and of the closest texture." Hooke clearly maintained that the "quantity of matter contained within the same space" is not the cause of gravity. It is "the modification of that matter and the receptivity it hath of the uniform power" of gravitation that makes bodies more or less heavy. Hooke's ether is not a simple and uniform fluid operating through traditional mechanical actions. It conveys one of the different motions originating from the internal parts of the earth. Each of these different internal motions is "communicated from the affecting to the affected body" by a distinct medium. "And so," Hooke adds, "I conceive also that the medium of gravity might be distinct and differing both from that of light, and from that of sound." According to Hooke, gravity, electricity, magnetism, light, and sound all consist in the motions of different mediums emitted from the internal constituents of bodies. Electricity, for instance, originates from "an internal vibrative motion of the parts of the electrick bodies," which a specific medium conveys to the bodies, whose internal texture is receptive to it.[136]

In *Micrographia* Hooke announced a new "theory of the magnet."[137] Hooke reiterated the announcement in 1666 and finally presented his hypotheses in 1682.[138]

---

[133] Royal Society Classified Papers, vol. XX, f. 117v; cf. Guicciardini (1998), 52–3; Id. (2011), 129.

[134] Hooke (1705), 29.

[135] Pugliese (2004), 955.

[136] Hooke (1705), 168, 181–4.

[137] Hooke (1665), 31.

[138] Birch (1756–57), vol. II, 85–6.

The delay is very relevant, because in 1665 Hooke's view of magnetical philosophy considerably changed. After 1666, Hooke no longer maintained that gravity was a magnetic force, and clearly distinguished between magnetic and gravitational attractions. Both these phenomena originate from the "internal motions of the earth." Gravitational attraction is universal; magnetism is not. The internal motions referred to by Hooke are of different kind, since they are conveyed by different mediums and affect different sorts of bodies. Yet they all originate from the motion of the earth "or any other celestial globulous body" around its axis. Since there are so many different subtle fluids, their effects depend on the congruity of their different motions with the textures of the bodies that they affect. Gravity, therefore, consists in the "receptivity" of bodies to the movement of gravitational ether only. Magnetism, on the other hand, is a "certain motion" of an "appropriate medium, that affects or moves certain bodies capable of receiving the impressions thereof according to determinate laws." What makes these bodies receptive to the "circular and vibrating motion" of the magnetic ether is their internal structure. According to Hooke, "all magnetic bodies have the constituent parts of them of equal magnitude and equal tone."[139] Although largely based on the action of a corporeal medium, Hooke's mature "theory of the magnet" was clearly different than some traditional mechanical views. In the late 1660s and early 1670s, Hooke took part in the discussions on Cartesian and atomistic hypotheses at the Royal Society and carried out experiments to test the alteration of magnetic attraction.[140] In *A Discourse of the Magnetical Variation*, read in 1686, he rejected the mechanical hypotheses based on "corpuscules," "magnetic effluvia" or "atoms." He also distanced his views from those held by scholars who reduced magnetism to a "spiritual" phenomenon due to an undefined "magnetic virtue," the "hylarchick spirit," or the "anima mundi."[141] One of these scholars was probably Martin Lister, whose paper on magnetism was read at the Royal Society in 1683. In the debate that followed, Hooke criticised Lister's ideas and conducted experiments to prove the corporeal nature of magnetic attraction.[142] The debate with Lister illustrates Hooke's position in Restoration natural philosophy. Far from defending a form of mechanical reductionism, Hooke's mature "theory of magnet" built on two apparently inconsistent principles: the motion of a corporeal fluid and congruity. This latter is as relevant as the former in defining the nature of Hooke's heterodox mechanical theory, and it can be considered inconsistent with a genuine mechanical philosophy only if one chooses to define mechanism in strictly Cartesian terms.[143]

Congruity was not the only innovative element that made Hooke's mechanical hypothesis heterodox. Hooke described the motion of the gravitational ether as an indefinitely short pulse that emanates from the interior of the earth. The source of

---

[139] Hooke (1705), 182, 192, 364.

[140] Birch (1756–57), vol. II, 85–6; vol. III, 124, 128–33.

[141] Hooke (1705), 484.

[142] Birch (1756–57), vol. IV, 265–6, 269–71.

[143] Cf. Pumfrey (1987), 9–10, 11, 14; Boantza and Ross (2015), 388–89.

the motion coincides with the point towards which earthly bodies are attracted; the direction of the gravitational ether is hereby opposed to that of the bodies on which it operates. This hypothesis is very different from more orthodox mechanical explanations. In 1690, for instance, Huygens updated Descartes's hypothesis on gravity. In Huygens' centrifugal model, the Cartesian *matière subtile* pushes the bodies towards the centre, whereas in Hooke's centripetal model the vibrating ether pulls them.[144] In both models the ether plays a mechanical role, since it is not an obstacle to gravity, but its medium.[145] Hooke's heterodox model was less intuitive and self-evident than Huygens'. Hooke was aware of the problems that his centripetal approach involved for a mechanical ethereal hypothesis. "It may," he noted, "perhaps seem a little strange how the propagation of a motion outward should be the cause of the motion of heavy bodies downwards." Unlike light and sound, gravity is due to "reversed" rather than direct vibrations of an ether.[146] To explain how, within a mechanical framework, such a reverse action could take place undoubtedly proved difficult.[147] Before Hooke, Gassendi undertook this task. Unlike ancient atomists, the French savant maintained that gravity was caused by an "attractive force" originating from the centre of the earth. "The motion of heavy bodies," he wrote, "is due to an external rather than an internal principle, namely the attraction of the earth." Weight depends on the quantity of matter of each body.[148] In Gassendi's view, the *tractorium instrumentum* consists in corpuscles emitted from the earth. But his writings do not provide a detailed explanation of how the motion of these corpuscles reverses once they impact earthly bodies. He seems to maintain that the *pondus* of atoms, their innate undirected motion, could solve the issue.[149] Like Gassendi, Hooke employed mechanical analogies rather than providing a clear explanation of the reverse action of the gravitational ether. "To make this more intelligible," Hooke compares it to something "very commonly known among tradesmen," the fixing of a hammer or axe in a helve:

> To do the easiest way, they commonly strike the end of the helve, holding the helve in their hands, and the axe or hammer at the lower end hanging downward, by which means they not only make the axe to go upon the helve, but make it ascend, if they continue striking, even to their hands.

Once again, Hooke employed an analogy derived from the work of craftsmen, and extended this mechanical model to every attractive motion, such as magnetism and electricity. Although it seems to be a mechanical reduction of forces at a distance, the analogy employed by Hooke offers only a partial explanation of his wider views. These, in fact, were based on the principle of congruity. Hooke, for instance, does not say which side of the helve is struck. The analogy with the action of the

---

[144] Huygens (1888–1950), vol. XXI, 455–7.

[145] Gaukroger (2002), 131–4.

[146] Hooke (1705), 171, 185.

[147] Freudenthal (1993), 163.

[148] Gassendi (1658), vol. I, 384–5, 389–90.

[149] Palmerino (2008), 151–6.

ether suggests that it is the side where the hammer is located. The helve is compared to the "medium of propagation," while the hammer is compared to the body subject to the direct motion of the medium, "so that at every stroke that is given by the globe of the earth to the propagating medium, one degree of velocity of descent is given to the grave body."[150] However, striking that side would produce the opposite effect to that described by Hooke. The hammer would move, but in the opposite direction to the hand. Where is the helve struck? Such a relevant detail is missing in Hooke's description. Nonetheless, it is a significant detail, since it is the element that could make the reverse motion possible. A corresponding element in Hooke's physical model is missing. Since in Hooke's natural philosophy there is no atomistic void, one can suspect that a non-gravitational ether plays this role. Moreover, the analogy with the striking hammer does not provide any explanation of the interaction between the gravitational ether and the internal textures of bodies. In Hooke's view, the weight of bodies does not depend on their quantity of matter. Rather it is due to their receptivity of the particular kind of motion propagated by the gravitational ether. Hooke seems to maintain that the internal texture is the cause of the congruity between bodies and ethereal motions.

Bibliography

Aiton, Eric. 1972. *The vortex theory of planetary motion*. London: Mcdonald.

Applebaum, Wilbur. 1996. Keplerian astronomy after Kepler. *History of Science* 34: 451–504.

Armitage, Angus. 1950. "Borell's hypothesis" and the rise of celestial mechanics. *Annals of Science* 6: 268–282.

Auzout, Adrien. 1665. *L'éphéméride du comète*. Paris.

Bacon, Francis. 1857–74. *Works*, 7 vols., ed. Robert L. Ellis, James Spedding, Douglas D. Heath. London: Longman.

Bennett, Jim. 1975. Hooke and Wren and the system of the world: Some point toward an historical account. *British Journal for the History of Science* 8: 32–61.

———. 1980. Robert Hooke as mechanic and natural philosopher. *Notes and Records of the Royal Society of London* 35: 33–48.

———. 1981. Cosmology and the magnetical philosophy, 1640–1680. *Journal for the History of Astronomy* 12: 165–177.

———. 1982. *The mathematical science of Christopher Wren*. Cambridge: Cambridge University Press.

———. 1989. Magnetical philosophy and astronomy from Wilkins to Hooke. In *Planetary astronomy from the Renaissance to the rise of astrophysics, Part A: Tycho Brahe to Newton*, ed. R. Taton and C. Wilson, 222–230. Cambridge: Cambridge University Press.

Bertoloni Meli, Domenico. 2006. *Thinking with objects: The transformation of mechanics in the seventeenth century*. Baltimore: Johns Hopkins University Press.

Birch, Thomas. 1756–57. *The history of the Royal Society of London*, 4 vols. London.

Boantza, Victor. 2011. From cohesion to pesanteur: the origins of the 1669 debate on the causes of gravity. In *Controversies within the Scientific Revolution*, eds. Marcelo Dascal and Victor Boantza, 77–100. Amsterdam and Philadelphia: John Benjamins Publishing Company.

---

[150] Hooke (1705), 185.

Boantza, Victor D., and Anna Marie Ross. 2015. Mineral Waters across the channel: Matter theory and natural history from Samuel Duclos' minerallogenesis to Martin Lister's chymical magnetism, ca. 1666–86. *Notes and Records of the Royal Society of London* 69: 373–394.

Boner, Patrick. 2006. Kepler on the origin of comets: Applying earthly knowledge to celestial bodies. *Nuncius* 21: 31–47.

———. 2013. *Kepler's cosmological synthesis: Astrology, mechanism and the soul*. Leiden: Brill.

Borelli, Giovanni Alfonso. 1666. *Theoricae mediceorum planetarum*. Florence.

Boyle, Robert. 1999. *The works of Robert Boyle*, 14 vols., ed. Michael Hunter and Edward Davies. London: Pickering and Chatto.

———. 2001. *The correspondence of Robert Boyle*, 6 vols., ed. Michael Hunter, Antonio Clericuzio and Laurence Principe. London: Pickering and Chatto.

Bucciantini, Massimo. 2003. *Galileo e Keplero: filosofia, cosmologia e teologia nell'eta' della controriforma*. Turin: Einaudi.

Charleton, Walter. 1654. *Physiologia epicuro-gassendo-charletoniana*. London.

Descartes, René. 1964–74. *Oeuvres*, 12 vols., ed. Charles Adam and Paul Tannery. Paris: Vrin.

———. 1983. *Principles of philosophy*. Trans. Valentine Rodger Miller and Rees P. Miller. Dordrecht: Reidel.

———. 1984–91. *The philosophical writings*, 3 vols., Trans. John Cottingham, Robert Stoothoff and Douglas Murdoch. Cambridge: Cambridge University Press.

Ducheyne, Steffen. 2012. *The main business of natural philosophy: Isaac Newton's natural philosophical methodology*. Dordrecht: Springer.

Freudenthal, Gideon. 1993. Clandestine Stoic concepts in mechanical philosophy: The problem of electrical attraction. In *Renaissance and revolution: Humanists, scholars, craftsmen and natural philosophers in early modern Europe*, ed. J.V. Field and Frank A. James, 161–172. Cambridge: Cambridge University Press.

Gal, Ofer. 2002. *Meanest foundations and nobler superstructures: Hooke, Newton and the "Compounding of the Celestial Motions of the Planetts"*. Dordrecht: Kluwer.

Gal, Ofer, and Raz Chen-Morris. 2005. The archaeology of the inverse square law: (1) Metaphysical images and mathematical practices. *History of Science* 43: 391–414.

Galilei, Galileo. 1890–1909. *Opere*, 20 vols, ed. Antonio Favaro. Florence: Barbera.

———. 1974. *Two new sciences, including Centres of gravity and Force of percussion*. Trans. Stillman Drake. Madison/London: University of Wisconsin Press.

Gassendi, Pierre. 1658. *Opera omnia*, 6 vols. Leiden.

Gaukroger, Stephen. 2002. *Descartes' system of natural philosophy*. Cambridge: Cambridge University Press.

Giglioni, Guido. 2013. How Bacon Became Baconian. In *The mechanization of natural philosophy*, ed. Daniel Garber and Sophie Roux, 27–54. Dordrecht: Springer.

Gilbert, William. 1600. *De magnete*. London.

———. 1900. *On the magnet*. Trans. Sylvanus Thompson. London: Chiswick Press.

Gouk, Penelope. 1980. The role of acoustic and music theory in the scientific work of Robert Hooke. *Annals of Science* 37: 573–605.

Granada, Miguel Angel. 2010. A quo movetur planetae? Kepler et la question de l'agent du mouvement planétaire après la disparition des orbes solides. *Galilaeana* 7: 111–141.

Guicciardini, Niccolò. 1998. *Newton: un filosofo della natura e il sistema del mondo*. Milan: Le Scienze.

Guicciardini, Niccoló. 2011. *Newton*. Rome: Carocci.

Heilbron, John. 1979. *Electricity in the 17th and 18th centuries*. Berkeley/Los Angeles/London: University of California Press.

Henry, John. 2011. Gravity and De gravitatione: The development of Newton's ideas on action at a distance. *Studies in History and Philosophy of Science* 42: 11–27.

Hooke, Robert. 1661. *An attempt for the explication of the phaenomena*. London.

———. 1665. *Micrographia*. London.

———. 1674a. *An attempt to prove the motion of the Earth by observations*. London.

————. 1674b. *Animadversions on the first part of the Machina coelestis*. London.

————. 1677. *Lampas*. London.

————. 1678. *Lectures and collections*. London.

————. 1705. *Posthumous works*, ed. Richard Waller. London.

————. 1726. *Philosophical experiments and observations*, ed. William Derham. London.

Horrocks, Jeremiah. 1673. *Opera posthuma*, ed. John Wallis. London.

Huygens, Christiaan. 1888–1950. *Oeuvres completes*, 22 vols., ed. Société hollandaise des sciences. The Hague: Martinus Nijhoff.

Johnston, Stephen. 2010. Wren, Hooke, and graphical practice. *Journal for the History of Astronomy* 41: 381–392.

Jullien, Vincent. 2006. *Philosophie naturelle et géométrie au XVIIe siècle*. Paris: Honoré Champion.

Kepler, Johannes. 1937–98. *Gesammelte Werke*, 20 vols., ed. Kepler-Kommission der Bayerischen Akademie der Wissenschaften. Munich: Beck.

————. 1992. *New astronomy*. Trans. William Donahue. Cambridge: Cambridge University Press.

————. 1995. *Epitome of Copernican astronomy and harmonies of the world*. Amherst: Prometheus Books.

Koyré, Alexandre. 1968. *Newtonian studies*. Chicago: University of Chicago Press.

————. 1973. *The astronomical revolution: Copernicus, Kepler, Borelli*. Trans. R. E. W. Maddison. London: Methuen.

Kuhn, Thomas. 1957. *The Copernican revolution: Planetary astronomy in the development of Western thought*. Cambridge, MA: Harvard University Press.

Lombardi, Anna Maria. 2008. *Keplero: una biografia scientifica*. Turin: Codice.

Mach, Ernst. 1919. *The science of mechanics: a critical and historical account of its development*. Trans. Thomas J. McCormack. Chicago/London: Open Court Publishing.

Manzo, Silvia. 2006. *Entre el atomismo y la alquimia: la teoria de la materia de Francis Bacon*. Buenos Aires: Editorial Biblos.

Milani, Nausicaa Elena. 2013. The Prodromus cometicum in the Académie des sciences and the Royal Society: The Hevelius-Azout controversy. In *Johannes Hevelius and his Gdańsk*, ed. Marian Turek, 195–208. Gdańsk: Societas Scientiarum Gedanensis.

Mormino, Gianfranco. 1993. *Penetralia motus: la fondazione relativistica della meccanica in Christiaan Huygens*. Florence: La Nuova Italia.

Nauenberg, Michael. 1994. Hooke, orbital motion and Newton's Principia. *American Journal of Physics* 62: 331–350.

Oldenburg, Henry. 1965–86. *The correspondence of Henry Oldenburg*, 13 vols., ed. Alfred Rupert Hall and Marie Boas Hall. Madison: University of Wisconsin Press.

Palmerino, Carla Rita. 2007. Bodies in water like planets in the skies: Uses and abuses of analogical reasoning in the study of planetary motion. In *Mechanic and cosmology in the medieval and early modern period*, ed. Massimo Bucciantini, Michele Camerota, and Sophie Roux, 145–168. Florence: Olschki.

————. 2008. Une force invisible à l'œuvre: le rôle de la vis attrahens dans la physique de Gassendi. In *Gassendi et la modernité*, ed. Sylvie Taussig, 141–176. Turnhout: Brepols.

Patterson, Louise Diehl. 1949. Hooke's gravitation theory and its influence on Newton I: Hooke's gravitation theory. *Isis* 40: 327–341.

————. 1952. Pendulums of Hooke and Wren. *Osiris* 10: 277–321.

Pepys, Samuel. 1995. *The diary of Samuel Pepys*, 11 vols., ed. Robert Latham and William Matthwes. London: Harper Collins.

Pugliese, Patri. 2004. Robert Hooke. In *Oxford dictionary of national biography*, ed. H.C.G. Matthew and Brian Harrison, vol. 27, 951–958. Oxford: Oxford University Press.

Pumfrey, Stephen. 1987. Mechanizing magnetism in restoration England: The decline of magnetic philosophy. *Annals of Science* 44: 1–22.

————. 1989. Magnetical philosophy and astronomy 1600–1650. In *Planetary astronomy from the Renaissance to the rise of astrophysics, Part A: Tycho Brahe to Newton*, ed. R. Taton and C. Wilson, 222–230. Cambridge: Cambridge University Press.

————. 1994. 'These 2 hundred years not the like published as Gellibrand has done de Magnete': The Hartlib circle and magnetic philosophy. In *Samuel Hartlib and universal reformation: Studies in intellectual communication*, ed. Mark Greengrass, Michael Leslie, and Timothy Raylor, 247–267. Cambridge: Cambridge University Press.

Purrington, Robert. 2009. *The first professional scientist: Robert Hooke and the Royal Society of London*. Berlin/Basel: Birkhäuser.

Rees, Graham. 1975. Francis Bacon's semi-paracelsian cosmology. *Ambix* 22: 81–101.

————. 1977. The fate of Bacon's cosmology in the seventeenth century. *Ambix* 24: 27–38.

Roberval, Gilles. 1644. *Aristarchi Samii de mundi systemate*. Paris.

Roux, Sophie. 2004. Cartesian mechanics. In *The reception of the Galileian science of motion in seventeenth-century Europe*, ed. Carla Rita Palmerino and J. M. M. H Thijssen, 25–66. Dordrecht: Kluwer.

Ruffner, James. 1971. The curved and the straight: Cometary theory from Kepler to Hevelius. *Journal for the History of Astronomy* 2: 178–194.

Shapin, Steven. 1994. *A social history of truth: Civility and science in seventeenth-century England*. Chicago/London: University of Chicago Press.

Sprat, Thomas. 1667. *History of the Royal Society*. London.

Vickers, Brian. 2007. Francis Bacon mirror of each age. In *Advancement of learning: Essays in honour of Paolo Rossi*, ed. John Heilbron, 15–57. Florence: Olschki.

Wallis, John. 1666. An essay of Dr. John Wallis exhibiting his hypothesis about the flux and reflux of the sea. *Philosophical Transactions* 263–289.

Wang, Xiaona. 2016. Francis Bacon and magnetical cosmology. *Isis* 107: 707–721.

Ward, Seth. 1653. *De cometis*. Oxford.

Ward, John. 1740. *The lives of the professors of Gresham College*. London.

Westfall, Richard. 1971. *Force in Newton's physics: The science of dynamics in the seventeenth century*. New York: American Elsevier.

————. 1972. Robert Hooke (1635–1703). In *Dictionary of scientific biography*, ed. Charles Gillispie, vol. VI, 481–488. New York: Scribner.

Wilkins, John. 1802. *The mathematical and philosophical works* , 2 vols. London.

Chapter 7
Matter and History

7.1 Physics and Metaphysics of the Earth

On 20 May 1663, Francis Glisson brought to the Royal Society a specimen of *lignum fossile*, "which was given to Mr. Hooke to have it cut even in order to see whether it would polish." When Hooke showed some thin polished slights of the specimen, Glisson noted that "the petrification of wood was occasioned by the passing of stony juices into the pores of wood throughout, and by filling them all up, and so coagulating there, without changing anything of the figure of the wood." Hooke's microscopic observations of the specimens supported Glisson's hypothesis.[1] Hooke's account was first published in Evelyn's *Sylva*, and then in his *Micrographia*. As the title suggests, however, the observation "of petrify'd wood and other petrify'd bodies" was not limited to the analysis of the specimen provided by Glisson. In *Micrographia* Hooke refined the hypothesis of corpuscular "transmutation," rejected the notion of "plastick virtue," and introduced the hypothesis of great transformations of the earth's surface. But Hooke still maintained that a "history of observations well rang'd, examin'd and digested" was needed in order to "perfectly and surely" know the "true original or production of all those kinds of stones."[2] He undertook this project soon after. After 1665, Hooke collected descriptions of "figured bodies" of all kinds, and observed several fossils through the microscope. In 1667, he reported to the Royal Society the observation of a sequence of horizontal and perpendicular "cliffs of stone near four miles together." Close to "a cliff in the isle of White" he found "shells of several sorts." In Hooke's view, only "some great earthquakes" could have produced those phenomena.[3] The first of Hooke's *Lectures and Discourses of Earthquakes*, written on September 15, 1668, includes five tables

---

[1] Birch (1756–57), vol. I, 245, 247–8, 260–2.

[2] Hooke (1665), 107–12.

[3] Birch (1756–57), vol. II, 183.

© Springer Nature Switzerland AG 2020

F. G. Sacco, *Real, Mechanical, Experimental*, International Archives
of the History of Ideas Archives internationales d'histoire des idées 231,
https://doi.org/10.1007/978-3-030-44451-8_7

of illustrations "design'd by Dr. Hooke himself," as Richard Waller noted. The "strictest survey" and diligent examination of "many hundreds of these figured bodies," along with the study of the "circumstances obvious enough about them," led Hooke to propose a consistent hypothesis on the nature of fossils and earth's history. In the lecture, Hooke reported detailed descriptions of the microscopic observation of many fossils. Although they "resemble" animals and vegetable bodies, "in all other proprieties of their substance, save their shape, are perfect stones, clays, or other earths, and seem to have nothing at all of figure in the inward parts of them." Such fossils are found below the surface of the earth and in places where the corresponding living species are not present. They are often "inclosed in some of the hardest rocks and thoughest metals." Through the comparative analysis of these "circumstances" about the "figured bodies" and their microscopic composition Hooke concluded that these are either remains of living organisms "converted into stones by having their pores fill'd up with some petrifying liquid substance," or the result of "impressions" made on these remains by fluid substances afterwards solidified. Hooke was aware of both the difficulties and the consequences of his conclusions. The corpuscular transmutations of organic bodies into stones are not ordinary processes, since "every kind of matter is not of it self apt to coagulate into a strong substance." These phenomena are "extraordinary" and mainly due to some "concurrent causes" originating from earthquakes and subterranean eruptions. Earthquakes raised "the superficial parts of the earth above their former level," created islands and mountains, but also "depressions and sinking of some part of the surface of the earth" leading to "vast vorages and abysses." During these transformations, "subversions, conversions, and transportations of the parts of the earth" originated the processes of "petrifaction" that transformed organic remains into mineral bodies. Hooke's hypothesis on the organic origin of fossils clearly entailed significant geomorphological transformations.

"A great part of the surface of the earth," he wrote, "hath been since the creation transformed and made another nature."[4] The presence of fossils of fishes in places far away from the sea is due to the alteration of geomorphology over time. These transformations account also for "the strange positions and intermixture of the veins of ores and minerals." Contrasting the ordinary force of gravity, earthquakes and subterraneous eruptions have "thrown up towards the surface of the earth" heavy mineral substances that are expected to be found, on the contrary, "under it to a sufficient depth." In Hooke's hands, the "figured bodies" became "monuments and records" of the history of nature:

> No coin can so well inform an antiquary that there has been such or such a place subject to such a prince, as these will certify a natural antiquary, that such and such places have been under water, that there have been such kind of animals, that there have been such and such preceding alterations and changes of the superficial parts of the earth.[5]

---

[4] Hooke (1705), 281, 288–91.

[5] Ibid., 305, 316–7, 321.

Hooke's innovative view of natural history did not emerge only from the diligent examinations of hundreds of "figured bodies," the microscopic observation of their substance, and the industrious collection of information on their "circumstances." In the late 1680s, after his views on fossils and earth's history were criticised by some members of the Royal Society, Hooke significantly appealed to Baconian ideas. He claimed that his "doctrines and conclusions" on "the causes and reasons of the present state and phenomena of the surface of the earth" were the result of "a methodical induction from the phenomena themselves." Those who did not accept this evidence were biased by some "Idola (as my Lord Verulam says) which pre-possess the Minds of some men, and molest them in the discovery and imbracing of the sciences."[6] Two decades earlier, on the contrary, in the first lecture on earthquakes Hooke called for the use of "some pre-design'd module and theory" in the study of fossils. Natural history should have "some end and aim," experiments and observations "some purpose."[7] Although he later compared the microscopic observations of fossils to what "the Lord Verulam called Experimenta Crucis," Hooke's use of the microscope in the study of fossils was not free of theoretical elements.[8] When Glisson presented the specimen of "petrified wood" to the Royal Society, Hooke was engaged in a series of microscopic observations. The whole project that led to *Micrographia* was built on some corpuscularian principles. It is not surprising, therefore, that in the report on the observations of Glisson's specimen, Hooke described the process of fossilization as the action of "stony and earthly particles" that were conveyed by some "fluid vehicle" into the "microscopical pores" of the wood. Once in these pores that are observable by means of a microscope, according to Hooke, these particles reached "also into the pores which may perhaps be even in the parts of the wood which through the microscope appears most solid."[9] A particular image of nature guided Hooke's inference from the observed porous structure of the petrified wood to the description of what his compound microscope showed still as solid and undivided. In the "laboratory of nature," simple and consistent "chymical operations" produce all mineral bodies.[10] "Metalline waters" crystalize and originate stones and metals, such as antimony and gold.[11] The "curious geometrical forms" of crystals are "very easily explicable mechanicanlly."[12] Hooke described all these chymical operations as the result of mechanical processes due to the central heat of the earth. A corpuscular approach to chemical change led Hooke to the belief that fossils were the product of the "transmutation" of the "texture or schematism" of vegetable and animal remains by some "stony waters." On this ground, Hooke

---

[6] Ibid., 433.

[7] Ibid., 280.

[8] Ibid., 339; Lawson (2016), 25–6.

[9] Birch (1756–57), vol. I, 243, 262.

[10] Hooke (1705), 296, 315.

[11] Birch (1756–57), vol. II, 193; Hooke (1665), 94–5.

[12] Hooke (1705), 281.

rejected "plastick virtues" as contrary to the "great wisdom of nature," and questioned the account of earth's history derived from the Bible.[13]

Galileian astronomical discoveries played a significant role in the emergence of Hooke's ideas on fossils.[14] It was in *Micrographia* that Hooke first referred to "some deluge, indundation, earthquake, or some such other means" as the causes of the fossils. In the same work, he described the inequalities of the moon's surfaces. "These seem to me," he claimed, "to have been the effect of some motions within the body of the moon, analogous to our earthquakes."[15] In 1668, he added that the "several smoaks, and clouds, and spots that appear on the surface of the sun" are signs of internal combustion.[16] Telescopic observations provided Hooke with powerful analogies that helped situate his ideas on earth's history within a wider view of nature.[17] Internal combustion, subterraneous eruptions, and earthquakes took place in all the main celestial bodies of the solar system. In Hooke's opinion, these alterations were signs of a general dynamic principle underlying all natural productions. This principle emerged, for the first time, in December, 1664, when Hooke began his study of comets and planetary motions. Hooke introduced the hypothesis that gravity might have been altered "either by time or place." This was consistent with the general principle "that all bodies and motions in the world seem to be subject to change." Signs of this principle could be found in the sun itself, as Galileo showed.[18] Hooke's new system of the world was a self-regulating and evolving mechanism.[19] Such a dynamic cosmology removed one of the biggest obstacles to the organic hypothesis on the nature of fossils. The fact that great geological transformations have changed the shape of earth's surface is "not only possible, but probable, and altogether consonant and agreeable to the rest of the works of nature." The opposite view was inconsistent with the natural order, because of the unavoidable effects of motion on matter. Since the nature and effects of physical principles change over time, the study of natural phenomena should not be limited to the description of their current condition, but focus on "the constant and more necessary effects, which are produced by the working power, before it produces its final and ultimate effect." The knowledge of the formation and alterations of natural bodies is like "a torch, drum, or light, by which we are guided in our pursuit of nature, and be inabled to distinguish in what steps, and which ways nature proceeds."[20] Descartes expressed a similar opinion in the *Principia*. The knowledge of the principles that produced everything, according to Descartes, "explain[s] their nature much better than if we were merely to describe them as they are now." But the new world that emerged

---

[13] Hooke (1665), 109, 112; cf. Ross (2011), 166.

[14] Galileo (1989), 39–50.

[15] Hooke (1665), 111, 243.

[16] Hooke (1705), 326–7.

[17] Pineda de Avila (2015), 37–8, 42.

[18] Birch (1756–57), vol. I, 506.

[19] Patterson (1950), 44; Drake (1996), 78–9.

[20] Hooke (1705), 45, 349, 450.

from the chaos of poets that Descartes depicted in *Le Monde* did not translate that knowledge into a historical view of the earth or the universe.[21] In Hooke's works, the epistemological relevance of the processes of formation and alteration of natural phenomena was a consequence of an image of nature as a mechanism in perpetual motion. This image guided Hooke in the observations of fossils specimens in the mid-1660s, leading to the acknowledgement of the historical dimension of nature.

7.2 Catastrophes or Catastrophism?

The morphological alterations of the earth had been ongoing "since the creation," but the greatest part took place in "the younger dayes of the world."[22] Early earthquakes "have heretofore not only been much more frequent and universal, but much more powerful." In the *Discourse of Earthquakes* these early phenomena are often described as "catastrophes of the past." Hooke even considered the possibility that some variations of earth's axial displacement in the early ages contributed to these catastrophes.[23] Thus, Hooke's geological ideas have often been considered as an early form of catastrophism, or rather "actualist catastrophism," for Hooke thought of early earthquakes as similar in nature but more powerful and frequent than those that could be observed in his time.[24] According to Stephen Toulmin and June Goodfield, Hooke's catastrophism was the result of the limiting effects of the biblical timeframe on an otherwise revolutionary view of earth's history. Instead of completing a revolution in geology by adopting an expanded chronology consistent with his dynamic view of earth's past, Hooke "compressed" geological history and introduced a conjectural distinction between early and late earthquakes.[25]

In contrast to Toulmin and Goodfield's account, Ellen Tan Drake lamented that "it is too facile, thus, to claim that Hooke was bound by the biblical chronology."[26] In several works, she claimed that Hooke's revolution in natural history was not incomplete, rather it anticipated James Hutton's ideas, Darwinian evolutionary theory, and continental drift.[27] Since these accomplishments presupposed a longer chronology, Drake did not hesitate to consider it as an implicit premise that Hooke never made explicit because of its religious implications. "Hooke," she wrote, "showed that he desperately needed the earth to be older than a few thousand years in order to accommodate his cyclicity of processes, his polar wandering and his

---

[21] Descartes (1964–74), vol. XI, 33–4; see also Roger (1982), 91, 101–2.

[22] Hooke (1665), 243.

[23] Hooke (1705), 322, 326.

[24] Hooykaas (1959), 4; Huggett (1997), 60.

[25] Toulmin and Goodfield (1967), 90; see also Westfall (1972), 486; Albritton jr (1980), 51–2.

[26] Drake (1996), 101.

[27] Drake (1981), 964–71; Drake (2005), 73–4, 91–2; Drake and Komar (1983), 14–16; see also Chapman (2005), 140; for a different view see Ranalli (1982), 322–5; Id. (1983), 283; Id. (1984), 187.

theory of evolution."[28] Drake's conclusions were built on historically questionable inferences. She claimed, for instance, that "it is inconceivable that the mathematically minded Hooke, would not have made some quick mental estimates and calculations" on the chronological consequences of his geological ideas.[29] In Drake's opinion, the supposedly irresistible internal logic of Hooke's ideas provides more compelling evidence than his several statements rejecting longer chronologies. By focusing only on some ideas abstracted from the rest of the *Discourse*, Drake described Hooke as a forerunner of modern uniformitarianism.

Catastrophism and uniformitarianism often form a misleading dichotomy; because of their own history they have become an obstacle to historical analysis.[30] The alternative between a long history of slow processes on one hand and a shorter history punctuated by violent transformations on the other emerged only after the acknowledgement of the legitimacy of a historical investigation of the earth's past.[31] Hooke significantly contributed to the emergence of the history of nature as an object of natural history. The introduction of the effects of time in a traditionally descriptive and classificatory discipline raised some questions on matters of evidence and proofs. As Halley noted in 1694, for the greatest majority of seventeenth-century historians, naturalists, and philosophers there were two sources of knowledge about the past, "revelation" and "induction from a convenient number of experiments or observations." Physical phenomena that were no longer observable could not, therefore, be considered as intelligible causes.[32] Hooke described early alterations as a series of catastrophes, but chose earthquakes because they were causes of phenomena still observable. However, neither in contemporary nor in ancient natural histories could Hooke find "any parallel instances that can countenance such mutations, changes, and catastrophes as are, and must be supposed to solve these phenomena." The absence of the "observations" that Halley considered unavoidable requisites of a correct historical "induction" was not an insurmountable difficulty for Hooke. "The experience and duration of a man," he noted, "whether he looks forward or backward, is very short in comparison of what seems requisite for this determination." Hooke knew that a direct observation of many phenomena was often impossible to achieve because of limited human perception, both of time and space. Optical instruments helped to reach parts of space that were too far or too small, but no instrument could help to see "events so far distant both before and behind himself in time, as if close by, and now present." When direct evidence is impossible to secure through "observations, experiments, and trials," analogies and comparisons help to build a hypothetical inference from the known to the unknown.[33] The observations and reports of earthquakes and volcanic eruptions provided Hooke

---

[28] Drake (2006), 144.

[29] Drake (2007), 26.

[30] Gould (1987), 9; Rappaport (1997), 5.

[31] Rossi (1984), VII-IX.

[32] Halley (1724–25), 126; cf. Schaffer (1977), 28.

[33] Hooke (1705), 343, 371, 449.

with instances of the effects that these causes could produce. Similar effects, though bigger in entity, could therefore have been produced by the same cause when "there have been much greater plenty of those kind of minerals which have been consumed." Such an image of science helped Hooke to shape and support his ideas on earth's history.[34] That the current morphology of the earth is the effect of earthquakes that "are much more considerable than those which are now produced" cannot be rejected simply "because we do not now find instances of effects of the same grandure produced in our present age."[35]

Catastrophism, uniformitarianism, and actualism, therefore, were not part of Hooke's philosophical world. The Bible did not prevent Hooke from acknowledging the historical dimension of nature. Nor did its short chronology compel Hooke to introduce a distinction between early catastrophes and late earthquakes.[36] The boundaries of biblical chronology were not rigid. It could in fact be stretched from six to almost nine thousand years.[37] The chymical processes employed by Hooke seemed to provide enough time to accept even a shorter version of the Biblical chronology. In 1672, Hooke told the fellows of the Royal Society "that he was credibly informed, that there was a ground in Bedforshire, which would in a twelvemonth's time turn wood and other matter, that was not stony, into stone, without vitiating the figure."[38] The processes of corpuscular transmutation seemed consistent with a framework of six thousand years. Longer chronologies were seen with scepticism. "Some heathen historians," Hooke noted, "have assigned space large enough and even beyond belief almost." Building on the work of Ulug Beg, Hooke reported "that the Chinese do make the world 88,640,000 years old."[39] These accounts could not be credited. The long chronology of the Egyptians, for instance, was due to "no other cause but to make the world believe they were preceding all others in antiquity." But their years might just have been lunar rather than solar years.[40]

A biblical chronological framework seems to be an unquestioned assumption in Hooke's natural history.[41] Hooke claimed that Moses died "in the 1552 year before Christ's nativity" and Noah's flood took place "between sixteenth and seventeenth hundred years according to the Hebrew."[42] It is significant that these two dates are close to different accounts of the biblical chronology. The earlier is four years away from the estimates of Johann Alsted, Gherardus Vossius, and Joseph Juste Scaliger.[43]

---

[34] Porter (1977), 72; Oldroyd (1972), 111–2.

[35] Hooke (1705), 427.

[36] Rudwick (2014), 46; cf. Levitin (2013), 324; Poole (2010), 41–3.

[37] Rappaport (1997), 190–2; Poole (2010), 41–3.

[38] Birch (1756–57), vol. III, 75.

[39] Hooke (1705), 395; cf. Poole (2006), 48.

[40] Hooke (1705), 328, 404, 408; cf. Grafton (1995), 27.

[41] For a different view see Drake and Komar (1983), 14; Drake (2007), 25–6; Jackson (2003), 73–4.

[42] Hooke (1705), 389, 414.

[43] Scaliger (1629), 371; Alsted (1650), 7; Vossius (1659), 39; cf. Bibliotheca Hookiana, 2, 22, 71.

The latter is close to the date established by Jacob Ussher.[44] This seems to suggest that Hooke was not interested in defining a clear chronology consistent with one of the several versions of the Bible that circulated in the seventeenth century. He rather aimed to reconstruct the sequence of great natural catastrophes to understand the principles governing the earth. "Nature itself," he wrote, "has preserved somewhat of the memory of them by the medals or indelible characters of shells or other petrify'd, or otherwise preserved substances, which any, that have senses and understanding, may read." Influenced by the works of antiquarians, Hooke described fossils as "medals, urnes, or monuments of nature." The "natural antiquary" can build on them a "natural chronology." But Hooke's aim was "not so properly to search and inquire into the history, as to find out what have been the natural or physical causes" of the present form of the earth. The reconstruction of the sequence of geological transformations should lead to the understanding of "the method of proceeding" of Nature, which in Hooke's eyes seems "very constant, uniform and regular." This aim could be achieved thanks to those "monuments and hieroglyphic characters of preceding transactions."[45]

As we have seen, on the Isle of White Hooke observed a sequence of vertical and horizontal "cliffs of stones," and explained it as the effects of a series of earthquakes.[46] Why, then, did he not provide us with a new and longer natural chronology which he seemed to refer to? In fact, Hooke's geological mechanism did not need additional time, nor did he find any reason to question the traditional biblical timeframe. The sequence of "cliffs of stones" observed in 1667 was not, in Hooke's eyes, an instance of superposition. Neither had he seen a sequence of faunal remains in the hundreds of fossils that he observed. The principles of superposition and faunal succession were not part of Hooke's natural history.[47]

Hooke was aware that "to raise a chronology" out of fossils was very difficult, yet "not impossible" thanks to the evidence provided by "other means and assistance of information." The chronology Hooke was interested in was a sequence of prodigious natural transformations that could help to grasp their underlying mechanism. Hooke, in other words, did not read fossils as proofs of a longer natural history, nor as means to revise civil history. In Hooke's programme, the literary and historical sources integrate the evidence that emerged from fossils. The works of ancient poets, philosophers, and historians offer elements for a "chronologie of the preceeding times" that consists in a sequence of earthquakes and inundations that altered the primeval earth. The natural historian should handle these additional sources with care. The information they can provide is conveyed in "hieroglyphick and mythologick characters." The "chronologick account couched" in ancient sources should be read against the aims of these works. Ancient pagans, for instance, mentioned several deluges. According to Hooke, these are different events, none of

[44] Ussher (1660), 44; cf. Bibliotheca Hookiana, 8.

[45] Hooke (1705), 321, 335, 341, 411, 432; cf. Schneer (1954), 257, 263.

[46] Birch (1756–57), vol. II, 183.

[47] For a different view see Albritton jr (1980), 50; Oldroyd (1996), 62.

which preceded Noah's flood. The floods of Ogygges and Deucalion took place after the biblical deluge. The ancient Egyptians seem to have had "the notion of a Deluge" that is comparable to the universal flood of the Bible. In their accounts it happened long before Noah's one. Hooke rejected this claim, but did not take a clear position about the relationship between the two accounts.[48]

The "Bedforshire's ground" provided Hooke with evidence of the short time in which corpuscular transmutations could occur, but it provided no evidence in support of the different intensity and frequency of the early earthquakes. Rather than an effect of chronological constrains, this element of Hooke's history of the earth is a consequence of his chymical theories.[49] The great catastrophes of the past contributed most to shape the surface of the earth. The "greater islands and continents" were produced by these earthquakes thanks to the "greater magazines, or stores of the material fitted for this purpose." These transformations did not take place "at one time," but during "many ages." In the following "ages of the world" minor earthquakes created the main mountain chains of Europe and smaller islands. According to Hooke, these processes of transformation continued, but with minor effects due to the reduction of the material responsible for the internal combustion of the earth.[50]

In 1668, Hooke claimed that a possible shifting of the earth's poles could have produced the early great changes of geomorphology. The relationship between astronomical and chymical causes in Hooke's work is far from clear.[51] The "alteration of the center of gravity" was initially presented as an alternative to early earthquakes. It could have produced morphological alterations "by the departing of the waters to another part or side of the earth." A significant consequence of this phenomenon could have been the alteration of the meridian direction and latitude. The observation of the variation of magnetic direction, therefore, was for Hooke a proof of it. Earthquakes remained, nonetheless, a more probable cause. "I confess," Hooke wrote, "I do incline to believe that what mutations there have been of the superficial parts, have been rather caus'd by earthquakes and eruptions." In early 1687, Hooke reiterated the hypothesis of axial displacement, claiming that this was not "the only cause" neither the "principal," but an "adjuvant cause" of geological alterations.[52] Informed by Halley on the content of Hooke's new lectures, John Wallis voiced a general scepticism when he noted that Hooke did not provide evidence in support of his hypothesis.[53] In reply to Wallis, Hooke defended his position by claiming that the shift of earth's poles was not the only cause of geomorphology. But in the following lectures, Hooke's position changed.[54] Geological alterations were now "most probable ascribable to another cause, which was earthquakes and subterranean

---

[48] Hooke (1705), 389, 401, 408, 411.

[49] Rudwick (1976), 73–4.

[50] Hooke (1705), 320, 417, 422, 440.

[51] Oldroyd (1989), 231; Carozzi (1970), 86.

[52] Hooke (1705), 291, 3290–2, 328, 342, 345, 347.

[53] Turner (1974), 168–70; cf. Ray (1693), 73; Chapman (1994), 172–3.

[54] Rappaport (1986), 144.

eruptions of fire."[55] Unable to provide astronomical evidence, he unburdened his general theory of earth's past from a controversial hypothesis. Hooke's belief, however, did not change. "Since I have found," he wrote in 1689, "that severall other consequences have proved true, I shall therefore not despair of this till experiment has evidenced the contrary."[56] The variation of the axis and the centre of gravity remained part of Hooke's dynamic cosmology.[57]

The main causes of geological transformations were in Hooke's view the effects of processes of internal combustion of the earth. This "subterranean inkindling" is mainly due to a substance that reminds Hooke of "gun-powder," and is stored in subterranean caverns. Hooke's description of geodynamics is based on his chymical ideas. While claiming that the "fuel or cause of the subterraneous fire may be much wasted and spent by preceeding conflagrations," in the first lecture of 1668 Hooke hinted at the possibility that the remaining quantity of "fuel" could be enough to produce significant changes in the planet. He noted, furthermore, that "there may be some causes that generate and renew the fuel."[58] The permanence of central fire in the earth was a debated topic in the chymical tradition.[59] But Hooke did not assume any of the solutions proposed by early modern chymists. In the later lectures his position became more definite. Although he still noted that the same effect might be produced by different causes, Hooke excluded the possibility that a "renovation of the foment" was possible. The spent volcanoes were, in his eyes, an eloquent proof of a planet that was "growing older in respect of time and duration," as much as in "its constitution and powers."[60] The depletion of earth's internal combustible was, perhaps, the main consequence of Hooke's chymical theory on his ideas on geodynamics. For this reason, he maintained that "in former ages before we had history" the internal combustible and the relative fluidity of the planet produced "eruptions and conflagrations which have infinitely surpassed any that have happened of later years."[61]

7.3   Decadence vs Evolution

Polar shift and earthquakes were more than two alternative physical hypotheses, for they offered two antithetic models of history of the earth. The earlier entailed a cyclic process of transformation, whereas the latter led to a directional Earth

---

[55] Hooke (1705), 372.

[56] Royal Society Classified Papers, vol. XX, f. 172r.

[57] A dynamic but not vitalistic cosmology; cf. Kubrin (1990), 68–9.

[58] Hooke (1705), 327, 420.

[59] Debus (1977), 90–2.

[60] Hooke (1705), 426; Journal Book of the Royal Society, vol. IX, 2–3.

[61] Hooke (1705), 426, 440.

history.[62] Despite the downgrading of the hypothesis of axial displacement to a secondary and less probable cause of geodynamics, in the *Discourse* Hooke often referred to cyclical processes. Wind and "streams of water" erode earth's surface. Unlike early earthquakes, these processes can still be observed. Their effects are also opposite to those of earthquakes and subterranean eruptions. The earlier produced "all manner of asperity and irregularity of surface of the earth," while the latter "indeavour[s] to reduce them back again to their pristine regularity." The balancing outcome "is indeed consonant to all other methods of nature, in working with opposite principles." "A continual circulation" is then produced.[63] Was Hooke embracing a cyclical view of geodynamics? Those who saw in Hooke's work an implicit longer chronology described Hooke's theory as a Christianised form of cyclical thinking.[64] The different effects of earthquakes and water in shaping current geomorphology suggest instead that Hooke saw earth's history as a directional process.[65] In Hooke's opinion earthquakes were the universal "active principle" that has altered the primordial plain surface of the earth since the creation. Atmospheric agents, on the contrary, were less effective and needed "length of time." The smooth earth of the golden age was dramatically altered by natural catastrophes. After the big earthquakes of the iron age, the depleted "fuel" has not allowed any further great alteration. The earth ages and "doth, as it were, wash and smooth its own face, and by degrees removes all the warts, furrows, wrinkles and holes of her skin." The consistent action of gravity did "really shrink and grow less" the more fluid and bigger "body of the earth" emerged from the creation.[66] Hooke's idea of an ageing and decaying planet was a consequence of his chymical explanations of earthquakes and subterranean eruptions. It is significant that it found support in Burnet's work and was criticised by John Evelyn and other fellows of the Royal Society.[67] In Hooke's opinion, it was "not only possible, but probable, nay necessary." This view of earth's history was consistent with "the general method of nature, which is always going forward and continually making progress of changing all things." Hooke seems to place this linear process within the biblical framework of creation and final dissolution. Fossils are a "symptom of old age," whereas abundance of "combustible or inflammable juices and moisture is a sign of youth." The depletion of the internal combustible after the iron age led to the prevailing action of atmospheric agents over the effects of earthquakes. This suggests that in its later ages the earth could recover the original smoothness of the "preceding estate." Along with the earth, the

---

[62] Oldroyd (1989), 226; Oldroyd (2006), 38.

[63] Hooke (1705), 304, 312.

[64] Davies (1968), 89; Drake and Komar (1983), 14–5; Drake (1996), 3, 84, 120–2; Drake (2007), 20; According to Yushi Ito, Hooke's commitment to a "modified cyclic theory" clearly emerged in his criticism of Thomas Burnet's "non-cyclic" theory, Ito (1988), 303, 305. As Stephen Gould has shown, Burnet rather harmonised a cyclic view of Earth's history with biblical tenants, Gould (1987), 21–44; see also Bettini (1997), 127.

[65] Rudwick (1976), 75; Porter (1977), 72; Rappaport (1986), 135.

[66] Hooke (1705), 311, 316, 348, 424, 459–61.

[67] Journal Book of the Royal Society, vol. IX, 3; Rossi (1984), 36–7.

planets, the stars, the sun, the moon, and "the heavens themselves" show "many expressions that denote a continual decay, and a tendency to a final dissolution."

In Hooke's theory the surface was not the only part of the earth affected by this process. Fossils also proved that "terrestrial beings" were involved in the general transformation of nature. After "the first creation" every species has produced a sequence of "states" and "appearances" that concluded in "the final dissolution." Like individuals, animal species grow, decay, and die.[68] This conclusion was not only a consequence of the organic hypothesis on the origins of fossils. Hooke's image of nature played a significant role in his interpretation of the fossil specimens found in the British Islands. It was by focusing on fossil remains of still living organisms that Niels Stensen and Agostino Scilla, instead, defended the organic origins of fossils first introduced by Fabio Colonna. "Every specimen," Scilla remarked, "that we find in stones is exactly identical to sea creatures of the same species."[69] The only discrepancies between fossils and living beings of a species that Stensen observed in Italy consisted in bigger dimensions. This was not, in Stensen's view, "so considerable" an objection to the organic hypothesis, since individuals of bigger dimensions than usual have always been observed.[70] Stensen's argument proved insufficient to explain the major differences between living organisms and fossils found in the British Islands. Among the "infinite number and great variety of shells" observed by Martin Lister, many showed features of "species or race yet to be found in being at this day."[71] These remains even resisted traditional classifications.[72] Following Lister's call for a focus on accurate descriptions rather than deceiving similitudes, Robert Plot concluded that the differences between living organisms and British fossils were too significant to conclude that they belonged to the same species. These fossils, therefore, could not be remains of non-existent organisms; like all other fossils, they were nothing else than *lapides sui generis*.[73]

When Hooke read to the Royal Society the first lecture on fossils, Stensen's *Prodromus* was just being published in Florence. From 1668 on, Hooke acknowledged the differences between fossils found in Italy and in the British Islands. Although the latter "retain such particular characteristicks as are sufficient to denote and show to what species they belong," many show differences in magnitude and figure from individuals of those species that live in the same place where the fossils are found. No objection to the organic hypothesis was "more pressingly urged than this." Consistent to his image of nature, Hooke concluded "there may have been divers species of things wholly destroyed and annihilated, and divers others changed and varied." Since "there are some kind of animals and vegetable peculiar to certain places, and not to be found elsewhere," it was possible for Hooke that after

[68] Hooke (1705), 425, 427, 435.

[69] Scilla (1670), 150; Id. (2016), 150.

[70] Stensen (1668), 62; Id. (1671), 88–9.

[71] Lister (1671), 2282–3.

[72] Lhwyd (1699), 44–5, 59, 86–7.

[73] Plot (1676), 104–5.

geological alterations the remains of these living organisms were carried by the surface on which they lived. The absence of living organisms corresponding to some fossils was due to the disappearance of some species and the formation of some others "which have not been from the beginning."[74]

Hooke's conclusions were built on the assumption that environment had a direct influence on animals' bodies. Discussing the "various and seemingly irregular" generations of insects in *Micrographia,* Hooke noted that these animals "change both their skin and shape" and "may be quite alter'd by the hew of their progenitors" when alterations occur in their environment, "like Mores translated into northern European climates."[75] Individuals of the same species might acquire significant physical variations if generated in "different places" and "in different times." Thus "new varieties" of the same species emerged when a great alteration of climate and nourishment produced "a very great alteration in those bodies that suffer it." This is why, Hooke concluded, there are "divers kinds of petrify'd shells, of which kind we have none now naturally produced."[76] Hooke's biological ideas, therefore, were not an anticipation of modern evolutionary theories. Rather than a forerunner of Darwin, as he has been described, Hooke was a seventeenth-century natural philosopher whose most innovative ideas were strictly linked to traditional beliefs.[77]

For Hooke, the hypothesis of disappearance and transformation of species was the main but not the only reason for the fossils found in the British Islands. He did not rule out that "a more full account of the production of the shores and oceans" could show the living organisms of some of the most controversial specimens. The works of travellers, explorers, and naturalists have shown many new species unknown in Europe. "We might with reason enough assert," Hooke claimed, that "there are many more yet latent, which time may make manifest."[78] But unlike Ray and John Wooodward, who made of this their only answer to the question raised by British fossils, Hooke did not "insist further on this way of defence" of the organic hypothesis.[79] Until the missing organisms are found, the disappearance and transformation of species remained for Hooke a "real and true" explanation.[80]

This was also the most radical solution to what Hooke saw as the most challenging question concerning the organic hypothesis. In the eyes of his critics, Hooke's hypothesis entailed a radical departure from the then current view of the *scala naturae.* According to Plot, it was inconsistent with the nature of the "providence which took so much care to secure the works of the creation in Noah's flood."[81] Others remarked that the extinction of species entailed that in the "first creation" there were

---

[74] Hooke (1705), 291, 327, 337, 342.

[75] Hooke (1665), 193–4, 206.

[76] Hooke (1705), 56, 327–8.

[77] Poole (2010), 127; cf. Lewis (2012), 64.

[78] Hooke (1705), 344–5.

[79] Ray (1693), sig. b2v; Woodward (1695), 25–6.

[80] Hooke (1705), 435.

[81] Plot (1676), 113–4.

more species "than what it was absolutely necessary to its present and future state, so would be a great derogation from the wisdom and power of the omnipotent creator." Hooke knew that the notion of providence that was challenged by his hypotheses was built on a view of nature as a static, balanced, and immutable system. It was, then, to this level that he moved the discussion with his critics. The transformations and decadence of both celestial bodies and terrestrial beings were not, according to Hooke, alterations of the ordinary course of nature, but the ordinary course itself. "We find nothingin Holy Writ," Hooke argued, "that seems to argue such a constancy of nature." The world created by God is perfect, but inexorably leading towards a final dissolution. Transformation and decadence are, for Hooke, fundamental features of Christian providence. According to these principles, nature has produced a chain of organisms in "a gradual transition from one to another." "Harmony, consent and uniformity" remained at the core of Hooke's *scala naturae*, but were presented as effects of a process rather than attributes of an unchanged and unchangeable creation.[82]

7.4   Ovid and Moses

In spite of the absence of any proof of early catastrophes in ancient natural histories, Hooke maintained that many ancients "had some knowledge of the catastrophys of some parts of the world."[83] Ancient poets and philosophers rather than natural historians recorded and passed down over centuries this knowledge. That ancient myths had allegorical meanings was a belief shared by many naturalists and historians in early modern Europe. Bacon described ancient myths "as sacred relics and light airs breathing out of better times" that the Greeks had conveyed to us.[84] For Bacon the literal meaning of many fables was too absurd. There had to be, therefore, an allegorical meaning behind these narrations. In *De sapientia veterum* and *De principiis*, Bacon used ancient myths to support his new ideas on the nature of matter and the reform of science.[85] Influenced by Bacon, Hooke maintained that the knowledge of these transformations of the mythical times passed to Egyptians, Chaldean, and Greeks of the historical time. Among these latter, Hooke elected Ovid's *Metamorphoses* as the summa of the ancient knowledge on nature.[86] In Hooke's opinion, the works of Ovid and other ancient poets and philosophers provided the evidence of early catastrophes that was absent in natural histories.[87]

---

[82] Hooke (1705), 341, 433, 435–6, 450.

[83] Ibid., 374.

[84] Bacon (1857–74), vol. VI, 627, 698.

[85] Rossi (1968), 88–93.

[86] Hooke (1705), 377.

[87] Birkett and Oldroyd (1991), 156; Rappaport (1997), 70–1, 96–7, 102–3; Levitin (2015), 215–8.

Hooke's geological hypotheses were criticised by many in the Royal Society. John Wallis, Robert Plot, and Martin Lister expressed scepticism and lamented the conjectural character of Hooke's conclusions.[88] From 1687 on, Hooke devoted a large parts of his geological lectures to ancient myths, as part of a strategy to address criticism.[89] Following Bacon's preference for the "light of nature" over the "darkness of antiquity," in 1680 Hooke opposed the straightforward study of nature to the "hunt for the causes of things among the worm-eaten volumes of antiquity."[90] But the criticism of some fellows demanded an answer, and a few years later Hooke appealed to mythology for additional evidence for his geological hypothesis. This strategy did not prove effective. Facing the increased scepticism of his audience regarding his use of myths and fables, Hooke acknowledged in 1693 that different interpretations were possible and that his geological hypothesis did not depend on ancient sources, but "must stand and fall with its fate." Nature itself is the only witness "that must as it were *viva voce* declare it."[91]

Nonetheless, the emphasis on ancient mythology in Hooke's works remained a relevant component of his view of earth's history.[92] Already in 1668 Hooke had employed ancient myths as sources of evidence alternative to the unsatisfactory natural histories of the past. In the first lecture on fossils, Hooke led the foundation of a view of the past that legitimized euhemerist interpretations. "It is not impossible," he wrote, "but that there may have been a preceding learned age wherein possibly as many things may have been known as now, and perhaps many more." The early catastrophes transformed the primeval earth and annihilated a golden age of learning. The knowledge of these catastrophes, along with the achievements in the arts and sciences reached before was nonetheless hidden behind the apparently meaningless myths of the ancient people who inherited it. On this ground, two decades later Hooke resumed and expanded his euhemeristic theory in defence of his hypothesis. He called for a closer and more detailed study of ancient sources, widened to include "mythologick history." A literal interpretation of pagan myths could make them "sufficiently ridiculous and impious." But Hooke thought that some of them had a different meaning, since they conveyed "a history of the production, ages, states, and changes that have formerly happened to the earth." Other myths hid also some of the science of the "learned age," such as atomistic philosophy, heliocentric astronomy, and Pythagorean cosmology.

According to Hooke, Ovid epitomised in the *Metamorphoses* the history of the early earthquakes and the "theories of the most ancient and most approv'd philosophers." Ovid and "those we now call the ancients" received "ruins and fragments" from the preceding learned age. The "secondary ancients" translated these incomplete materials in a new language. The use of these "extravagant marks" of

---

[88] Oldenburg (1965–86), vol. VIII, 213–5; Plot (1676), 11–2; Turner (1974), 168–70.

[89] Rappaport (1986), 134–6.

[90] Bacon (1857–74), vol. III: 605; Farrington (1964), 87; Hooke (1705), 106.

[91] Royal Society Classified Papers, vol. XX, f. 181r.

[92] Poole (2006), 48.

mythology aimed to help the transmission of knowledge by raising "extraordinary attention and wonder." The difficulty of the interpretation of this puzzling language was increased by the fact that myths and fables were indifferently used to convey different forms of knowledge: "physical," "historical," and "moral." Ovid converted "true histories" into myths "by personating things and powers." In order to identify the philosophical meaning conveyed by each mythological figure or fable, the reader should compare different versions of the same story and refer to the physical knowledge that moderns could achieve.[93] These sources offer only "hints" that the reader should follow "in the consulting of the nature itself." And only a few fables conveyed some "hints" of real knowledge of the early ages of the world[94]:

> Nor that I do here undertake for the truth of history in every fable, for I conceive that there are as various kind of fables as there are of histories. Some are repeated and believed fable which are true histories, others are believed true, but are really fables; some are believed fables and are really so, and others are believed true and are really so. But of this fourth head I fear is the smallest number.

Ovid's account of the creation was in Hooke's eyes a summa of the ideas of Plato, Pythagoras, the ancient Egyptians and Chaldeans on the formation of the world. All these stories echoed the Genesis narrative.[95] Like these myths, the literal meaning of the "expressions in the Scripture relating to physical matters" resisted a literal interpretation. The language of those parts of the Bible was "accommodated generally to the most common and believ'd opinions of men concerning them."[96] Hooke, therefore, called for a physical interpretation of the Genesis narrative and the deluge according to the real natural philosophy.[97] This view was not uncommon among the fellows of the Royal Society. Wren underlined the importance of mathematics and astronomy for a better understanding of the Genesis narrative and Noah's flood.[98] Building on the works of Galileo and Campanella, Wilkins rejected the literal interpretations of the Bible and considered some pagan fables of Phaeton and Ogyges as indirect accounts of the biblical deluge.[99] Despite scepticism towards the knowledge of the gentiles, even Edward Stillingfleet acknowledged that "some kind of tradition" originating from the Bible had been preserved in pagan stories.[100] The orthodox nature of Hooke's exegetical principles, however, did not attenuate the effects of his euhemeristic reading of the Bible. By comparing the narratives of Ovid and Moses, the distance between pagan myths and the holy writings was

---

[93] Hooke (1705), 307–8, 323–4, 328, 372, 376, 378–9, 381, 384, 392, 394, 396, 402.

[94] Royal Society Classified Papers, vol. XX, f. 181r.

[95] Hooke (1705), 396–7, 409, 413.

[96] Id. (1726), 228.

[97] Id. (1705), 423.

[98] Ward (1740), 31–2.

[99] Wilkins (1802), vol. I, 138, 140, 146–7, 149–58, 156–7, 188.

[100] Stillingfleet (1662), 577–8; cf. Hutton (1993), 108.

significantly reduced.[101] The interpretation of both depended now on natural science.[102] Hooke did not leave unnoticed the similitude between Ovid's "short history of the formation of the earth" and "Descartes theory, and that of the ingenious Dr. Burnet in his *Theoria Sacra*."[103] He significantly did not mention the Cartesian interpretation of his friend Francis Lodwick, but the Genesis narrative was one of the most debated topics during the informal gatherings of what has been described as Hooke's circle.[104]

Like Lodwick, Hooke offered a description of the formation of the world in which the role of God was limited to the creation of matter and motion. The laws of mechanics did the rest.[105] The decisive role played by light and gravity in Hooke's account suggests that Hooke's interest in a physical explanation of the formation of the world neither originated from, nor was limited to, the discussion of pagan myths. Hooke undertook a physical description of the "history of creation" in *Cometa*, published in 1678. Another account was included in a lecture on comets read to the Royal Society in 1682. According to the second principle of Hooke's system of the world published in 1674, "all bodies whatsoever that are put into a direct and simple motion" continue to move in a straight line unless deflected into a curve "by other effectual powers."[106] Planets are masses of matter united by a gravitating principle; the attraction of the sun on these masses has transformed their inertial rectilinear motion into a curvilinear one. How did this dynamic system emerge? Hooke's accounts of the formation of the world might perhaps provide some elements to answer this question. According to Hooke, light – a straight and vibrative motion in his opinion – shaped a chaotic dark matter. "The second general rule of natural motion," i.e., gravity, completed the separation of fluids and solids and the formation of bodies with a "conglobating property." "These two powers," Hooke noted, "seem to constitute the souls of the greater bodies of the world."[107] Rectilinear inertial tendency and centripetal attraction were, therefore, not just the principles governing the motions of celestial bodies, but the physical powers that created Hooke's dynamic system of world.

Hooke's estimates on the beginning of the great alterations of the earth were different. He often claimed that they began "since the creation," but he also stated that the primeval earth remained unaltered till the deluge. According to Hooke the afflux of water that caused the flood was not the effect of a miracle but of an "extraordinary earthquake."[108] The analogies between Hooke's accounts of the Genesis narrative and the deluge on the one hand, and those of Burnet and the so-called theorists

---

[101] Harrison (1998), 124–5; Poole (2006), 47.

[102] Westfall (1992), 86.

[103] Hooke (1705), 378.

[104] Lodwick (2011), 252, 257; Poole (2005), 249–50, 261.

[105] Hooke (1705) 413.

[106] Id. (1674), 28.

[107] Id. (1678), 230–1; Id. (1705), 174–5.

[108] Id. (1705) 313–4, 328, 413–6.

on the other, should not overshadow significant differences. Hooke's use of the Bible was opportunistic and instrumental to his scientific interests.[109] The physical interpretation of the formation of the world was not meant as a contribution to biblical exegesis, but as a proof that his dynamic system of the world was "perfectly consonant" to the "Holy Writ," inasmuch as it agreed "to reason, and the nature of things themselves."[110] Although his lectures never assumed an anti-biblical tone, Hooke's history of the earth was alternative to biblical history.

7.5   God, Science, and the Bible

In seventeenth-century natural history, the deluge was a watershed between contrasting views of the earth's past and the study of nature. In a meeting of the Royal Society in July, 1668, the topic of fossils caused a clear divergence of opinions among the fellows. The presence of fossils on European mountains and in England "was conceived by some not possible to be solved but by earthquakes, though others thought it might be by the deluge."[111] The latter hypothesis soon became prevalent among British naturalists. In Burnet's *Telluris Theoria Sacra* the deluge was the only event that changed the earth after the creation. Fossils, mountains, oceans, and the disposition of minerals in the underground were all effects of that great transformation. According to Burnet, since the deluge the morphology of the earth had been affected only by a slow, albeit consistent, process of levelling.[112] Unlike Burnet, Woodward believed that, thanks to the continuous involvement of divine providence, earth's post-diluvial morphology was stable and unaltered. As part of this system, earthquakes balanced the continual erosion due to atmospheric agents. Hooke's hypotheses, in Woodward's view, "have not due warrant from observation, but are clearly repugnant thereunto." The deluge, on the contrary, was "the most horrible and portentous catastrophe that nature ever yet saw." This event, though, was part of "the great design of providence," since it created a stable order that favoured human life.[113] Even William Whiston rejected Hooke's hypotheses and drew a view of earth's past as stable and unaltered after the deluge.[114] Despite significant differences in their interpretations of the Bible, Burnet, Woodward, and Whiston maintained the uniqueness of the biblical deluge in shaping the earth's current morphology. In their diverse accounts, there was no place for the history of alterations that Hooke championed.[115]

---

[109] Poole (2010), 110–1.

[110] Hooke (1705), 175.

[111] Birch (1756-57), vol. II, 307.

[112] Burnet (1681), 21–2, 53–7, 93–5.

[113] Woodward (1695), 40, 46–9, 82–3, 243.

[114] Whiston (1696), 164–6, 208, 259–61.

[115] Poole (2010), 95; Harrison (2000), 169, 178, 180–1.

Hooke did not build a theory of the earth based on the deluge. In contrast to the theorists' projects, Hooke introduced a historical dimension into the history of the earth. The resistance to this radically alternative view of nature was strong. The new history of earth "was likewise opposed and thought improbable." "For so long time as our history will reach backwards," Hooke noted, "it was affirmed there had happened no such change" and that "no such history could be produced"[116] But the difference between Hooke's geological ideas and seventeenth-century theories of the earth should be measured not by the closeness to modern historical geology, rather by the distance from biblical history.[117] Hooke did not deny the universal nature of the deluge. In his opinion, that "total deluge" did not last enough to account for the amount and wider distribution of fossils; neither was the only catastrophe that took place on earth. Many other catastrophes contributed to the production of fossils and earth's surface.[118]

Along with these catastrophes, Noah's flood and the Genesis narrative could be described as wholly natural processes following the creation of matter and motion by God. In Hooke's account, the formation and the succeeding history of the earth were part of a wider dynamic system of the world. Like the earth, celestial bodies have been altered by natural phenomena comparable to earthquakes. The irregular surface of the moon suggested to Hooke that 'moonquakes' took place on the satellite of the earth. The production of light by the sun and the stars was the result of a process of dissolution of a sulphuric body by an aerial nitrous substance. The globular form and the positions of celestial bodies could be explained as the mechanical effects of the vibrative motions of light and gravity on an originally chaotic and shapeless matter. Even the origin and nature of puzzling phenomena like comets could be understood against the background of the emergence of Hooke's system of the world.[119]

The absence of any supernatural intervention in this process could hardly be accepted by many of Hooke's contemporaries. Among several hypotheses "repugnant" to the "plain assertion of Moses," Stllingfleet listed the opinion of those who described the origin of the world "merely by the mechanical laws of the motion of matter."[120] According to Robert Boyle, corpuscular philosophy offered the best description of natural phenomena, but could not venture in the explanation of how the world was formed. Any mechanical hypothesis on the formation of the world would be necessarily insufficient and impious, "for many things can be perform'd by matter variously figur'd and mov'd, which yet would never be perform'd by it, if it had been still left to it self." In Boyle's view, God created matter and motion, established natural laws, and directed the formation of the world. When this process was concluded, his active involvement in the ordinary phenomena could not be

---

[116] Hooke (1705), 404.

[117] Redwood (1976), 131; for a different view see Huggett (1989), 51.

[118] Hooke (1705), 328, 341, 408.

[119] Hooke (1678), 249.

[120] Stillingfleet (1662), 422–3; cf. Hutton (1993), 113–4.

excluded.[121] On this topic, Newton agreed with Boyle. Behind Newton's refusal of any physical explanation of sacred history and current geology was the belief that the origin and following course of the world was due to the action of an intelligent being.[122] Burnet, Woodward, and Whiston undertook a physical description of biblical history, about which Newton was sceptical. But Whiston and Woodward maintained Newton's assumption that the process could not be satisfactorily explained by ruling out the direct action of God. Gravity offered the means to reconcile their accounts with Newton's position.[123] "As to that affection of bodies which is called their gravity," Woodward noted, "it clearly surpasses all the powers of meer nature, and all the mechanism of matter." In Woodward's view, gravity was due to the "direct concourse of the power of the author of nature," who used it to manage and support "this stupendous fabric of the universe."[124] Even for Whiston the action of God in the creation and the following course of nature was constant. Gravity was "the most mechanical affection of bodies;" it seemed also "the most natural," but in reality it depended "entirely on the constant and efficacious, and if you will the supernatural and miraculous influence of Almighty God."[125]

In Hooke's opinion, Whiston's explanation of gravity "is well begun and I conceive he need to say or solve noe more for this will solve all, and all the other solutions will be insignificant and needless."[126] Among the latter, Hooke probably included also his own vibrative hypothesis. As William Whewell noted, by "referring gravity to the will of the Deity as the First Cause" Newtonian philosophers "assumed a superiority over those whose philosophy rested in second causes."[127] Hooke was one of them. In 1677 he opposed More's "hylarchick or matter-governing spirit" because gravity, colours and all other natural phenomena could be "clearly solved by the common rules of mechanicks."[128] In the *Enchiridion Metaphysicum*, More had few years earlier claimed that Hooke's hypothesis of colours and Descartes' theory of gravity were insufficient explanations.[129] In More's view, the action of a plastic immaterial spirit was a physical necessity, for passive matter could not operate according to the laws of nature unless directed by an active agent.[130] The creation and succeeding course of nature would have been impossible without this divine agent. More's argument was based on the assumption of a fundamental tenant of the mechanical philosophy that he aimed to amend, for he

[121] Boyle (1999), vol. III, 242, 253, 259; vol. V, 306, 353–4.

[122] Newton (1959–77), vol. II, 334; Kubrin (1967), 326; Roger (1982), 108; Mandelbrote (1994), 157–8, 163; Ducheyne (2012), 261–2.

[123] Henry (1994), 123, 128–9.

[124] Woodward (1695), 52–3.

[125] Whiston (1696), 6, 218.

[126] Royal Society Classified Papers, vol. XX, f. 187r.

[127] Whewell (1837), vol. II, 197.

[128] Hooke (1677), 31–3.

[129] More (1679), vol. II, 139, 190–2.

[130] Hall (1990), 115; Bondì (2001), 134–5.

favoured Cartesian matter theory over atomism and other non-Cartesian corpuscular philosophies.[131] In Hooke's eyes, the action of immaterial agents on matter was not needed, especially if these were expected to operate according to mechanical laws. More's plastic spirit was supposed to perform "the effects which do clearly belong to mechanical motions and powers." It was, therefore, redundant, since natural phenomena "are performed and regulated exactly according to quantity and quality of matter, and according to the general and universal rules of motion, and not otherwise."[132]

In Hooke's system of the world there was neither need nor room for supernatural causes and immaterial agents. More's introduction of a "hylarchick spirit" and Whiston's appeal to the "supernatural and miraculous influence of Almighty" to account for a natural phenomenon like gravity seemed to Hooke "subterfuges of ignorance." There was no reason why "the immediate, extraordinary and divine power" had to act according to the already existing and self-sufficient laws of mechanics rather than "a singular and particular determination of that power" itself. The intervention of any superior and immaterial principle proved inconsistent with the existence of a natural order. The universal laws of mechanics were sufficient to explain all natural phenomena. In Hooke's view, the real philosophy could only be mechanical, but not atheistic, since matter, motion, and natural laws were created by God. There was no need, therefore, to "suppose new created causes" or "to believe that everything effected supernaturally, of which we cannot find out the natural cause." For this reason, the limitation of natural knowledge within the boundaries of mechanical laws did not entail irreligious consequences, or "any ways detract from the omnipotency and power of God."[133]

The distinction between secondary natural causes and the primary divine cause was a common solution to some of the theological questions risen by the new mechanical philosophy.[134] Descartes considered the laws of nature as "the secondary and particular causes" of natural phenomena.[135] The permanence of the natural order due to these secondary causes was, however, due to the primary divine cause in Descartes' system.[136] A significant relationship between the divine first cause and the corporeal secondary causes was established also by Gassendi. Although physical phenomena could be explained only by physical causes, these depend on the "first cause."[137] On this ground, Wilkins associated the "usual course of nature" with the "usual way of providence."[138]

---

[131] More (1679) 178–9, 192–3.

[132] Hooke (1677), 34.

[133] Id. (1705), 165, 392, 423.

[134] Westfall (1970), 6; Shea (2007), 469.

[135] Descartes (1964–74), vol. VII.1, 124; Id. (1983), 59.

[136] Garber (1992), 264–5; Osler (1994), 102–1.

[137] Gassendi (1658), vol. I, 133, 287, 333, 334, 335.

[138] Wilkins (1802), vol. I, 233.

"It is the same omnipotent power," Hooke wrote, "which does influence the remote causes as well as the proximate." Like Wilkins, Hooke maintained that providence operates through natural causes. "The universal providence," he remarked, "that ordereth all the effects, doth also determine and appoint all the causes and means conducing thereunto." The natural order is an effect of the first cause because God created matter and motion, and established the universal laws governing their interactions. For Hooke this view did not question nor limit God's power, for this "is not less wonderful in producing the causes of things, than in producing and disposing things more immediately." Despite secondary causes being a direct effect of the first one, the study of nature can achieve a detailed and complete knowledge of natural laws only. If a naturalist asks "what is the inlivening power that orders, disposes, governs and performs all these wonderful effects, there he finds the *Ne plus ultra*, there is the miracle that he may truly admire but cannot understand."

"We should go as far as we can if we cannot go further," Hooke concluded, quoting Horace's *Epistles*. The study of nature shows "more powerfully" than "all the other methods of contemplation or argumentation whatsoever" that the divine providence "rules and regulates the things of this world." But it cannot say anything else concerning the nature of God.[139] Hooke's position on natural theology echoed Bacon's ideas. The new natural and experimental history was described in the *Parasceve* as "the book of God's works and another kind of Holy Writ."[140] According to Bacon it is providence that "by a fatal and necessary law" produced "all the order and the beauty of the universe."[141] But natural theology shows only the power of God, not his will. For this reason, it might serve to prove atheism wrong, but it should not be confused with science.[142]

The microscope showed to Hooke evidence of "the omnipotency and infinite perfections of the great creatour."[143] This design clearly appeared in the organs of the animals. Everything in the "animated" creatures seemed to Hooke "contrived on purpose and with a design."[144] The structures of inorganic bodies showed as well marks of a design. "We shall in all things," Hooke noted, "find that nature does not onely work mechanically, but by such excellent and most compendious, as well as stupendious contrivances." The perfection, uniformity, and order of nature proved that it could not be a "product of chance." Thus, Hooke directed this arguments against a very common target of seventeenth-century natural theology, i.e., "Epicurus and his followers."[145] In Hooke's lectures after *Micrographia* there is not "far less invocation of the Deity, and little evidence of serious commitment to the natural

---

[139] Hooke (1705), 392, 423–4.

[140] Bacon (2004), 468–9.

[141] Bacon (1857–74), vol. VI, 657, 731.

[142] Rossi (1989), 65; Gaukroger (2001), 94–5; Henry (2002), 85–6; Matthews (2007), 68–9, 73.

[143] Hooke (1665), 8.

[144] Id. (1705), 120–1.

[145] Id. (1665), 171–2, 177; cf. Hunter (1990), 441, 444; Lolordo (2011), 661–3.

theological enterprise," as Steven Shapin has claimed.[146] Despite the refusal of biblical natural history, in the lectures on fossils, for instance, natural theology played a relevant role in favour of the organic origin of fossils and the transformations of the earth.[147] To those who described fossils as *lapides sui* generis produced by a plastic virtue, Hooke opposed the view of nature as a mechanical, uniform, and ordered system. "It is certain," he claimed, "that nature doth nothing *frustra*, but manifestly with an admirable and wise design."[148] Why should nature produce stones similar to living being? It was more consistent with the order of nature to believe that fossils were produced by the living beings whose figures they reproduce. Far from being unremarkable, in the *Cutlerian lectures* the natural theological principles of order and uniformity of nature assumed a dynamic feature. As the fossils showed, the earth was the result of a continuous series of transformations directed by some fundamental and universals laws of mechanics created by God, rather than a stable and unaltered system.

Bibliography

Albritton, Claude, Jr. 1980. *The dark abyss of time: Chancing conceptions of the Earth's antiquity after the sixteenth century*. San Francisco: Freeman, Cooper and Co.

Alsted, Johann Heinrich. 1650. *Thesaurus chronologiae*. Herborn.

Bacon, Francis. 1857–74. *Works*, 7 vols., eds. Robert L. Ellis, James Spedding, Douglas D. Heath. London: Longman.

———. 2004. *The Instauratio magna. Part II: Novum organum*, ed. Graham Rees with Maria Wakely. Oxford: Oxford University Press.

Bettini, Amalia. 1997. *Cosmo e apocalisse: Teorie del millennio e storia della terra nell'Inghilterra del seicento*. Florence: Olschki.

Birch, Thomas. 1756–57. *The history of the Royal Society of London*, 4 vols. London.

Birkett, Kristen, and David Oldroyd. 1991. Robert Hooke's physico-mythology: Knowledge of the world of the ancient and knowledge of the ancient world. In *The uses of antiquity*, ed. Stephen Gaukroger, 145–170. Dordrecht: Kluwer.

Bondì, Robert. 2001. *L'onnipresenza di Dio: Saggio su Henry More*. Soveria Mannelli: Rubbettino.

Boyle, Robert. 1999. *The works of Robert Boyle*, 14 vols., ed. Michael Hunter and Edward Davies. London: Pickering and Chatto.

Burnet, Thomas. 1681. *Telluris theoria sacra*. London.

Carozzi, Albert. 1970. Robert Hooke, Rudolf Enrich Raspe and the concept of earthquakes. *Isis* 61: 85–91.

Chapman, Alan. 1994. Edmond Halley's use of historical evidence in the advancement of science. *Notes and Records of the Royal Society of London* 48: 167–191.

———. 2005. *England's Leonardo: Robert Hooke and the seventeenth-century scientific revolution*. Bristol/Philadelphia: Institute of Physics Publishing.

Davies, Gordon. 1968. *The earth in decay: A history of British geomorphology 1578–1878*. London: Mcdomald.

Debus, Allen George. 1977. *The chemical philosophy*. New York: Science History Publishing.

---

[146] Shapin (1989), 278.

[147] Rudwick (1976), 55–6.

[148] Hooke (1705), 341.

Descartes, René. 1964–74. *Oeuvres*, 12 vols., eds. Charles Adam and Paul Tannery. Paris: Vrin.
———. 1983. *Principles of philosophy.* Trans. Valentine Rodger Miller and Rees P. Miller. Dordrecht: Reidel.
Drake, Ellen Tan. 1981. The Hooke imprint on Huttonian theory. *American Journal of Science* 281: 963–973.
———. 1996. *Restless genius: Robert Hooke and his earthly thoughts.* Oxford: Oxford University Press.
———. 2005. Hooke's concept of the earth in space. In *Robert Hooke and the English Renaissance,* ed. Paul Kent and Alan Chapman, 75–94. Leominster: Gracewing.
———. 2006. Hooke's ideas of the terraqueous globe and a theory of evolution. In *Robert Hooke: Tercentennial studies,* ed. Michael Cooper and Michael Hunter, 135–149. Aldershot: Ashgate.
———. 2007. The geological observations of Robert Hooke (1635–1703) on the Isle of Wight. In *Four centuries of geological travel: The search for knowledge on foot, bicycle, sledge and camel,* ed. Patrick N. Wyse Jackson, 19–30. London: The Geological Society.
Drake, Ellen Tan, and Paul Komar. 1983. Speculations about the earth: The role of Robert Hooke and others in the 17th century. *Earth Sciences History* 2: 11–16.
Ducheyne, Steffen. 2012. *The main business of natural philosophy: Isaac Newton's natural philosophical methodology.* Dordrecht: Springer.
Farrington, Benjamin. 1964. *The philosophy of Francis Bacon.* Liverpool: Liverpool University Press.
Galileo, Galilei. 1989. *The sidereal messenger.* Trans. Albert Van Helden. Chicago/London: University of Chicago Press.
Gassendi, Pierre. 1658. *Opera omnia,* 6 vols. Leiden.
Garber, Daniel. 1992. *Descartes' metaphysical physics.* Chicago/London: University of Chicago Press.
Gaukroger, Stephen. 2001. *Francis Bacon and the transformation of early modern philosophy.* Cambridge: Cambridge University Press.
Gould, Stephen Jay. 1987. *Time's arrow, time's cycle: Myth and metaphor in the discovery of geological time.* Cambridge, MA: Harvard University Press.
Grafton, Anthony. 1995. Tradition and technique in historiographical chronology. In *Ancient history and the antiquarian: Essays in memory of Arnaldo Momigliano,* ed. Michael H. Crowford and Christopher Ligota, 15–32. London: The Warburg Institute.
Hall, Alfred Rupert. 1990. *Henry More: Magic, religion and experiment.* Oxford: Blackwell.
Halley, Edmund. 1724–25. Some considerations about the cause of the universal deluge. *Philosophical Transactions* 33: 118–123.
Harrison, Peter. 1998. *The Bible, protestantism and the rise of natural science.* Cambridge: Cambridge University Press.
———. 2000. The influence of Cartesian cosmology in England. In *Descartes' natural philosophy,* ed. Stephen Gaukroger, John Schuster, and John Sutton, 168–192. London: Routledge.
Henry, John. 1994. "Pray do not ascribe that notion to me": God and Newton's gravity. In *The books of nature and scripture: Recent essays on natural philosophy, theology, and biblical criticism in the Netherland of Spinoza's time and the British Isles of Newton's time,* ed. James Force and Richard Popkin, 123–147. Dordrecht: Kluwer.
———. 2002. *Knowledge is power: Francis Bacon and the method of science.* Cambridge: Icon Books.
Hooke, Robert. 1665. *Micrographia.* London.
———. 1674. *An attempt to prove the motion of the earth by observations.* London.
———. 1677. *Lampas.* London.
———. 1678. *Lectures and collections.* London.
———. 1705. *Posthumous works,* ed. Richard Waller. London.
———. 1726. *Philosophical experiments and observations,* ed. William Derham. London.
Hooykaas, Reyer. 1959. *Natural law and divine miracle: A historical-critical study of the principles of uniformity in geology, biology and theology.* Leiden: Brill.

Huggett, Richard. 1989. *Cataclysms and earth history: The development of diluvialism*. Oxford: Clarendon Press.

———. 1997. *Catastrophism: Asteroids, comets and other dynamic event in Earth history*. London: Verso.

Hunter, Michael. 1990. Science and heterodoxy: An early modern problem reconsidered. In *Reappraisal of the scientific revolution*, ed. David Lindberg and Robert Westman, 437–460. Cambridge: Cambridge University Press.

Hutton, Sarah. 1993. Science, philosophy, and atheism: Edward Stillingfleet's defence of religion. In *Scepticism and irreligion in the seventeenth and eighteenth centuries*, ed. Richard H. Popkin and Arjo Vanderjagt, 102–120. Leiden: Brill.

Ito, Yushi. 1988. Hooke's cyclic theory of the earth in the context of seventeenth-century England. *The British Journal for the History of Science* 21: 295–314.

Jackson, Patrick Wyse. 2003. *The chronologers' quest: The search for the age of the earth*. Cambridge: Cambridge University Press.

Kubrin, David. 1967. Newton and the cyclical cosmos: Providence and the mechanical philosophy. *Journal of the History of Ideas* 28: 325–346.

———. 1990. "Such an impertinently litigious lady": Hooke's "great pretending" vs Newton's Principia and Newton's and Halley's theory of Comets. In *Standing on the shoulders of the giants*, ed. Norman Thrower, 55–90. Berkeley/Los Angeles/London: University of California Press.

Lawson, Ian. 2016. Crafting the microworld: How Robert Hooke constructed knowledge about small thigs. *Notes and Records of the Royal Society of London* 70: 23–44.

Levitin, Dmitri. 2013. Edmund Halley and the eternity of the world revisited. *Notes and Records of the Royal Society of London* 67: 315–329.

———. 2015. *Ancient wisdom in the age of the new science: Histories of philosophy in England, c. 1640–1700*. Cambridge: Cambridge University Press.

Lewis, Rhodri. 2012. *William Petty on the order of nature: An unpublished manuscript treatise*. Tempe: Arizona Center for Medieval and Renaissance Studies.

Lhwyd, Edward. 1699. *Lithophylacii britannici ichonographia*. London.

Lister, Martin. 1671. A letter of Martin Lister. *Philosophical Transactions* 6: 2281–2284.

Lodwick, Francis. 2011. *On language, theology, and utopia*, ed. Felicity Henderson and William Poole. Oxford: Clarendon Press.

Lolordo, Antonia. 2011. Epicureanism and early modern naturalism. *The British Journal for the History of Philosophy* 19: 647–664.

Mandelbrote, Scott. 1994. Isaac Newton and Thomas Burnet: Biblical criticism and the crisis of late seventeenth-century England. In *The books of nature and scripture: Recent essays on natural philosophy, theology, and biblical criticism in the Netherland of Spinoza's time and the British Isles of Newton's time*, ed. Richard Popkin and James Force, 149–178. Dordrecht: Kluwer.

Matthews, Steven. 2007. Reading the two books with Francis Bacon: Interpreting God's will and power. In *The word and the world: Biblical exegesis and early modern science*, ed. Kevin Killen and Peter Forshaw, 61–77. London: Palgrave Macmillan.

More, Henry. 1679. *Opera omnia*, 2 vols. London.

Newton, Isaac. 1959–77. *The correspondence of Isaac Newton*, 7 vols., ed. H. W. Turnbull, J. F. Scott, A. R. Hall and L. Tilling. Cambridge: Cambridge University Press.

Oldenburg, Henry. 1965–86. *The correspondence of Henry Oldenburg*, 13 vols., ed. Alfred Rupert Hall and Marie Boas Hall. Madison: University of Wisconsin Press.

Oldroyd, David. 1972. Robert Hooke's methodology of science as exemplified in his Discourse of Earthquakes. *The British Journal for the History of Science* 6: 109–130.

———. 1989. Geological controversy in the seventeenth century: Hooke vs. Wallis and its aftermath. In *Robert Hooke: New studies*, ed. Michael Hunter and Simon Schaffer, 207–233. Woodbridge: Boydell Press.

———. 1996. *Thinking about the earth*. Cambridge, MA: Harvard University Press.

———. 2006. *Earth cycles: A historical perspective*. Westport: Greenwood Press.

Osler, Margaret. 1994. *Divine will and the mechanical philosophy: Gassendi and Descartes on contingency and necessity in the created world*. Cambridge: Cambridge University Press.

Patterson, Louise Diehl. 1950. Hooke's gravitation theory and its influence on Newton II: The insufficiency of the traditional estimate. *Isis* 41: 32–45.

Pineda de Avila, Nydia. 2015. Crater-pear-vale: Earth-moon analogies in Robert Hooke's Micrographia. In *Newberry essays in medieval and early modern studies*, ed. Karen Christianson and Andrew K. Epps, vol. 9, 29–45. Chicago: Newberry Library.

Plot, Robert. 1676. *The natural history of Oxfordshire*. Oxford.

Poole, William. 2005. Francis Lodwick's creation: Theology and natural philosophy. *Journal of the History of Ideas* 66: 245–263.

———. 2006. The Genesis narrative in the circle of Robert Hooke and Francis Lodwick. In *Scripture and scholarship in early modern England*, ed. Ariel Hessayon and Nicholas Keene, 41–57. Aldershot: Ashgate.

———. 2010. *The world makers: Scientists of the restoration and the search for the origins of the Earth*. Oxford: Peter Lang.

Porter, Roy. 1977. *The making of geology: Earth science in Britain 1660–1815*. Cambridge: Cambridge University Press.

Rappaport, Rhoda. 1986. Hooke on earthquakes: Lectures, strategy and audience. *The British Journal for the History of Science* 19: 129–146.

———. 1997. *When geologists were historians, 1665–1750*. Ithaca/London: Cornell University Press.

Ranalli, Giorgio. 1982. Robert Hooke and the Huttonian theory. *Journal of Geology* 90: 319–325.

———. 1983. Robert Hooke and the Huttonian theory: A reply. *Journal of Geology* 91: 233–234.

———. 1984. Speculation about the earth: The role of Robert Hooke and others in the 17th century: A discussion. *Earth Sciences History* 3: 187.

Ray, John. 1693. *Three physico-theological discourses*. London.

Redwood, John. 1976. *Reason, ridicule and religion: The age of Enlightenment in England, 1660–1750*. London: Thames and Hudson.

Roger, Jacques. 1982. The Cartesian model and its role in eighteenth-century "Theory of the Earth". In *Problems of Cartesianism*, ed. Thomas Lennon, John Nicholas, and John Davis, 95–111. Kingston/Montreal: McGill-Queen's University Press.

Ross, Anna Marie. 2011. Salient theories in the fossil debate in the early Royal Society: The influence of Johann Van Helmont. In *Controversies within the scientific revolution*, ed. Marcelo Dascal and Victor Boantza, 151–170. Amsterdam/Philadelphia: John Benjamins Publishing Company.

Rossi, Paolo. 1968. *Francis Bacon: Magic to science*. London: Routledge and Kegan Paul.

———. 1984. *The dark abyss of time: The history of earth and the history of nations from Hooke to Vico*. Trans. Lydia Cochrane. Chicago/London: Chicago University Press.

———. 1989. *Aspetti della rivoluzione scientifica*. Turin: Bollati Boringhieri.

Rudwick, Martin. 1976. *The meaning of fossils*. Chicago/London: Chicago University Press.

———. 2014. *Earth's deep history: How it was discovered and why it matters*. Chicago/London: Chicago University Press.

Scaliger, Joseph Juste. 1629. *De emendatione temporum*. Geneve.

Schaffer, Simon. 1977. Halley's atheism and the end of the world. *Notes and Records of the Royal Society of London* 32: 17–40.

Schneer, Cecil. 1954. The rise of historical geology in the seventeenth century. *Isis* 45: 256–268.

Scilla, Agostino. 1670. *La vana speculazione disingannata dal senso*. Neaples.

———. 2016. Vain speculation undeceived by sense, trans. Rodney Palmer, Rosemary Williams and Ilaria Bernocchi. Cambridge: Sedgwick Museum of Earth Science.

Shapin, Steven. 1989. Who was Robert Hooke? In *Robert Hooke: New studies*, ed. Michael Hunter and Simon Schaffer, 253–285. Woodbridge: Boydell Press.

Shea, William. 2007. The scientific revolution really occurred. *European Review* 15: 459–471.

Stensen, Niels. 1668. *De solido intra solidum naturaliter content dissertationis prodromus*. Florence.

———. 1671. *Dissertation concerning solids naturally contained within solids*. London.

Stillingfleet, Edward. 1662. *Origines sacrae*. London.

Toulmin, Stephan, and Jane Goodfield. 1967. *The discovery of time*. London: Penguin.

Turner, Anthony J. 1974. Hooke's theory of earth's axial displacement: Some contemporary opinions. *The British Journal for the History of Science* 7: 166–170.

Ussher, James. 1660. *Chronologia sacra*. Oxford.

Vossius, Gerard. 1659. *Chronologiae sacrae isagoge*. The Hague.

Ward, John. 1740. *The lives of the professors of Gresham College*. London.

Westfall, Richard. 1970. *Science and religion in seventeenth-century England*. Hamden: Archon Books.

———. 1972. Robert Hooke (1635–1703). In *Dictionary of scientific biography*, ed. Charles Gillispie, vol. VI, 481–488. New York: Scribner.

———. 1992. The scientific revolution of the seventeenth century: The reconstruction of a new world view. In *The concept of nature*, ed. John Torrance, 63–93. Oxford: Clarendon Press.

Whewell, William. 1837. *History of inductive sciences*, 3 vols. London.

Whiston, William. 1696. *A new theory of the earth*. London.

Wilkins, John. 1802. *The mathematical and philosophical works*, 2 vols. London.

Woodward, John. 1695. *An essay towards a natural history of the earth*. London.

Chapter 8
Beyond Priority

8.1 Old Disputes and New Questions

On August 28, 1686, the manuscript of Newton's *Principia* was presented to the Royal Society. The book was dedicated to the Society, and the fellows "were so very sensible of the Great Honour" that they considered printing the book at the Society's expenses.[1] According to Halley's account, one fellow was less enthusiast than others about that "incomparable treatise." "Mr Hooke," Halley wrote to Newton, "has some pretentions upon the inventions of ye rule of the decrease of Gravity, being the square of the distances from the Center."[2] A controversy between the two men soon sparked. Hooke, on the one hand, acknowledged that only Newton successfully demonstrated how gravity leads to the elliptic orbits of planets. He claimed that Newton adopted "the notion" of the inverse square law from him, and expected to be mentioned in the preface of that "incomparable treatise." Newton, on the other hand, refused that claim altogether. He countered that he knew the inverse square law long before he discussed it with Hooke, who probably had it from Wren and Borelli. For this reason, his name was mentioned in the *System of the World* along those of the other two scholars.[3] Newton would concede nothing more.

Since then scholars have taken positions on both sides, claiming for Hooke or for Newton the priority of the discovery of the inverse square law. Despite Hooke's claims being supported by some of the early fellows of the Royal Society, after his death a Newonian narrative dominated for three centuries. In the mid-twentieth century, Louise Diehl Patterson first questioned the myth of Newton's *annus mirabilis*. Focusing on the letters that Newton and Hooke exchanged in 1679 and 1680, Patterson maintained that in 1666 Newton's ideas on celestial mechanics were far

---

[1] Guicciardini (2018), 147.

[2] Newton (1959–1977), vol. II, 431.

[3] Ibid., vol. II, 433–7; Id. (1999), 452; Id, (1779–85), vol. I, 51.

© Springer Nature Switzerland AG 2020

F. G. Sacco, *Real, Mechanical, Experimental*, International Archives
of the History of Ideas Archives internationales d'histoire des idées 231,
https://doi.org/10.1007/978-3-030-44451-8_8

from close to those expressed in the *Principia*. In her view, it was Hooke who discovered the inverse square law and communicated it to Newton.[4] Following Patterson's focus on the epistolary exchange between the two claimants, Alexandre Koyré advanced a more balanced reconstruction. In Koyré's view, Hooke embraced the inverse square law only in 1679, and through the discussion with Newton on the trajectory of falling bodies. Inasmuch as this latter was decisive in this process, so were Hooke's hypotheses for the evolution of Newton's views on the physical nature of gravity.[5] Koyré's narrative has recently been rejected by advocates of both contenders. According to Michael Nauenberg a manuscript kept at the Wren Library of the University of Cambridge proves that Hooke's contribution went further than Newton and other scholars have hitherto conceded.[6] In *The Laws of Circular Motion*, Hooke demonstrated that the orbit of a body in a central field of force is an approximate ellipse centred at the origin of the field itself.[7] This graphic demonstration was drafted in September, 1685. Less than one year earlier, Halley had informed the Royal Society that Newton had completed a manuscript *de motu corporum in gyrum*, and agreed to send "the curious treatise" to the Society "to be entered upon their register."[8] One can suspect, therefore, that Hooke built on Newton's instantaneous impulse construction available in *De motu*.[9]

Scholars' persistent interest in questions of priority concerning the inverse square law has proven an impediment to a full understanding of the historical debate on celestial dynamics. It seems, furthermore, irrelevant, since in the second half of the seventeenth century many natural philosophers reached it through different ways and maintained it for different reasons.[10] Hooke was aware of the law long before he asked Newton about the "proprietys of a curve line (not circular nor concentricall) made by a central attractive power which makes the velocitys of descent from the tangent line or equall straight motion at all distances in a duplicate proportion to the distances reciprocally taken."[11] The analogy between the ethereal propagation of light and gravity offered Hooke an insight into the quantitative determination of the force, although he did not provide any demonstration of it before 1685.[12] Newton pointed out that the analogy employed by Hooke derived from the idea of Ismaël Boulliau that "all force respecting ye sun as its center & depending on matter must be reciprocally in duplicate ratio of ye distance from ye center."[13] In *Astronomia philolaica* the French mathematician rejected Kepler's geometrical distinction

---

[4] Patterson (1949), 334; Id. (1950), 33–4.

[5] Koyré (1968), 234–5, 253–4; see also Fiocca (1998), 52–7.

[6] Trinity College Library, Cambridge, MS. 0.11a.1/16; cf. Pugliese (1989), 200.

[7] Nauenberg (1994), 332; Id. (1998), 92; Id. (2005), 523; Id. (2006), 7–8.

[8] Birch (1756–57), vol. IV, 347.

[9] Erlichson (1997), 170–1, 182–3; Gal and Chen Morris (2006), 55.

[10] Bennett (1989), 229; Lohne (1960), 19; Gal (2002), 9.

[11] Newton (1756–77), vol. II, 313.

[12] Hooke (1674), 28; Id. (1705), 185.

[13] Newton (1956–77), vol. II, 437.

between the spherical (*orbiculariter*) diffusion of light and the attractive force of the sun operating along the plane (*circulariter*) of the ecliptic.[14] As Kepler knew, since light spreads in concentric spheres around the source, the quantity of it that reaches any given point is inversely proportional to the square of the distance from the centre because the surface of a sphere is directly proportional to the square of its radius.[15] Boulliau concluded that this basic principle of geometry applies also to the quantification of solar attractive force. Borelli, whom Newton listed as a source of Hooke, referred to Boulliau's principle.[16] And so did Hooke, who maintained that light and gravity were not immaterial virtues or a "geometrical medium between a body and a spirit"; rather they consisted in two different ethereal motions originating from the sun and spherically distributed.[17]

Hooke's references to Bouillau's principle suggest that *Astronomia philolaica* was his main way to the inverse square law.[18] Other routes to the same law were also available. As Newton showed, the determination of the centrifugal force combined with Kepler's third law could lead to the same conclusion. In the second half of 1660s, Newton found that the *conatus recedendi a centro* of a body revolving in a uniform motion is proportional to the square of its speed and inverse to the distance from the centre.[19] In the mid-1660s the young Lucasian professor believed that gravity was the effect of an ethereal motion. Far from being an impediment to the development of Newtonian ideas, his early commitment to vortexes played a major role in the definition of the inverse square law.[20] Contrary to what Newton later claimed, this fundamental law was not inconsistent with Cartesian celestial mechanics.[21] After the *Principia*, Huygens proved that a centrifugal approach to gravity could provide a physical explanation of the inverse square law. In Huygens's opinion, the attractive force reciprocal to the square of the distance between the centres of two bodies whatsoever is "a very remarkable property of weight whose cause is worth inquiring." This cause lies, according to Huygens, in the centrifugal force rather than the centripetal one employed by Newton, for the weight of a body consists in the "quantity of fluid matter" needed to replace it in a vortex.[22]

It is, perhaps, time to abandon this old priority dispute and focus instead on the different views that Newton and Hooke held conccerning gravity, and on the philosophical exchange between them.[23]

---

[14] Bouillau (1645), 3–7, 21–4; Applebaum (1996), 460; Wilson (1970), 107.

[15] Gal (2002), 169–70, 200–1; Id. (2005), 532; Gal and Chen Morris (2006), 50–4.

[16] Borelli (1666), 32.

[17] Hooke (1705), 114, 132.

[18] Gal (2006), 45; Gal and Chen Morris (2005), 391–2, 397.

[19] Newton (1956–77), vol. I: 297–303; Gal (2002), 171–2; Whiteside (1970), 10–1; Westfall (1971), 352, 358–60.

[20] Guicciardini (2011), 62–4; Whiteside (1991), 20; Casini (1997), 204.

[21] Newton (1777–85), vol. I, 453, 454–7; vol. II, 633–5, 791; Id. (1999), 784,786–8, 815–6, 939; see also Bertoloni Meli (2006a), 324–9; Westfall (1972a), 186–7; Ducheyne (2012), 120, 125.

[22] Huygens (1888–1950), vol. XXI, 458, 471.

[23] Gal (2002), 18; Guicciardini (2005), 512.

8.2   Congruity and Sociableness

In Hooke's cosmology, Bouillau's principle coexisted with a dynamic approach based on centripetal attraction. This approach, as we have shown in Chap. 6, originated from the magnetic philosophy. It was, perhaps, because of this that Hooke's definition of the force quantified by the inverse square law showed a persistent ambiguity.[24] Hooke aimed to provide a general theory "which would solve all the unequal motions of the planets," yet in 1687 Hooke seemed to believe in the existence of some differences among the "gravitating powers" of sun, moon, and earth. "Tho' they may in most particulars be consonant," there can be "something specifick in each of them."[25] Hooke had no doubts about the extension of these powers. Each of them operates within a limited sphere of action, as the paths of comets show. Even the sun's gravitating power operates within a limited sphere of action, which is large enough to affect the motions of primary and secondary planets and thus create a solar system.[26] Beyond this system other stars, such as the fixed stars, exert their gravitational powers over other celestial bodies within their respective "sphere[s] of activity or expansion proportionate to their solidity and activity."[27] For this reason, some scholars have warned that Hooke's was not a theory of universal gravitation. From a post-Newtonian point of view, the presence of innovative elements such as the inverse square law and a centripetal celestial dynamic seem to entail a universal notion of gravity. In spite of these, Hooke maintained a view of gravity linked to the magnetic notion of *orbis virtutis*, which led him towards a system of many different gravities.[28] Such a view, however, has not been shared by all scholars. According to Koyré, for instance, Hooke's emphasis on the "spheres of action" of "gravitational powers" was not prejudicial to the acknowledgment of a universal force operating on all celestial bodies and identical with terrestrial gravity.[29]

The notion of a sphere of activity does not seem inconsistent with the inverse square law, for it can be seen as the physical description of a force depending on distance and magnitude. But the existence of physically different forces suggested by Hooke in 1687 clearly contrasts with the notion of universal gravitation entailed by that law. If these forces are physically different, how could they operate according to the same law? Hooke's hypothesis on the "gravitational powers" of sun, moon, and earth, however, was not carried further. When he first acknowledged the existence of "a principle of gravitation" in the moon, Hooke linked it to the presence of an "internal elastical body" similar to that of the earth. "And to make this

---

[24] Bertoloni Meli (2006a), 219–20; Westfall (1972b), 485.

[25] Birch (1756–57), vol. II, 188; Hooke (1705), 546.

[26] Hooke (1678), 228, 247.

[27] Id. (1674), 6.

[28] Bertoloni Meli (2006a), 219–20; Westfall (1972b), 458; Westfall (1971), 268, 270–1; Ducheyne (2012), 147.

[29] Koyré (1968), 232–3; Aiton (1972), 97.

probable," Hooke added, "I think we need no better argument then the roundness or global figure of the body of the moon itself, which we may perceive very plainly by the telescope." By showing the earthly nature of the moon and the alterations that take place on the surface of the sun, the telescope decisively contributed to the notion of a system of celestial bodies whose shape and motions depended on the same force, gravity.[30] The link established between globular form and gravity supports Hooke's commitment to a universal notion of gravity. Since the same force maintains the spherical form of celestial bodies and attracts all others in a ratio defined by the inverse square law, there cannot be any difference in nature among the "gravitating powers" of each celestial body. "I take this roundness," Hooke claims, "to be as convincing an argument as any to prove that there is the like power in every globular celestial body, as there is in the earth." All "globous bodies of the universe" are produced by the same force. A uniform "method of nature" operates in all globular productions, regardless of their dimensions.[31] At the root of this "method" there was a more fundamental principle in Hooke's natural philosophy, congruity.

As Halley reported to Newton, Hooke saw his ideas about gravity as "a small part of an excellent system of nature."[32] The construction of this system did not take place by means of mathematics. The quest for a mathematical proof that a force operating according to the inverse square law produces elliptical orbits was not part of Hooke's programme. He was rather interested in graphical geometrical models and in mechanical devices that could offer insight into celestial motions.[33] When he resolved to provide a geometrical construction of these motions, the outcome was determined by his assumption that gravity "is a continuall impulse expanded from the centre of the earth indefinitely by a conicall expansion."[34] This very physical idea was rooted in his general "system of nature" through the notion of congruity.

It is not surprising that the young Newton focused most on this notion while reading Hooke's *Micrographia*, which can be considered one of the various sources of heterodox mechanical ideas of Newton. Far from being limited to the work of Boyle, as an old but still popular view maintains,[35] the young Newton read the works of many British natural philosophers whose ideas contributed to the emergence of a mechanical philosophy critical of Cartesianism, and open to active principles.[36]

In his notes to *Micrographia*, Newton describes the cause of attractions and repulsions among bodies as "ye agreement or disagreement in their motion (caused by their various bulkes, densitys or figures)." Following Hooke, Newton maintains that congruity may be a "coefficient" in "almost all ye phaenomena of nature."

---

[30] Van Helden (1974), 56–7; see Hooke (1665), 246; Id. (1705), 85–6, 91–2.

[31] Hooke (1705), 88, 177.

[32] Newton (1956–77), vol. II, 446.

[33] Guicciardini (2009), 27.

[34] Trinity College Library, Cambridge, Ms. 0.11a.1/16.

[35] For example Boas (1952), 521 and Janiak (2008), 6–7.

[36] Henry (1992), 202–3; Machamer, McGuire, and Kochiras (2012), 374.

Gravity might just be due to the action of an aether whose motion is "incongruous to all other bodys," which are forced "to retire from those places where it is in greatest plenty, towards ye earth where it is in least plenty."[37] A similar ethereal mechanism can be found in a manuscript likely written in 1666 or 1667.[38] There Newton suggests that magnetism can be explained by the motions of two different "streams" that are "unsociable to one another." Two years after *Micrographia*, Newton turned Hooke's congruity into "ye [un]sociableness of matter."[39] This new principle had a long story ahead in Newton's natural philosophy.[40] A "secret principle of unsociableness" was employed in the *Hypothesis* sent to the Royal Society in 1675 and in the system outlined in a letter to Boyle in 1679.[41] It likely influenced Newton's assumption that there are short-range forces operating at microscopical level.

In *Micrographia*, Newton also found that "aire is the menstruum or universal dissolvent of all sulphurious bodys." This property of air is due to the presence of "parts such as are fixed in salt peeter wch parts are true dissolving bodys."[42] Boyle's works had already shown to the young Newton that air is an elastic fluid.[43] In *Micrographia* first, and in the works of Mayow later, Newton found a more advanced description of the elastic component of air. He could then retrace the sources of this ideas in the works of alchemists, such as Michael Sendivogius.[44] It would be a mistake, nonetheless, to ignore the influence of the corpuscular chymistry of Hooke and Mayow.[45] This evidently appears in the outlines of the system of nature that Newton drew between 1675 and 1679.[46] This system is based on the assumption that "there is an aetheriall medium of the same constitution with air, but far rarer, subtiler & more strongly elastic." Various "aethereal spririts" are included in this medium. The most relevant is a spirit "very thinly & subtly diffused through it, perhaps of an unctuous or Gummy, tenacious & springy nature" similar to "the vitall aereall Spirit requisite for the conservation of flame & vitall motions." Gravity can result from the continual process of condensation of this aethereal spirit by the earth, as this might cause a very quick motion towards the centre. In this "descent," Newton adds, "it may beare downe with it the bodys it pervades with force proportionall to the superficies of all their parts it acts upon."[47]

---

[37] Newton (1962), 400–1.

[38] Westfall (1971), 413; Home (1985), 107.

[39] Cambridge University Library MS 3970, f. 473r.

[40] Henry (1989), 156; Westfall (1983), 219.

[41] Newton (1956–77), vol. I, 368; vol. II, 229.

[42] Id. (1962), 407.

[43] Henry (2011), 20.

[44] Hall (1998), 56–7.

[45] Henry (1986), 344.

[46] Westfall (1970), 91; Id. (1971), 364.

[47] Newton (1956–77), vol. I, 364–6.

8.3 Anni Mirabiles

Even after the inclusion of congruity and nitro-aerial "spirits," Newtonian cosmology in 1679 still maintained some fundamental Cartesian principles.[48] In November of that year, Hooke began an epistolary exchange that proved decisive for the evolution of Newton's ideas.[49] As newly elected secretary, Hooke asked Newton to maintain the correspondence with the Society. Despite the previous disagreement on lights and colours, or, perhaps, because of it, Hooke specifically asked for Newton's "objections" to his hypotheses. He was especially interested in Newton's "thoughts of that [hypothesis] of compounding the celestiall motions of planetts of a direct motion by the tangent & and an attractive motion towards the centrall body." Hooke published this hypothesis for the first time in 1674. Newton claimed that he did not read Hooke's *Attempt to prove the motion of the Earth* and that he heard of the principles of Hooke's system of the world for the first time in November 1679. In a following letter, Hooke clarified that gravitational attraction is directed to the centre of planets and operates according to the inverse square law. Hooke's hypothesis was clearly different from what Newton expressed in a letter to Boyle earlier the same year.

After Hooke's letters, Newton began to reshape his ideas and assumed a new approach that led to the celestial dynamic of the *Principia*. Decisive in this intellectual revolution was Hooke's new approach to circular motion.[50] As Derek Whiteside has pointed out, during the exchange with Hooke, Newton's position already shifted towards a "Borellian viewpoint."[51] In a reply to Newton's description of the celestial motions as the result of a dynamic balance between planets' *vis centrifuga* and a "supposed uniform" attraction from the centre, Hooke reminded Newton that "attraction always is in a duplicate proportion to the distance from the center reciprocall."[52] After Hooke's reply, Newton did not continue the epistolary exchange. After what is now known as Halley's Comet appeared in November, 1680, Newton focused on the motion of comets. Influenced by Hooke's innovative ideas, Newton turned to this latter's description of the cometary motions published in 1678.[53] In February, 1681, he rejected Flamsteed's view that cometary curvilinear motions are due to the combination of the sun's magnetic attraction and the motion of the vortexes that carry them. "I can easily allow," Newton wrote, "an attractive power in the sun whereby the planets are kept in their courses about him from going away in tangent lines." But this force cannot be magnetic, i.e., attractive and repulsive. The retrograde motion of Halley's comet with respect to the planets entailed

---

[48] Bertoloni Meli (2006b), 321, Guicciardini (2011), 129; Ducheyne (2012), 38.

[49] Kochiras (2008), 39–41; Bertoloni Meli (2005), 539; Whiteside (1991), 21–2.

[50] Westfall (1970), 93; Id. (1971), 426–30; Guicciardini (1998), 32.

[51] Whiteside (1970), 13.

[52] Newton (1956–77), vol. II, 307, 309.

[53] Cambridge University Library MS 4004, f. 103r-v.

that the comet moves in a direction opposite to that of Flamsteed's vortex.[54] In Newton's eyes the recent observations of the comet supported Hooke's hypothesis. A few years later, Newton definitively embraced Hooke's views and described the motions of comets as the result of their inertial tendency and the attraction "towards the centre of the sun and each of the planets."[55]

Once set in motion, Newton's philosophical evolution significantly diverged from Hooke's views. In what can be described as the real *annus mirabilis*,[56] Newton completed a new analysis *de motu corporum in gyrum* that went further than Hooke's physical principles and geometrical instruments. In the *Principia* Newton rejected Cartesian vortexes and claimed that no corporeal medium directly acting on bodies could account for gravity. Regardless of how one interprets Newton's position about action at distance, it seems unquestionable that in his mature cosmology there was no room for purely mechanical agents.[57] But Hooke's model was not entirely rejected. Newton employed ethereal substances whose operations were not entirely mechanical because of principles such as sociableness. Although these substances were described in different ways to fill different explanatory functions,[58] Newton's aethers maintained throughout the years some features of the elastic and congruous fluids often employed by Hooke.[59] In 1706, Newton introduced a "medium exceedingly more rare and subtle than air, and exceedingly more elastic and active," whose resistance to the motions of bodies was negligible.[60] These were also the properties of a substance responsible for some physiological functions of animals, such as motion and vision.[61] From 1692, Newton identified in the chymical principles of Hooke and Mayow a source of active principles operating in dissolutions, fermentations, and combustions.[62] In the query 31 of *Opticks*, Newton suggested that bodies have some "powers, virtues, or forces" by means of which they act at distance "but also one another." A "great part of the phaenomena of nature" may depend on these active principles. Fermentation and combustion, for instance, are the effects of the action of these forces on the interaction between the nitrous component of air and the sulphuric components of some bodies.[63]

In 1717, when this dynamic mechanical philosophy was advanced,[64] Newton considered capillarity as an example of these interactions at a microscopic level. This was not Newton's view before 1706, when Francis Hauksbee published in the

---

[54] Newton (1955–77), vol. II, 337–8, 341; Guicciardini (2011), 129.

[55] Cambridge University Library MS 3965, f. 613r; cf. Ruffner (2000), 262–3.

[56] Whiteside (1970), 14.

[57] Ducheyne (2014), 692–3; Kochiras (2009), 276.

[58] Hall (1998), 59; Home (1993), 197–9.

[59] Henry (1986), 348; Kochiras (2008), 134 n. 304.

[60] Newton (1779–85), vol. IV, 224.

[61] Ibid., 226–7; cf. Mamiani and Trucco (1991), 87–95.

[62] Newton (1692), 256–7.

[63] Newton (1779–85), vol. IV, 242–51.

[64] Westfall (1971), 378.

Philosophical Transactions an account of an experiment "shewing that the seemingly spontaneous ascention of water in small tubes open at both ends is the same in vacuo as in open air."[65] For many years, Newton maintained that capillarity was due to the "unsociableness" between air and the glass of the pipes.[66] In the notes to *Micrographia*, Newton observed that liquors in a vessel ascend thin pipes "because ye vessel hath more or less congruity wth ye Aire yn wth those liquors."[67] But Hauksbee's experiment confirmed that these phaenomena take place even in vacuo, i.e., in absence of air. Thus, Newton turned to attractive short-range forces between the liquids and the glass, the same forces that caused cohesion of bodies.[68] Richard Westfall has described this theoretical shift in Newton as Hooke's posthumous "revenge" through Hauksbee.[69]

It is significant that Hauksbee expressed scepticism towards Hooke's explanation based on congruity. When he entered the Royal Society as curator of experiments, Hauksbee had already developed a corpuscular view of nature mainly influenced by Boyle's works.[70] As new curator, Hauksbee was instrumental in the realization of the experimental agenda of the new president of the Society, Sir Isaac Newton.[71] The experiment on capillarity, however, shows that the relationship between the curator and the president was of mutual influence. Hauksbee's effluvial theories have been credited as the main source of Newton's late electrical spirit.[72] Newtonian dynamic mechanical philosophy, on the other hand, influenced the evolution of Hauksbee's views on matter and forces. After a series of ingenious experiments on electricity, the new curator published in 1709 his *Physico-Mechanical Experiments on Various Subjects*. An expanded edition was published ten years later. While in 1709 Hauksbee explained electrical phaenomena in terms of "fine effluvia," in the appendix added in 1719 he emphasized the role of the "considerable force and activity" of those bodies that produce electricity by friction.[73]

The influence of Newtonian philosophy was already present in 1709, when Hauksbee expanded the outcome of some experiments on capillarity carried out few years later. In the "matter of fact" added to the experiment originally published in the *Transactions* of 1706, Hauksbee concluded that air pressure was not involved in the phenomenon. The levels that liquids, regardless of atmospheric pressure, reach seem to depend just on the dimension of the holes rather than the thickness of tubes. In spite of this, Hauksbee considered capillarity as one of the phenomena of attraction and repulsion that he had been investigating in his experiments on electricity.

---

[65] Hauksbee (1706–07), 2223.

[66] Westfall (1971), 348; Id. (1983), 746.

[67] Newton (1962), 400.

[68] Id. (1779–85), vol. IV, 251–63.

[69] Westfall (1971), 384, 412 n.142.

[70] Home (1981), 24, 32; Schofield (1970), 67.

[71] Guerlac (1977), 108.

[72] Id. (1967), 46–8; Id. (1977), 110, 117; Home (1985), 102.

[73] Hauksbee (1719), 55–7, 238–9; cf. Heilbron (1979), 234, 237–8.

As such, it could easily be explained by some general principles of nature, like attraction. Hooke's congruity, on the contrary, seemed to Hauksbee both experimentally and theoretically inadequate. Compared to the Newtonian gravitational attraction, Hooke's congruity became in Hauksbee's eyes an unintelligible principle. Since Newton introduced a uniform principle of attraction, Hauksbee considered it his duty to show that capillarity "may be handsomly accounted for by it, without being forc'd upon any of those obscure precarious suppositions." The similitudes between capillarity and magnetism seemed enough to suggest "that the phaenomena of the load-stone, and of small tubes, depend upon one and the same principle in general." The attraction between the internal parts of the small tubes and the liquids is the main cause of the liquids' rise. The superficial particles of glass exert some small-range attractive forces that pulls the liquids upwards.[74] By claiming the superiority of Newton's dynamic mechanical philosophy over Hooke's, the new curator showed in fact how Newton built on some of Hooke's most heterodox ideas and how far he was able to see a new order in nature by standing also on Hooke's shoulders.[75]

Bibliography

Aiton, Eric. 1972. *The vortex theory of planetary motion*. London: Mcdonald.

Applebaum, Wilbur. 1996. Keplerian astronomy after Kepler. *History of Science* 34: 451–504.

Bennett, Jim. 1989. Magnetical philosophy and astronomy from Wilkins to Hooke. In *Planetary astronomy from the renaissance to the rise of astrophysics, part A: Tycho Brahe to Newton*, ed. R. Taton and C. Wilson, 222–230. Cambridge: Cambridge University Press.

Bertoloni Meli, Domenico. 2005. Who is afraid of centrifugal force? *Early Science and Medicine* 10: 535–541.

———. 2006a. *Thinking with objects: The transformation of mechanics in the seventeenth century*. Baltimore: Johns Hopkins University Press.

———. 2006b. Inherent and centrifugal forces in Newton. *Archive for History of Exact Sciences* 60: 319–335.

Birch, Thomas. 1756–1957. *The history of the Royal Society of London*, 4 vols. London.

Boas, Marie. 1952. The establishment of the mechanical philosophy. *Osiris* 10: 412–541.

Borelli, Giovanni Alfonso. 1666. *Theoricae mediceorum planetarum*. Florence: Ex Typographia S.M.D.

Boulliau, Ismaël. 1645. *Astronomia philolaica*. Paris: Simeonis Piget.

Casini, Paolo. 1997. "Magis amica veritas": Newton e Descartes. *Rivista di filosofia* 88: 197–221.

Ducheyne, Steffen. 2012. *The main business of natural philosophy: Isaac Newton's natural philosophical methodology*. Dordrecht: Springer.

———. 2014. Newton on action at a distance. *Journal of the History of Philosophy* 52: 675–701.

Erlichson, Herman. 1997. Hooke's September 1685 ellipse vertices construction and Newton's instantaneous impulse construction. *Historia Mathematica* 24: 167–184.

Fiocca, Alessandra. 1998. The southern deviation of freely falling bodies: From Robert Hooke's hypothesis to Edwin Hall's experiment (1679–1902). *Physis* 35: 51–83.

Gal, Ofer. 2002. *Meanest foundations and nobler superstructures: Hooke, Newton and the "compounding of the celestial motions of the planetts"*. Dordrecht: Kluwer.

---

[74] Hauksbee (1719), 99–100, 200–1, 202, 204, 208.

[75] Cf. Newton (1959–77), vol. I, 416.

————. 2005. The invention of celestial mechanics. *Early Science and Medicine* 10: 529–534.

Gal, Ofer, and Raz Chen-Morris. 2005. The archaeology of the inverse square law: (1) metaphysical images and mathematical practices. *History of Science* 43: 391–414.

————. 2006. The archaeology of the inverse square law: (2) the use and non-use of mathematics. *History of Science* 44: 49–67.

Guerlac, Henry. 1967. Newton's optical aether. *Notes and Records of the Royal Society of London* 22: 45–57.

————. 1977. *Essays and papers in the history of modern science.* Baltimore: Johns Hopkins University Press.

Guicciardini, Niccolò. 1998. *Newton: un filosofo della natura e il sistema del mondo.* Milan: Le Scienze.

————. 2005. Reconsidering the Hooke-Newton debate on gravitation: Recent results. *Early Science and Medicine* 10: 511–517.

————. 2009. *Isaac Newton on mathematical certainty and method.* Cambridge, MA/London: MIT Press.

Guicciardini, Niccolò. 2011. *Newton.* Rome: Carocci.

————. 2018. *Isaac Newton and natural philosophy.* London: Reaktion Books.

Hall, Alfred Rupert. 1998. Isaac Newton and the aerial nitre. *Notes and Records of the Royal Society* 52: 51–61.

Hauksbee, Francis. 1706–07. An experiment made at Gresham College. Philosophical Transactions 25: 2223–2224.

————. 1709. *Physico-mechanical experiments on various subjects.* London: Printed by R. Brugis.

————. 1719. *Physico-mechanical experiments on various subjects.* London: Printed for J. Senex; and W. Taylor.

Heilbron, John. 1979. *Electricity in the 17th and 18th centuries.* Berkeley/Los Angeles/London: University of California Press.

Henry, John. 1986. Occult qualities and the experimental philosophy: Active principles in pre-Newtonian matter theory. *History of Science* 24: 335–381.

————. 1989. Robert Hooke, the incongruous mechanist. In *Robert Hooke: New studies*, ed. Michael Hunter and Simon Schaffer, 149–180. Woodbridge: Boydell Press.

————. 1992. The scientific revolution in England. In *The scientific revolution in national context*, ed. Roy Porter and Mikuláš Teich, 178–209. Cambridge: Cambridge University Press.

————. 2011. Gravity and De gravitatione: The development of Newton's ideas on action at a distance. *Studies in History and Philosophy of Science* 42: 11–27.

Home, Roderick. 1981. *The effluvial theory of electricity.* New York: Arno Press.

————. 1985. Force, electricity, and the powers of the living matter in Newton's mature philosophy of nature. In *Religion, science and worldview: Essays in honor of Richard S. Westfall*, ed. Margaret Osler and Paul Lawrence Farber, 95–117. Cambridge: Cambridge University Press.

————. 1993. Newton's subtle matter: The Opticks queries and the mechanical philosophy. In *Renaissance and revolution: Humanists, scholars, craftsmen and natural philosophers in early modern Europe*, ed. J.V. Field and Frank A. James, 193–202. Cambridge: Cambridge University Press.

Hooke, Robert. 1674. *An attempt to prove the motion of the earth by observations.* London.

————. 1705. *Posthumous works*, ed. Richard Waller. London.

Huygens, Christiaan. 1888–1950. *Oeuvres completes*, 22 vols., ed. Société hollandaise des sciences. The Hague: Martinus Nijhoff.

Janiak, Andrew. 2008. *Newton as philosopher.* Cambridge: Cambridge University Press.

Kochiras, Hylarie. 2008. *Force, matter, and metaphysics in Newton's natural philosophy.* Phd Dissertation, University of North Carolina at Chapel Hill.

————. 2009. Gravity and Newton's substance counting problem. *Studies in History and Philosophy of Science* 40: 267–280.

Koyré, Alexandre. 1968. *Newtonian studies.* Chicago: University of Chicago Press.

Lohne, Johs. 1960. Hooke versus Newton: An analysis of the documents in the case of free fall and planetary motion. *Centaurus* 7: 6–52.

Machamer, Peter, J.E. McGuire, and Hylarie Kochiras. 2012. Newton and the mechanical phi-losophy: Gravitation as the balance of the heavens. *The Southern Journal of Philosophy* 50: 370–388.

Mamiani, Maurizio, and Emanuela Trucco. 1991. Newton e i fenomeni della vita. *Nuncius* 6: 69–96.

Nauenberg, Michael. 1994. Hooke, orbital motion and Newton's Principia. *American Journal of Physics* 62: 331–350.

———. 1998. On Hooke's 1685 manuscript on orbital mechanics. *Historia Mathematica* 25: 89–93.

———. 2005. Hooke's and Newton's contributions to the early development of orbital dynamics and the theory of universal gravitation. *Early Science and Medicine* 10: 518–528.

———. 2006. Robert Hooke's seminal contribution to orbital dynamics. In *Robert Hooke: Tercentennial studies*, ed. Michael Cooper and Michael Hunter, 3–32. Aldershot: Ashgate.

Newton, Isaac. 1779–85. *Opera quae extant omnia*, 5 vols., ed. Samuel Horsley, London.

———. 1959–77. *The correspondence of Isaac Newton*, 7 vols., ed. H. W. Turnbull, J. F. Scott, A. R. Hall and L. Tilling. Cambridge: Cambridge University Press.

———. 1962. In *Unpublished scientific papers*, ed. Alfred Rupert Hall and Marie Boas Hall. Cambridge: Cambridge University Press.

———. 1999. In *The Principia: Mathematical principles of natural philosophy*, ed. I. Bernard Cohen and Anne Whitman. Berkeley/Los Angeles/London: University of California Press.

Patterson, Louise Diehl. 1949. Hooke's gravitation theory and its influence on Newton I: Hooke's gravitation theory. *Isis* 40: 327–341.

———. 1950. Hooke's gravitation theory and its influence on Newton II: The insufficiency of the traditional estimate. *Isis* 41: 32–45.

Pugliese, Patri. 1989. Robert Hooke and the dynamics of motion in a curved path. In *Robert Hooke: New studies*, ed. Michael Hunter and Simon Schaffer, 181–205. Woodbridge: Boydell Press.

Ruffner, James. 2000. Newton's propositions on comets: Steps in transition, 1681–1684. *Archive for History of Exact Sciences* 54: 259–277.

Schofield, Robert. 1970. *Mechanism and materialism: British natural philosophy in the age of reason*. Princeton: Princeton University Press.

Van Helden, Albert. 1974. The telescope in the seventeenth century. *Isis* 65: 38–58.

Westfall, Richard. 1970. Uneasy fitful reflections on fits of easy transmission. In *The annus mira-bilis of Isaac Newton 1666–1966*, ed. Robert Palter, 88–104. Cambridge, MA: The MIT Press.

———. 1971. *Force in Newton's physics: The science of dynamics in the seventeenth century*. New York: American Elsevier.

———. 1972a. Circular motion in the seventeenth mechanics. *Isis* 63: 184–190.

———. 1972b. Robert Hooke (1635–1703). In *Dictionary of scientific biography*, ed. Charles Gillispie, vol. VI, 481–488. New York: Scribner.

———. 1983. *Never at rest: A biography of Isaac Newton*. Cambridge: Cambridge University Press.

Whiteside, Derek. 1970. Before the Principia: The maturing of Newon's thoughts on dynamical astronomy, 1664–1684. *Journal for the History of Astronomy* 1: 5–19.

———. 1991. The prehistory of the Principia from 1664 to 1686. *Notes and Records of the Royal Society of London* 54: 11–61.

Wilson, Curtis. 1970. From Kepler's laws, so-called, to universal gravitation: Empirical factors. *Archive for History of Exact Sciences* 6: 89–170.

Chapter 9
Conclusion: Did Hooke Have a Natural Philosophy?

9.1 Professional Barriers and Social Boundaries

By now it should be evident that over more than three decades Robert Hooke produced an innovative natural philosophy. Neither the image of the staunch mechanic enemy of Newton nor that of the undisciplined forerunner of modern ideas seem adequate descriptions of Hooke's philosophical persona.[1] This latter, however, remains controversial. Throughout his life, Hooke was an assistant of Willis and Boyle, a professor of geometry at Gresham College, an architect, a curator of experiments and a secretary of the Royal Society.[2] The complex dynamics among these offices and occupations led Hooke towards different topics and resulted in a sophisticated natural philosophy. His contemporaries discussed, praised and opposed his ideas. But how did they see him? Was he, in their eyes, a natural philosopher after all?

Hooke's status is far from clear. As curator of experiments and secretary, Hooke played a major role in the early Royal Society. In the first decade of activity, he provided a significant number of experiments discussed at meetings, a long series of microscopic observations that he turned into a book, and a continuous flow of ideas for the fellows. When his experimental contributions reduced, the Society declined.[3] As secretary from 1677 to 1684, Hooke aimed to increase the scientific activities of the fellows.[4] But after his tenure, the decline of the Society continued until Newton was elected president. Hooke's failure is significant of his ambivalent position within the Society. Far from being a modern scientific professional,[5] Hooke was at

---

[1] For example Nakajima (2001), 162.

[2] Cooper (2003), 95–190; Inwood (2002), 19–20.

[3] Espinasse (1974), 335; Boas Hall (1991), 32–3, 88–90.

[4] Mulligan and Mulligan (1981), 331.

[5] As Purrington (2009), 86 claims.

© Springer Nature Switzerland AG 2020
F. G. Sacco, *Real, Mechanical, Experimental*, International Archives
of the History of Ideas Archives internationales d'histoire des idées 231,
https://doi.org/10.1007/978-3-030-44451-8_9

once an employee and a fellow.[6] However relevant his influence could have been, the Society remained a complex social and intellectual entity.[7] The historical vicissitudes of the Society and the personality of Hooke contributed to shape the unprecedented role of curator.[8] As he wrote in 1663, the young Hooke still "belonged" to Boyle when he was appointed by the Royal Society.[9] A year after his first book was published, Hooke was proposed as curator by Robert Moray. When the fellows invited Hooke to "come and sit among them," he was still working for Boyle, who received the "thanks of the society for dispensing with him for their use." After some fellows managed to have him elected professor of geometry at Gresham College, Hooke left Boyle's house in Pall Mall and moved into his new lodgings in the City, where the repository of the Society was initially located.[10] Thanks to the election of Hooke, the fellows could continue to hold their meetings at the college. The project of the merchant John Cutler "of giving fifty pounds a year to Mr. Hooke during his life for the reading of the history of trades at Gresham college" offered the Society a further opportunity to finance the newly instituted office of curator.[11]

From his first nomination, Hooke's position in the Royal Society proved contradictory. In 1663 he was elected fellow, but exempt from the annual fees that other fellows were expected to pay. He was employed by the Society, but his salary was reduced when he obtained the position at Gresham College. There he gave lectures in geometry and prepared experiments for the weekly meetings of the Society. But even when he was employed by Boyle, Hooke was not an invisible technician.[12] As an assistant and protégé of Boyle, in the early 1660s Hooke designed and realized one of the most important scientific instruments of his time, the air pump.[13] After the publication of Boyle's *New Experiments*, he contributed to the debate on the nature of elasticity of air and drew an innovative and heterodox system of natural philosophy around the principle of congruity. As a mechanic, he designed new instruments and looked for protection to individuals like Boyle or groups like the Society. As a natural philosopher, he openly debated scientific questions among his peers.[14] Although subordinate to the fellows, the office of curator was not similar to that of the invisible technicians employed by early modern virtuosi. Since Hooke assumed it in 1664, the Society employed operators and assistants.[15] One of them, Denis Papin, became curator and was elected fellow. Another fellow, Richard Lower,

---

[6] Boas Hall (1991), 31; Shapin (1989a), 256, 285.

[7] Hunter (1982b), 460.

[8] Pumfrey (1991), 2, 3.

[9] Boyle (2001), vol. II, 97.

[10] Thomas (2009), 18–20; Id. (2011), 2; Swann (2001), 87; cf. Feingold (1998), 172; Hunter (1985), 159, 166.

[11] Birch (1756–57), vol. I, 123–4, 250, 442, 473, 479, 484–5, 496.

[12] Cf. Shapin (1989b), 554–6, 560.

[13] Van Helden (1991), 158.

[14] Bennett (1980), 34.

[15] Shapin (1988), 382.

declined the office of anatomy curator because of his more lucrative medical profession. The physicians Nehemiah Grew and Edward Tyson, on the contrary, became curators after their election to the fellowship.

Hooke's hybrid status as fellow remained, and the Society did not hesitate to withdraw the curator's salary when in 1683 he failed to prepare the experiments required.[16] Whereas in the early years of activity many fellows proposed experiments and contributed to their realization, the increase in the number of amateur aristocrats contributed to a shift towards a major role for curators. The non-scientist fellows expected to be entertained by that part of the scientific work in which hypotheses are confuted rather than being instructed by long and repetitive experimental activities. For this reason, curators prepared a specimen of experiments and observations, and operators reproduced them in the weekly meetings. When the momentum of the early years disappeared, the Society lost the ability to sustain a continuous interest in planned experimental researches.[17] Thus, the weekly experiments realized by operators have been compared to *mises en scène* aiming to secure the assent of the members of the community to some fundamental matters of fact resulting from a longer, day to day activity.[18] In this view, the presence of aristocrats was a source of legitimacy for the whole experimental activity, because of the disinterestedness associated with their social status.[19] If the moral economy of Restoration science reproduced wider social standards, mechanics and merchants could not be considered peers, albeit in the restricted experimental community, of new aristocrat natural philosophers such as Boyle. For this reason, Hooke has been described as a subordinate figure whose aim to be part of the scientific community was constantly frustrated because of his "artisanal rather than gentlemanly moral economy."[20]

As Jan Golinski acknowledged, such sociological explanations tend to replace the historical quest for empirical evidence with an analytic approach.[21] The image of the early Royal Society as a community of gentlemen scholars stands on the Foucaultian notions of episteme and regimes of truth.[22] The a-historical nature of these concepts evidently appears in many such sociological reconstructions, despite these often employing skilfully crafted arguments supported by a wide range of sources.[23] Nevertheless, when powerful philosophical tools are compared against complex historical backgrounds, a more nuanced picture emerges.[24] Not all fellows of the Royal Society, for instance, were gentlemen. Most considered themselves as

---

[16] Birch (1756–57), vol. III: 42, 47, 491; vol. IV, 188, 207–8, 277.

[17] Heilbron (1983), 13, 15; Boas Hall (1991), 24, 33, 42; Lawson (2015), 192.

[18] Licoppe (1996), 11, 54.

[19] Dear (1992), 627; Shapin (1994), 42, 45–6, 65, 83, 86–7.

[20] Shapin (1989a), 254–6, 262–3; Biagioli (1996), 199–200; cf. Shapin (1991), 295; Id. (2006), 181–2, 190; Daston (1995), 3–6, 12.

[21] Golinski (2005), 12.

[22] Burke (2004), 54; Gimelli Martin (2002), 27.

[23] Feingold (1996) 132–8; see also Hacking (1991), 235, 239.

[24] Feingold (2006), 203–4.

virtuosi, learned and ingenious individuals interested in natural and mechanical knowledge.[25] Along with Hooke, many other virtuosi inhabited more than one social space in the multi-layered social world of Restoration. For the courtier-inventor Samuel Morland, the apparently inconsistent categories of mechanic and gentleman coexisted.[26] Like Hooke, Jonas Moore inhabited both the world of gentlemen and that of mechanics.[27] In the coffee-houses people of different social status could mix. It is significant that Hooke used to meet his closest associates there, and planned a new philosophical club in which the aristocrat Christopher Wren took part as well as the merchant Francis Lodwick.[28] Although it was not a concrete realization of an egalitarian utopia, the experimental community of seventeenth century England provided a space where mechanics, merchants and aristocrats could interact following some new epistemological rules that added to traditional codes of behaviour. In this space, for instance, social status was one but not the only factor involved in assessing witnesses.[29] The criteria adopted for the evaluation of testimony in natural philosophy were also influenced by the outcome of Renaissance dialectic and legal theory. The value of testimony depended also on the skills and competences of the observers. For this reason, for instance, the Society used his curator's astronomical observations to judge those of the Polish aristocrat Hevelius. Since the difference between the observations of Hevelius and Auzout was about matters of fact, Lord Broucker noted, "'tis the authority, number and reputation of other observers that must cast the balance."[30] "Even of honest and sincere witnesses," Boyle added, "the testimony may be insufficient if the matters of fact require skill in the Relator."[31] A mechanic but skilled observer, therefore, was worth some trust even though he was not considered as disinterested as the noble Boyle. Interest, in fact, was to a certain extent seen as a fundamental component of the new experimental enterprise, not an obstacle.[32]

9.2   A Philosophical Servant

Boyle, Steven Shapin wrote, was the "instantiation of the Christian experimental natural philosopher."[33] In this view, his nobility and disinterestedness were the foundations of a clear demarcation between matters of fact and conjecture in the new

---

[25] Yeo (2014), 6–7; cf. Roger (1984), 304; Schuster and Taylor (1997), 512–3.

[26] Ratclif (2007), 161–2.

[27] Stewart (1998), 138–9.

[28] Hooke (1935), 199–200.

[29] Serjeantson (1999), 196–7, 202; Shapiro (2000), 25; cf. Shepard (2015), 1–5, 32, 303–13.

[30] Boyle (2001), vol. II, 610.

[31] Royal Society, Boyle Papers, vol. IX, f. 70r.

[32] Keller (2015), 219.

[33] Shapin (1991), 299.

experimental philosophy. The inferior social position of Hooke, on the contrary, was accompanied by a consistent concern for his interests as inventor and discoverer. His inclination to formulate and stand by controversial hypotheses, therefore, seems nothing else than an expression of a different moral economy for which there was no place in the community of gentlemen-scholars.[34] Hooke's philosophical ambitions were frustrated by this experimental form of life.[35] He began as Boyle's servant and remained a subordinate figure.

Although this image of Boyle and Hooke explains their positions in the early 1660s, it fails to take into account the evolution that took place over more than 30 years.[36] "So apt are we to be mis-led," Boyle wrote in the *New Experiments*, "even by experiments themselves, into mistakes, when either we consider not that most effects may proceed from various causes, or minde onely those circumstances of our experiment, which seem to comply with our preconceiv'd hypothesis or conjecture." Boyle criticised those who claimed more than they could really know; he largely used expressions of doubts and caution in his early works on pneumatics. Some of these, such as the *Defence of the Doctrine touching the Spring and the Weight of Air,* published in 1662, were not written to "establish a theory and principles, but to devise experiments and to enrich the history of nature with observations faithfully made and deliver'd."[37] But Boyle did not rule out hypotheses as detrimental to experimental philosophy, rather he considered matters of fact as valuable insofar as they were given interpretative shape.[38] In the *Experimental History of Colours*, for instance, the "theoretical part of the enquiry" was "interwoven with the historical" one. The task of a naturalist was "to devise hypotheses and experiments."[39] In a manuscript written between 1668 and 1670, Boyle listed sense, reason and authority as the main foundations of human knowledge.[40] "The organs of senses are but the instruments of reason in the investigation of truth." Hypotheses and deductions cannot replace experiments, but the inquiry into nature would not be successful without the "instruments of knowledge" devised by human reason.[41]

Not less problematic than the demarcation between matters of fact and hypotheses is the distinction between the gentleman philosopher and his mechanic servants and invisible technicians. Influenced by Erasmian ideals, the young Boyle saw industry and commitment to material pursuits as virtues worthy of an aristocrat. He engaged in a form of natural philosophy modelled on the work of chymists, mechanics and practitioners.[42] Despite the permanence of social barriers between him and

---

[34] Id. (1989a), 256, 270–2.

[35] Shapin and Schaffer (1985), 44–55.

[36] Clericuzio (1997), 109–10.

[37] Boyle (1999), vol. I, 210; vol. III, 12.

[38] Hunter (2000), 9; Sargent (1995), 211; Rogers (1972), 254.

[39] Boyle (1999), vol. IV, 26; vol. V, 287.

[40] Ricciardo (2010), 396, 409.

[41] Royal Society, Boyle Papers, vol. IX, ff. 6r, 12r, 69r, 125r.

[42] Oster (1992), 260, 270.

his servant, Boyle acknowledged Hooke's intellectual abilities and philosophical status. Hooke read Euclid at Westminster School in London and Descartes for the first time at Oxford in 1656.[43] "He was there," Aubrey noted, "assistant to Dr. Thomas Willis in his cymistry; who afterwards recommended him to the hon'ble Robert Boyle, esqre, to be usefull to him in his chymicall operations. Mr. Hooke then read to him (R. B. esqre) Euclid's Elements, and made him understand Des Cartes' Philosophy."[44] Boyle acknowledged Hooke's philosophical expertise when he assigned to him a defence of mechanical explanations of the spring of air from the criticism of Francis Linus. The text was published in the *Defence* of 1662. Even when he was no longer his assistant, it was to Hooke again that Boyle in 1677 submitted a draft of a manuscript on *Final Causes* dealing with Cartesian ideas.[45] This evidence supports the view that the relationship between the two virtuosi was not only shaped by differences in social status.[46]

That in Boyle's eyes Hooke was also a legitimate member of the philosophical community is suggested also by the links between their early projects of reform of natural history composed in the mid-1660s. Even before he left London for Oxford in 1668, Boyle's involvement in the activities of the Royal Society was episodic. Rather than a community to shape according to a new experimental form of life, since its foundation the Society's Baconian agenda influenced Boyle's philosophical style.[47] To implement Bacon's project, the fellows designed various formats of historical inquiry for correspondents and travellers around the world.[48] As Sprat stated in 1667, the fellows' collective enterprise relied on "plain, diligent, and laborious observers; such, who, though they bring not much knowledge, yet bring their hands, and their eyes uncorrupted."[49] To direct them in the observation of unknown parts of the globe, the fellows provided guidelines based on Bacon's topics for natural history. Influenced by these works, Boyle composed a *Design about Natural History,* in which one can find the Baconian doctrine which formed thereafter the centrepiece of his works.[50] In Boyle's *Design* there is room for hypotheses and theories. Boyle does not maintain Bacon's tripartite structure; rather he adopts a wider list of topics.[51] A related manuscript included evidence that Hooke's early lectures on natural history, read to Gresham College, influenced Boyle's *Design.*[52] The aim and structure of these works are indeed similar. In January, 1666, Oldenburg informed Boyle that "Hooke has also ready (having shewed it me and others) a

---

[43] Hooke (1665), 44; Hunter (2003), 131–2.

[44] Aubrey (1898), vol. I, 410–1.

[45] Davies (1994), 158–9.

[46] Feingold (2006), 204.

[47] Hunter (2009), 145, 152, 164, 169.

[48] Daston (2011), 89–90; Raj (2007), 28–9.

[49] Sprat (1667), 72–3.

[50] Hunter (2007), 3–5.

[51] Boyle (2001), vol. III, 170–5.

[52] Anstey and Hunter (2008), 91, 117–8; Hunter (2000), 124–5.

method for writing a Naturall History."[53] Hooke's lectures provided a list of topics of inquiry, and considered hypotheses as a legitimate component of natural history.[54] In some notes originally written in the mid-1660s, Boyle drew a scheme of a philosophical history of nature modelled on Hooke's. Even Hooke's philosophical algebra was included[55]:

> One of the noblest and usefullest things that may be reduced to the title of loos experiments is a philosophical or physical algebra, whereby divers of the practices of symbolical arithmetick may be applyed to natural and experimental things as for example, to resolve a question, or perform an operation we may reckon up and digest into the best order, all the phenomena that we know, and other means that are in our power already, and look upon these as our Data, then by considering the nature and tenor of the proposition, we may find out whether the Data we have be sufficient or no and if insufficient what other Data we want. We may also give symbolical marks to our Data and other particulars and by adding, subtracting etc. in a way suitable to the nature of this physical algebra, we may frame now propositions, whence will oftentimes result new truths and which will at least frequently suggest new inquirys and experiments.[56]

9.3 The Swinging Pendulum of Historiography

Like all mono-causal explanations, those founded on social status alone leave many questions with no adequate answer.[57] The emphasis on concrete particular events and the public character of the inquiry played a significant role in the emergence of the modern notion of scientific fact.[58] But the experimental philosophy was not limited to matters of fact guaranteed by aristocratic witnesses. When the young Isaac Newton sent his *New Theory of Light and Colours* to the Royal Society, his text clearly did not follow the so-called literary technology that has been attributed to Boyle.[59] In spite of the abstract nature of Newton's account, Hooke was able to reproduce the experiment with two prisms and confirm its outcome. But no further agreement was reached, and the contenders kept discussing alternative hypotheses until Newton idiosyncrasies prevailed.[60]

Identifying Restoration experimental philosophy with a codified set of rules to produce consensus and avoid theoretical disagreements risks reviving a positivist image of inductive science.[61] In this evergreen view, Hooke is at the margins of the

---

[53] Boyle (2001), vol. III, 46.

[54] Guildhall Library, London MS 1757.11, f. 103r; Oldroyd (1987), 159; Hooke (1705), 21–2; cf. Boas Hall (1981), 183; Vickers (1987), 4; Jardine (1974), 88 n.2.

[55] Anstey and Hunter (2008), 91, 117–8.

[56] Royal Society, Boyle Papers, vol. IX, ff. 72r-73r.

[57] Lux (1991), 187.

[58] Shapiro (2000), 167.

[59] Schaffer (1989), 68.

[60] Illife (2017), 126–7, 318–28.

[61] Principe (1998), 24–5; Newman (2006), 11; Shea (2007), 460–1; Boschiero (2007), 23, 31.

experimental community because of his use of hypotheses and supposedly conservative mechanical beliefs.[62] This image significantly reproduces Newton's disdain for Hooke's natural philosophy.[63] Despite the persistance of this Newtonian privileged point of view among historians, neither Glanvill's moderate scepticism nor Wilkin's natural religion provided a unifying epistemology to the Royal Society.[64] Moreover, as we have seen in the previous chapters, Hooke's experimental and mechanical philosophy aimed at a demonstrative form of knowledge by means of a combination of deductions and experiments.

9.4 Hooke's Philosophical Character

Morderchai Feingold had recently vindicated Hooke's philosophical reputation against what he described as a "merciless – not to say spurious – representation grounded on misrepresentation of the evidence and on gross misunderstanding of the nature of seventeenth-century science and its cultural milieu." Although not an aristocrat, Hooke was, in Feinglod's words, "a gentleman of science."[65] His affiliation with and role in the experimental community were nonetheless the result of a process that was influenced both by social norms and his individual motives. As his diary shows, Hooke was seldom a passive object of external social forces. He navigated the various social codes of the Restoration and shaped for himself a distinctive role in the experimental community.[66]

The process of publication of *Micrographia* shows that since the beginning this process proved neither easy nor immediate. Hooke's early involvement in the Royal Society included carrying out a project of microscopical research begun by Wren and Wilkins in Oxford. On March 25, 1663, the fellows solicited the new curator to "prosecute his microscopical observations in order to be published." After the manuscript of Power's Experimental Philosophy reached the Society, pressure on the curator increased, and he was then expected to read an account of a new observation at every meeting. Once Hooke's manuscript was ready, the fellows decided that "if Mr. Hooke's observations should be printed by order of the society, they might be perused and examined by some members."[67] In a letter to Boyle in early November, 1664, Hooke lamented that the examination by "several of the members of the Society" was delaying the publication. The preface, Hooke added, "has been very long in the hands of some, who were to read it." Although troubled by the delays and

---

[62] Chapman (2005), 39.

[63] Koyré (1968), 235 n.2; Ducheyne (2012), 60–2.

[64] Cf. Illife (2004), 437–8; see also Van Leeuwen (1963), 34, 80–90; Shapiro (2000), 4–5, 11–4, 105–10, 113–5, 144–9, 153–4, 162; Burns (1981), 20–37.

[65] Feingold (2006), 203.

[66] Golinski (2005), 58.

[67] Birch (1756–57), vol. I, 213, 397, 443.

the "great expectations" of the fellows, Hooke was still confident "to prevail with the printer" and see the book published by mid-November.[68] But on November 23, 1664, when the entire book was already printed, the fellows demanded:

That Mr. Hooke give notice in the dedication of that work to the Society, that though they have licensed it, yet they own no theory, nor will be thought to do so; and that several hypotheses and theories laid down by him therein, are not delivered as certainties, but as conjectures; and that he intends not at all to obtrude or expose them to the world as the opinion of the society.[69]

Hooke abided, and added a dedication to the fellows. By "avoiding dogmatising, and the espousal of any hypothesis not sufficiently grounded and confirm'd by experiments," the Society has set some principles that "may preserve both philosophy and natural history from its former corruptions." In the book, on the contrary, "there may perhaps be some expressions, which may seem more positive then your prescription will permit." These "conjectures and quaeries," Hooke concluded, were his own.[70]

The conflict between Hooke and the council over the licence of *Micrographia* might appear as the effect of a violation of the clear social boundary between matters of fact and hypotheses. In fact, it hardly supports the claim that the Society was a gentlemen's club with a codified experimental form of life.[71] Hooke did not downgrade the theoretical elements included in *Micrographia* to private conjectures because of the demarcation between matters of fact and hypotheses. It is difficult to believe that if these rules were in place, the curator would have been in a position to express his philosophical views along with the description of his microscopical observations. By including causal explanations both in the weekly reports to the fellows and in the final manuscript in the first place, Hooke claimed a role as author that did not seem to be consistent with the Society's early microscopical project. The conflict inevitably originated not from the supposed methodological rules of the Society, as a first look might suggest, but rather because Hooke's ideas contrasted with those of other fellows.

Questions of credibility and authorship emerged in the same period also in other European scientific academies, where the social and epistemological dynamics of Restoration science did not apply.[72] Despite these differences, the solutions adopted by early modern experimenters were not very different. The accounts of experiments published in 1667 by the secretary of the Florentine Accademia del Cimento, for instance, raised questions of authorship not far from those raised by Hooke's *Micrographia*.[73] Like the fellows, the Florentine *accademici* published a series of

---

[68] Boyle (2001), vol. II, 412; cf. Wood (1980), 20.

[69] Birch (1756–67), vol. I, 490.

[70] Hooke (1665), sig. A2v.

[71] Shapin and Schaffer (1985), 321–2; Dear (1985), 146–7, 157; cf. Dennis (1989), 324–5, 347, 350; Wilding (2006), 125–7.

[72] Shapiro (2000), 143; cf. Wilding (2014), 2–3; Hahn (1990), 5, 8.

[73] Boschiero (2007), 3; cf. Id. (2010), 75; Holmes (1991), 164, 170–1, 179.

experiments and observations carried out and discussed during their meetings. Their discordant explanations were reported in the journal book of the Accademia, but were banned from the *Saggi* edited by the secretary Lorenzo Magalotti.[74] As Magalotti noted in the preface, those hints of speculation that survived his severe screening of the journal book were just ideas of individual members.[75] In the short life of the Cimento, consensus on common principles was rarely achieved.[76]

Disagreement emerged also among the English virtuosi over the explanations of some observations reported by Hooke. Consistent to his early views on the philosophical history of nature, in the weekly reports to the Society Hooke did not omit hints to possible explanations of what the microscope showed. His hypotheses on the formation of fossils, in particular, were widely discussed and criticised by many members of the Royal Society in 1664. When Hooke's observations were already in print, the fellows "approved of the modesty used in his assertions, but advised him to omit what he had delivered concerning the ends of such petrifications." Hooke did not follow the advice, and the fellows soon realised how inextricable Hooke's philosophical ideas, observations and experiments were. Unable to limit the book to a bare report of what Hooke saw and described to them, by the end of 1664 the fellows could only ask that Hooke's authorship of the book became clearer if possible. Like each fellow who saw Hooke's early drawings and listened to his weekly reports on microscopic observations, the reader had to be informed that hypotheses and conjectures included in the book were not agreed upon by all members of the Society. Unlike Borelli, a vocal critic of Magalotti's editing of the *Saggi*, Hooke was the sole author of the experiments and observations collected in *Micrographia*. For this reason, perhaps, his claim over the phenomena described and explained in the book were successful. Thus, rather than a tangible proof of the industrious activities of the newly formed Royal Society, *Micrographia* became the manifesto of Hooke's own real, mechanical, experimental philosophy.

Even though their disagreement involved methodology, neither the council nor Hooke advocated an unmitigated ban of theoretical elements from their experimental activities. The fellows, on the one hand, asked Hooke to clarify that "the several hypotheses and theories laid down by him therein, are not delivered as certainties, but as conjectures."[77] Hooke, on the other, described the rules of the Society as aiming to avoid "dogmatizing," not the search for causes by means of conjectures and questions, which the Society's "method does not altogether disallow."[78]

As their Horatian motto suggested, the fellows rejected *jurare in verba magistri*, rather than *in verba naturae*.[79] In the early Royal Society, a Baconian rhetoric provided a unifying public image covering a plurality of different and sometimes

---

[74] Knowles Middleton (1971), 54, 68; Beretta (2000), 141.

[75] Magalotti (1667), sig. *3v.

[76] Galluzzi (1981), 804–5; Bertoloni Meli (2001), 88, 94.

[77] Birch (1756–57), vol. I, 247–8, 260–2, 463, 491.

[78] Hooke (1665), sig. A2v.

[79] Sutton (1994), 57–9.

contrasting natural philosophies.[80] Baconianism, in short, had different meanings for different fellows. The Society absorbed all of them under a general programme of natural and experimental history that allowed each fellow to carry on his own philosophical agenda. The consequent theoretical neutrality of the institution, however, proved insufficient or detrimental to its philosophical goals. Along with the precarious financial conditions, the decline of members' involvement in scientific activities was the main source of dissatisfaction among the most active fellows. Talks about reform began within a decade of its foundation. After the publication of *Micrographia* and the beginning of the Cutler lectures, Hooke was able to express his criticism of the way the Society was run by the council, the president Brounker and the secretary Henry Oldenburg. The way these managed the priority dispute with Huygens over the application of springs to mechanical watches caused resentment in Hooke. The criticism of his early ideas on fossils and the history of nature by some fellows led Hooke to suggest a new "method" and new "rules" to replace those to which he abided instead in 1664. "Though this honourable society," he wrote in 1668, "have hitherto seem'd to avoid and prohibit pre-conceived theories and deductions from particular and seemingly accidental experiments; yet I humbly conceive, that such, if knowingly and judiciously made, are matter of greatest importance." Towards the end of the 1660s, Hooke saw himself as a respected philosopher engaged in the reform of natural knowledge along the lines set in *Micrographia* and the early Cutler lectures.[81] The implementation of his plan for a philosophical history was inconsistent with the then current arrangement of the Society. A reform of the Society, therefore, was for Hooke instrumental to the implementation of his philosophical programme. In a project probably written in the early 1670s, Hooke emphasised the "necessity of a Society" wherein the "joynt unanimous and regulated labour of a multitude" takes places.[82] The contemporary Royal Society, on the contrary, lacked the institutional setting to sustain a planned research agenda. Financial instability led to increasing the number of members. Those who could pay the annual fees proved disinterested in the everyday work of an early scientific academy, whose activities inevitably declined.[83]

Hooke's project aimed to restrict the membership only to those "known to be naturally and zealously inclined to the prosecution of the Designe of the Society." To favour the participation of craftsmen and tradesmen, he stressed the importance of rules safeguarding the secrecy of the Society's activities and suggested a series of prizes and rewards for inventors and contributors.[84] To attract learned as well as active members, Hooke noted in a later manuscript, the Society should dispose of the best instruments and show tangible proofs of the results that a long and con-

---

[80] Hoppen (1976) 276; Snider (1991), 120; Hunter (1995), 102–3.

[81] Hooke (1705), 20–1, 280; Oldroyd (1987), 151–2.

[82] Hunter and Wood (1986), 87.

[83] Birch (1756–57), vol. III, 136–7, 158–9; vol. IV, 7; cf. Hunter (1990), 18; Id. (1982a), 12–3.

[84] Hunter and Wood (1986), 88; British Library, Sloane MS 1039, f. 112; cf. Mulligan and Mulligan (1981), 332, 346, 348; Vermeir (2012), 166, 170.

sistent experimental activity can achieve.[85] These results were both theoretical and practical. Inventions and discoveries were expected to make the nation less dependent on "foreign productions whether natural or artificial." Even if trade was to be encouraged, the progress of natural knowledge could "advance the interest, stock and power" of the nation.[86] Well directed interest, therefore, was not an obstacle to the new experimental philosophy.[87] Neither was it inconsistent with a theoretical inquiry. In Hooke's view a clear philosophical agenda involving hypotheses and deductions was consistent with the inclusion of craftsmen and merchants. Instead of an obstacle to an honest inquiry into nature, the interest associated with these social figures was for Hooke the sign of an active involvement typical of the new experimental philosopher.[88] Hooke, who spent most of his time with both instrument makers and virtuosi, saw the workshops of the former as leading to the laboratories of the latter. His lodgings in Gresham College were a crossroad for mechanics, merchants and natural philosophers.[89]

As he noted in *Micrographia*, along with the "presence of very many of the chief nobility and gentry" among the fellows of the Society, the esteem of tradesmen was the sign of its success. What moves merchants, Hooke claimed, is "*meum et tuum*, that great rudder of human affairs."[90] Rather than disinterested gentlemen looking for entertainment, Hooke aimed to increase the participation of merchants and craftsmen. The belief in the existence of a close relationship between these latter and the experimental philosophers was shared also by the aristocrat Wren. In Wren's eyes Restoration London was like ancient Alexandria, a place where merchants, craftsmen, mathematicians and natural philosophers benefitted from each other's influence, where science could progress, and the common wealth increase.[91] Sprat considered the "noble and inquisitive genius" of English merchants a decisive component of the new experimental community.[92] Rather than the expression of an inferior moral economy, therefore, Hooke's ideas on the reform of the Royal Society seem an important stage in the emergence and development of his philosophical persona which, at the same time, contributed to shaping the model of the new experimental philosopher in Restoration London.[93]

---

[85] Royal Society Classified Papers, vol. XX, f. 172r; see also Hooke (1705), 420.

[86] Royal Society Classified Papers, vol. XX, f. 178r.

[87] Hooke (1665), sig. 1gr; cf. Jacob (1997), 18; Mokyr (2005), 304; Keller (2015), 219, 223; Mokyr (2016), 161;.

[88] Serjeantson (2013), 27; Bennett (2006), 65; Stewart (1992), 15; cf. Heilbron (2007), 485; Smith (2006), 293, 296–7; Long (2011), 1–3, 9, 37, 60–1, 95–6; Young (2017), 536, 542–3.

[89] Cf. Johnson (1957), 330; Feingold (1984)174–85; Turner (1985), 219.

[90] Hooke (1665), sig. g1v.

[91] Ward (1740), 36–7; Cf. Wolfe (2004), 6–57; Harkness (2007), 7–8, 10, 119–20, 141.

[92] Sprat (1667), 88.

[93] Henderson (2013).

Bibliography

Anstey, Peter, and Michael Hunter. 2008. Robert Boyle's 'Design about natural history'. *Early Science and Medicine* 13: 83–126.

Aubrey, John. 1898. *Brief lives*, 2 vols. Oxford: Clarendon Press.

Bennett, Jim. 1980. Robert Hooke as mechanic and natural philosopher. *Notes and Records of the Royal Society of London* 35: 33–48.

———. 2006. Instruments and ingenuity. In *Robert Hooke: Tercentennial studies*, ed. Michael Cooper and Michael Hunter, 65–76. Aldershot: Ashgate.

Beretta, Marco. 2000. At the source of western science: The organization of experimentalism at the Accademia del Cimento (1657–1667). *Notes and Records of the Royal Society of London* 54: 131–151.

Bertoloni Meli, Domenico. 2001. Authorship and teamwork around the Cimento academy: Mathematics, anatomy, experimental Philosophy. *Early Science and Medicine* 6: 65–95.

Biagioli, Mario. 1996. Etiquette, interdependence, and sociability in Seventeenth-Century science. *Critical Inquiry* 22: 193–238.

Birch, Thomas. 1756–57. *The history of the Royal Society of London*, 4 vols. London.

Boas Hall, Mary. 1981. Solomon's house emergent: The early Royal Society and cooperative research. In *The analytic spirit: Essays in the history of science in honor of Henry Guerlac*, ed. Harry Woolf, 177–194. Ithaca/London: Cornell University Press.

———. 1991. *Promoting experimental learning: Experiment and the Royal Society 1660–1727*. Cambridge: Cambridge University Press.

Boschiero, Luciano. 2007. *Experiment and natural philosophy in seventeenth-century Tuscany: The history of the Accademia del Cimento*. Dordrecht: Springer.

———. 2010. Translation, experimentation and the spring of the air: Richard Waller's Essayes of natural experiments. *Notes and Records of the Royal Society of London* 64: 67–83.

Boyle, Robert. 1999. *The works of Robert Boyle*, 14 vols., eds. Michael Hunter and Edward Davies. London: Pickering and Chatto.

———. 2001. *The correspondence of Robert Boyle*, 6 vols., eds. Michael Hunter, Antonio Clericuzio and Laurence Principe. London: Pickering and Chatto.

Burke, Peter. 2004. *What is cultural history?* Cambridge: Polity Press.

Burns, Robert. 1981. *The great debate on miracles: From Joseph Glanvill to David Hume*. London/Toronto: Bucknell University Press.

Chapman, Alan. 2005. *England's Leonardo: Robert Hooke and the seventeenth-century scientific revolution*. Bristol/Philadelphia: Institute of Physics Publishing.

Clericuzio, Antonio. 1997. Notes on corpuscular philosophy and pneumatical experiments in Robert Boyle's New experiments physico mechanical, touching the spring of the air. In *Die Schwere der Luft in der Diskussion des 17. Jahrhundertsw*, ed. Hrsg Wim Klever, 109–116. Wiesbaden: Harrasowitz Verlag.

Cooper, Michael. 2003. *A more beautiful city. Robert Hooke and the rebuilding of London after the great fire*. Stroud: Sutton Publishing.

Daston, Lorraine. 1995. The moral economy of science. *Osiris* 10: 2–24.

———. 2011. The empire of observation, 1600–1800. In *Histories of scientific observation*, ed. Loraine Daston and Elisabeth Lunbeck, 81–113. Chicago/London: The University of Chicago Press.

Davies, Edward. 1994. 'Parcere nominibus': Boyle, Hooke and the rhetorical interpretation of Descartes. In *Robert Boyle reconsidered*, ed. Michael Hunter, 157–175. Cambridge: Cambridge University Press.

Dear, Peter. 1985. Totius in verba: Rhetoric and authority in the early Royal Society. *Isis* 76: 145–161.

———. 1992. From truth to disinterestedness in the seventeenth century. *Social Studies of Science* 22: 619–631.

Dennis, Michael Aaron. 1989. Graphic understanding: instruments and interpretation in Robert Hooke's Micrographia. *Science in Context* 3: 309–364.

Ducheyne, Steffen. 2012. *The main business of natural philosophy: Isaac Newton's natural philosophical methodology*. Dordrecht: Springer.

Espinasse, Margaret. 1974. The decline and fall of Restoration science. In *The intellectual revolution of seventeenth century*, ed. Charles Webster, 347–368. London: Routledge & Kegan Paul.

Feingold, Mordechai. 1984. *The mathematicians' apprenticeship: Science, university and society in England, 1560–1640*. Cambridge: Cambridge University Press.

———. 1996. When facts matter. *Isis* 87: 131–139.

———. 1998. Of records and grandeur: The archives of the Royal Society. In *Archives of the scientific revolution: The formation and exchange of ideas in seventeenth-century Europe*, ed. Michael Hunter, 171–184. Woodbridge: Boydell Press.

———. 2006. Robert Hooke gentleman of science. In *Robert Hooke: Tercentennial studies*, ed. Michael Cooper and Michael Hunter, 203–217. Aldershot: Ashgate.

Galluzzi, Paolo. 1981. L'Accademia del Cimento: «gusti» del principe, filosofia e ideologia dell'esperimento. *Quaderni Storici* 48: 788–844.

Gimelli Martin, Catherine. 2002. The ahistoricism of the new historicism: Knowledge as power versus power as knowledge in Bacon's New Atlantis. In *Fault lines and controversies in the study of seventeenth-century English literature*, ed. Claude J. Summers and Ted-Larry Pebworth, 22–49. Columbia/London: University of Missouri Press.

Golinski, Jan 2005. Making natural knowledge: Constructivism and the history of science. Chicago/London: University of Chicago Press.

Hacking, Ian. 1991. Artificial phenomena. *The British Journal for the History of Science* 24: 235–241.

Hahn, Roger. 1990. The age of academies. In *Solomon's house revisited: The organization and institutionalization of science*, ed. Tore Frängsmyr, 3–12. Canton: Science History Publication.

Harkness, Deborah. 2007. *The jewel house: Elizabethan London and the scientific revolution*. New Haven/London: Yale University Press.

Heilbron, John. 1983. *Physics at the Royal Society during Newton's presidency*. Los Angeles: William Andrews Clark Memorial Library.

———. 2007. Coming to term with scientific revolution. *European Review* 15: 473–489.

Henderson, Felicity. 2013. *Robert Hooke and the construction of the scientific self* (Unpublished manuscript).

Holmes, Frederic. 1991. Argument and narrative in scientific writing. In *The literary structure of scientific argument: Historical studies*, ed. Peter Dear, 164–181. Philadelphia: University of Pennsylvania Press.

Hoppen, Theodore. 1976. The nature of the early Royal Society. Part II. *The British Journal for the History of Science* 9: 243–273.

Hooke, Robert. 1665. *Micrographia*. London.

———. 1705. *The Posthumous Works of Robert Hooke*, ed. Richard Waller. London.

———. 1935. *The diary of Robert Hooke 1672–1680*, eds. Henry W. Robinson and Walter Adams. London: Taylor and Francis.

Hunter, Michael. 1982a. *The Royal Society and its fellows: The morphology of an early scientific institution*. Oxford: The British Society for the History of Science.

———. 1982b. Reconstructing restoration science: Problems and pitfalls in institutional history. *Social Studies of Science* 12: 451–466.

———. 1985. The cabinet institutionalised: The Royal Society's 'repository' and its background. In *The origins of museums: The cabinet of curiosities in sixteenth- and seventeenth-century Europe*, ed. Olivier Impey and Arthur Macgregor, 159–168. Oxford: Clarendon Press.

———. 1990. First steps in institutionalization: The role of the Royal Society of London. In *Solomon's house revisited: The organization and institutionalization of science*, ed. Tore Frängsmyr, 13–29. Canton: Science History Publication.

————. 1995. *Science and the shape of orthodoxy: Intellectual change in late seventeenth-century Britain*. Woodbridge: Boydell Press.

————. 2000. *Robert Boyle (1627–91): Scrupulosity and science*. Woodbridge: Boydell Press.

————. 2003. Hooke the natural philosopher. In *London's Leonardo. The life and work of Robert Hooke*, ed. Jim Bennett, Michael Cooper, Michael Hunter, and Lisa Jardine, 105–162. Oxford: Oxford University Press.

————. 2007. Robert Boyle and the Royal Society: A reciprocal exchange in the making of Baconian science. *The British Journal for the History of Science* 40: 1–23.

————. 2009. *Boyle: Between god and science*. New Haven/London: Yale University Press.

Hunter, Michael, and Paul Wood. 1986. Towards Solomon's house: Rival strategies for reforming the early Royal Society. *The British Journal for the History of Science* 24: 49–108.

Illife, Rob. 2004. Abstract considerations: Disciplines and the incoherence of Newton's natural philosophy. *Studies in History and Philosophy of Science* 35: 427–474.

————. 2017. *Priest of nature: The religious worlds of Isaac Newton*. Oxford: Oxford University Press.

Inwood, Stephen. 2002. *The man who knew too much: The strange and inventive life of Robert Hooke*. London: Macmillan.

Jacob, Margaret. 1997. *Scientific culture and the making of the industrial West*. Oxford: Oxford University Press.

Jardine, Lisa. 1974. *Francis bacon: Discovery and the art of discourse*. Cambridge: Cambridge University Press.

Johnson, Francis. 1957. Gresham college: Precursor of the royal society. In *Roots of scientific thought: A cultural perspective*, ed. Philip Wiener and Aaron Noland, 328–353. New York: Basic Book.

Keller, Vera. 2015. *Knowledge and the public interest, 1575–1725*. Cambridge: Cambridge University Press.

Knowles Middleton, William. 1971. *The Experimeters: A study of the Accademia del Cimento*. Baltimore: The John Hopkins University Press.

Koyré, Alexandre. 1968. *Newtonian studies*. Chicago: University of Chicago Press.

Lawson, Ian. 2015. *Robert Hooke's microscope: The epistemology of a scientific instrument*. PhD diss., University of Sidney.

Licoppe, Christian. 1996. *La formation de la pratique scientifique: le discours de l'expérience en France et en Angleterre (1630–1820)*. Paris: Édition de la découverte.

Long, Pamela. 2011. *Artisan/practitioners and the rise of the new sciences, 1400–1600*. Corvallis: Oregon State University Press.

Lux, David. 1991. The reorganization of science 1450–1700. In *Patronage and institutions: Science, technology, and medicine at the European court 1500–1700*, ed. Bruce Moran, 185–194. Woodbrige: Boydell Press.

Magalotti, Lorenzo. 1667. Saggi di naturali esperienze. Florence.

Mokyr, Joel. 2005. The intellectual origins of modern economic growth. *The Journal of Economic History* 65: 285–351.

————. 2016. *A culture of growth: The origins of the modern economy*. Princeton: Princeton University Press.

Mulligan, Lotte, and Glenn Mulligan. 1981. Reconstructing Restoration science: Styles of leadership and social composition of the early Royal Society. *Social Studies of Science* 11: 327–365.

Nakajima, Hideto. 2001. Astronomical aspects of Robert Hooke's scientific research. In *Optics and astronomy: Proceedings of the XX International congress of history of science*, ed. Gérard Simon and Suzanne Débarbat, 161–166. Turnhout: Brepols.

Newman, William. 2006. *Atoms and alchemy: Chemistry and the experimental origins of the scientific revolution*. Chicago/London: University of Chicago Press.

Oldroyd, David. 1987. Some writings of Robert Hooke on procedures for the prosecution of scientific inquiry, including his 'Lectures of Things Requisite to a Ntral History'. *Notes and Records of the Royal Society of London* 41: 145–167.

Oster, Malcolm. 1992. The scholar and the craftsman revisited: Robert Boyle as aristocrat and artisan. *Annals of Science* 49: 255–276.

Principe, Lawrence. 1998. *The aspiring adept: Robert Boyle and his alchemical quest.* Princeton: Princeton University Press.

Pumfrey, Stephen. 1991. Ideas above his station: A social history of Hooke's curatorship of experiments. *History of Science* 29: 1–44.

Purrington, Robert. 2009. *The first professional scientist: Robert and the Royal Society of London.* Berlin/Basel: Birkhäuser.

Raj, Kapil. 2007. *Relocating modern science: Circulation and construction of knowledge in South-Asia and Europe, 1650–1900.* Basingstoke: Palgrave Macmillan.

Ratcliff, J.R. 2007. Samuel Morland and his calculating machines c.1666: The early career of a courtier-inventor in restoration London. *The British Journal for the History of Science* 40: 157–179.

Ricciardo, Salvatore. 2010. *Medicina, chimica, teologia: Robert Boyle e le origini della filosofia sperimentale.* PhD diss., University of Bergamo.

Roger, Jacques. 1984. Per una storia storica delle scienze. *Giornale critico della filosofia italiana* 63: 285–313.

Rogers, Graham Allen. 1972. Descartes and the method of English science. *Annals of Science* 29: 237–255.

Sargent, Rose-Mary. 1995. *The diffident naturalist: Robert Boyle and the philosophy of experiment.* Chicago/London: University of Chicago Press.

Schaffer, Simon. 1989. Glass works: Newton's prism and the uses of experiment. In *The uses of experiment: Studies in natural sciences,* ed. David Gooding, Trevor Pinch, and Simon Schaffer, 67–104. Cambridge: Cambridge University Press.

Schuster, John, and Alan Taylor. 1997. Blind trust: The gentlemanly origins of experimental science. *Social Studies of Science* 27: 503–536.

Serjeantson, Richard. 1999. Testimony and proof in early modern England. *Studies in History and Philosophy of Science* 30: 195-236.

———. 2013. Becoming a philosopher in seventeenth-century Britain. In *The Oxford handbook of British philosophy in the seventeenth century,* ed. Peter Anstey, 9–38. Oxford: Oxford University Press.

Shapin, Steven. 1988. The House of experiments in seventeenth-century England. *Isis* 79: 373–404.

———. 1989a. Who was Robert Hooke? In *Robert Hooke: New studies,* ed. Michael Hunter and Simon Schaffer, 253–285. Woodbridge: Boydell Press.

———. 1989b. The invisible technician. *American Scientist* 77: 554–563.

———. 1991. "A scholar and a gentleman": The problematic identity of the scientific practitioner in early modern England. *History of Science* 29: 279–327.

———. 1994. *A social history of truth: Civility and science in seventeenth-century England.* Chicago/London: University of Chicago Press.

———. 2006. The man of science. In *The Cambridge history of science vol. 3 early modern science,* ed. Katharine Park and Lorraine Daston, 179–191. Cambridge: Cambridge University Press.

Shapin, Steven, and Simon Schaffer. 1985. *The Leviathan and the air pump: Hobbes, Boyle and the experimental life.* Princeton: Princeton University Press.

Shapiro, Barbara. 2000. *A culture of fact. England 1550–1720.* Ithaca/London: Cornell University Press.

Shea, William. 2007. The scientific revolution really occurred. *European Review* 15: 459–471.

Shepard, Alexandra. 2015. *Accounting for oneself: worth, status, and the social order of early modern England.* Oxford: Oxford University Press.

Smith, Pamela. 2006. Laboratories. In *The Cambridge history of science vol. 3 early modern science,* ed. Katharine Park and Lorraine Daston, 290–305. Cambridge: Cambridge University Press.

Snider, Alvin. 1991. Bacon, legitimation and the origin of Restoration science. *Eighteenth Century* 32: 119–138.

Sprat, Thomas. 1667. *The history of the Royal Society of London*. London.

Stewart, Larry. 1992. *The rise of public science: Rhetoric, technology, and natural philosophy in Newtonian Britain, 1660–1750*. Cambridge: Cambridge University Press.

———. 1998. Other centres of calculation, or, where the Royal Society didn't count: Commerce, coffee-houses and natural philosophy in early modern London. *The British Journal for the History of Science* 32: 133–153.

Sutton, Clive. 1994. 'Nullius in verba' and 'nihil in verbis': Public understanding of the role of language in science. *The British Journal for the History of Science* 27: 55–64.

Swann, Marjorie. 2001. *Curiosities and texts: The culture of collecting in early modern England*. Philadelphia: University of Pennsylvania Press.

Thomas, Jennifer. 2009. *A philosophical storehouse: The life and afterlife of the Royal Society's repository*. PhD diss. Queen Mary University, London.

———. 2011. Compiling 'God's great book [of] universal nature': The Royal Society's collecting strategies. *Journal of the History of Collections* 23: 1–13.

Turner, Gerard l'E. 1985. The cabinet of experimental philosophy. In *The origins of museums: The cabinet of curiosities in sixteenth- and seventeenth-century Europe*, ed. Olivier Impey and Arthur Macgregor, 214–222. Oxford: Clarendon Press.

Van Helden, Anne. 1991. The age of the air-pump. *Tractrix* 3: 149–172.

Van Leeuwen, Henry. 1963. *The problem of certainty in English thought 1630–1690*. The Hague: Martinus Nijhoff.

Vermeir, Koen. 2012. Openness versus secrecy? Historical and historiographical remarks. *The British Journal for the History of Science* 45: 165–188.

Vickers, Brian, ed. 1987. *English science: Bacon to Newton*. Farnham: Ashgate.

Wilding, Nick. 2006. Graphic technologies. In *Robert Hooke: Tercentennial studies*, ed. Michael Cooper and Michael Hunter, 123–134. Aldershot: Ashgate.

———. 2014. *Galileo's idol: Gianfrancesco Sagredo and the politics of knowledge*. Chicago/London: University of Chicago Press.

Ward, John. 1740. *The lives of the professors of Gresham College*. London.

Wolfe, Jessica. 2004. *Humanism, machinery, and renaissance literature*. Cambridge: Cambridge University Press.

Wood, Paul. 1980. Methodology and apologetics: Thomas Sprat's "History of the Royal Society." *The British Journal for the History of Science* 13: 1–26.

Yeo, Richard. 2014. *Notebooks, English virtuosi, and early modern science*. Chicago/London: The University of Chicago Press.

Young, Mark Thomas. 2017. Nature as spectacle; experience and empiricism in early modern experimental practice. *Centaurus* 59: 72–96.

Appendix

Once a manuscript is transcribed in a mechanised letter form, it loses part of its original character. Because of the inevitable modifications of the original form, such a transcription is by its nature an edition.[1] The following transcriptions, however, are not meant as a definitive edition of Hooke's unpublished scientific papers, as they provide only a selection of those papers, chosen for their relevance to the topics discussed in this book. They are provided here to offer the reader easy access to some of the unpublished documents often referred to in the previous chapters. Some of the algebraic lectures and a lecture on "the uses and advantage of microscopes" have previously been transcribed only in two doctoral dissertations.[2] The remaining texts have never been transcribed and published before. In some cases, they have remained unnoticed both by archivists and scholars.

The first two texts are part of the "collection of papers dating from 1641 to 1695 made by Robert Hooke" kept at the Guildhall Library of London. A note at the end of the folder suggests that William Derham separated these papers from those kept at the Royal Society.[3] Derham probably did not consider them worth publishing among the *Philosophical Experiments and Observations of Robert Hooke,* which he edited in 1726. The first text is a transcription of MS 1757.11 (ff. 103–110), a "Lecture about the Improvement of Natural History" delivered at the Royal Society. The manuscript does not bear a date. It follows another undated lecture on "philosophicall history" kept at the Royal Society.[4] The pagination of MS 1757.11 often does not follow the original order of folios; the earlier pagination has been followed in the present transcription.

The second text includes transcriptions of two lectures on algebra delivered at Gresham College in 1665, along with a short introduction. All three documents are

---

[1] Hunter (2007), 59, 76–8, 82, 84.

[2] Pugliese (1982), 656–63; Lawson (2015), 251–9.

[3] Guildhall Library, London MS 1757.

[4] Oldroyd (1987), 151–9.

© Springer Nature Switzerland AG 2020
F. G. Sacco, *Real, Mechanical, Experimental*, International Archives of the
History of Ideas   Archives internationales d'histoire des idées 231,
https://doi.org/10.1007/978-3-030-44451-8

in Latin and undated; the third is not in Hooke's hand.[5] Although f. 112v reads "First Latin Lecture about Algebra," f. 112r does not seem a text of a lecture delivered by Hooke, but a short introduction to the two following lectures. The title seems the result of an intervention in different ink, which added the words "First Latin" and deleted the final letter of "Lectures." The original title, therefore, was "Lectures about Algebra."[6] The text of Hooke's first lecture on algebra is the second document included in MS 1757.11. It is the Latin version, with significant alterations, of "Mr Hooke first Algebraic Lecture" kept at the Royal Society and dated June 10, 1665.[7] Consequently, the third document included in MS 1757.11 is the text of the "Second Latin Lecture of Algebra."[8]

A transcription of Hooke's "first Algebraic Lecture" read at Gresham College follows here the Latin lectures kept at the Guildhall Library. Although both are in Hooke's hand, the two versions of the lecture carry significant differences that have been recorded in the notes and discussed in chapter one.

The fourth text is a transcription of one of Hooke's many lectures and papers on the question of penetration of bodies, discussed in chapter four. The manuscript is kept in the archives of the Royal Society (Classified Papers, vol. XX, ff. 171r–174r). Hooke probably composed or began to work on this lecture on July 3, 1689. This date is on the first folio, in Hooke's hand. According to the journal book of the Royal Society, in 1689 Hooke read lectures on this topic on May 22, 29, and July 26, but not on July 3 or 24. This last date has been reported by Hooke on the last folio and added, in different ink, to the date "3 July 1689."[9]

The fifth text is a transcription of a lecture kept in the archives of the Royal Society (Classified Papers, vol. XX, ff. 178r–179v). Hooke read it at the meeting held on December 18, 1694. This document does not bear a title.

The sixth text is a transcription of a lecture on "the uses and advantage of microscopes" read on November 29, 1693. The manuscript, kept in the archives of the Royal Society (Classified Papers, vol. XX, ff. 183r–184v), has been heavily edited. Among Hooke's papers edited by Derham in 1726 there is "Mr. Waller's Observations upon Dr. Hooke's Discourses, concerning Telescopes and Microscopes." In the text, Waller provides a summary of two lectures, on microscopes and telescopes, that were read at the Royal Society on November 29 and December 6, 1693. Derham chose to publish Waller's summary instead of Hooke's lectures because these seemed too long.[10] Although heavily edited, either by Waller or Derham, the first lecture is still kept at the Royal Society. The second, on the contrary, seems lost. It

[5] An English translation of the introduction and the second Latin lecture on algebra is available in Pugliese (1982), 652–3, 670–7.

[6] Guildhall Library, London MS 1757.11, f. 112v.

[7] Royal Society Classified Papers, vol. XX, ff. 65r–66v.

[8] Guildhall Library, London MS 1757.11, f. 121v.

[9] Royal Society Classified Papers, vol. XX, ff. 171r, 174r; Journal Book of the Royal Society, vol. VIII, 262–3, 267–8, 271.

[10] Hooke (1726), 270–3; cf. Journal Book of the Royal Society, vol. IX, 143–5.

is not even among Hooke's papers kept at the Guildhall Library, where some of the documents Derham did not consider worth publishing are currently kept.

The last two texts are transcriptions of manuscripts kept at the British Library (Sloane MS 1039, ff. 112r–113v, 114r). Neither of the manuscripts are titled. In the list of contents of "Robert Hooke, FRS: Collection of scientific papers and letters," the first manuscript is listed under "Royal Society: Design of, to print papers of advertisements every week or fortnight, and soliciting communications relative to the variation of the magnetic needle: late 17th cent." The second is not mentioned.[11] Both are texts of lectures read at the Royal Society. The second bears the date January 29, 1680,[12] the first is undated.

The following transcriptions are literal, retaining the original spelling, punctuation and capitalisation. Abbreviations, however, whether by way of contraction or suspension, in Latin or in English, have been silently expanded. Alterations have been recorded in different ways. Deletions, replacements, and corrections are recorded in the notes, while insertions are maintained within the text but denoted by ‹ ›. Editorial additions are indicated by square brackets. Damaged or illegible words are denoted by (…), either in the text or in the notes. Italics are employed to denote underlining. The original foliation has been indicated in bold by the number of the folio between inverted commas at the beginning of each recto or verso. Old style dates are retained, but with the year beginning on January 1, according to Hooke's use.

Guildhall Library, London MS 1757.11, ff. 103r–110v

'**f. 103r**' I indeavourd the last Day to show what Philosophicall History was And what ‹care and diligence›[13] ought to[14] [be] used[15] for the compiling of It.[16] And what method seemd best to regulate and direct ‹an undertaker therein›[17]; And that was first to Indeavour to range all things ‹under›[18] certaine heads[,] which heads ‹I shewd› might properly enough be Distinguisht according to the Natu[re] of the severall bodys that[19] were to be found in the world. Namely Coelestiall Elementary, minerall, vegetable, Animall, Artificiall or factitious. I shewd likewise into what heads each of these might be Divided. ‹the› Next ‹thing› was to indeavour to propound to

---

[11] British Library, MS Sloane 1039.

[12] Cf. Birch (1756–57), vol. IV, 66–7.

[13] Corrected: method

[14] Deleted: the words.

[15] Deleted: of.

[16] Royal Society Classified Papers, vol. XX, ff. 106r–109r.

[17] Corrected: (…) therein.

[18] Corrected: to.

[19] Deleted: are.

ones self what was requisite to be considerd and examind in ‹each of these›[20] bodys
to find out the true nature of them that is to consider well[21] what informations were
Desirable to know. The[22] proprietys of Any body. And accordingly[23] to sett downe in
writing as many querys[24] as can be thought of, the answering of which would give
one a full account of the Nature of the body to be examind. And though perhaps
many of them may not be thought solvable by any man,[25] seeming perhaps far
beyond the capacitys of men[']s senses,[26] yet ‹if they seem resistant to be answerd
before the nature of the body be known we ought› to propound them; for we know
not yet to what height the power and abilitys of man[']s ‹ratiocination› may be
increasd and improv'd and what help there may be found to advance them far
beyond their usuall[27] limitts. And ‹though to some› perhaps, some quaerys may
seem[28] difficult ‹enough› even for An angell to answer, yet there be many things
which were heretofore thought such and yet can now be answerd by the Industry
and diligence of men. my meaning is not to[29] make[30] these quaerys[31] Extravagancy
of Phancy; but rather to make them consist of such as upon a serious Consideration
of the nature and conditions of the body to be examind seem those necessary to be
answerd before the[32] constitution of that body can be known. And having thus set
downe upon the body[33] intended to ‹be› examined a Sufficient number of considerable quaerys. The third thing will be to consider what way each of these quaerys
may be Answerd.

'f. 104r' And here upon the examination of this It will[34] be of great use to be well
versed in all sorts of Naturall historians, and in the various ways of trying examining
and working upon matters[35] and ‹some may perhaps be 2ly answerd by authors, and
are to be found in books› in the principles of mechanics and mathematics. For others
by tradesmen and such as have been conversant about[36] those bodys that are under
examination and soe may have learnt more by experience causaltys and accidents,

---

[20] Corrected: (…).

[21] Deleted: with ones self.

[22] Deleted: true nature and.

[23] Deleted: thereunto.

[24] Deleted: ‹or questions›.

[25] Deleted: (…).

[26] Deleted: (…).

[27] Deleted: Station.

[28] Deleted: (…).

[29] Deleted: ca.

[30] Deleted: all.

[31] Deleted: a company of.

[32] Deleted: Nature and.

[33] Deleted: you.

[34] Deleted: (…).

[35] Deleted: (…).

[36] Deleted: (…).

others by praesent tryall and examination others then can be praesently attaind by one or two tryalls. Other quaerys may be solvd and answerd by by comparing praesent tryalls with such as have been made heretofore others perhaps will only admitt of A Solution by similitude ‹and comparison› As ther[e] may be a propriety of a body which I would willingly know, but I cannot think of any way of making tryall on that body how to finde it, but[37] knowing[38] an other body which has divers other things common with it and seems to have the same with that[39] on which I can make tryall ‹by this meanes I say 'tis possible many quaerys may be fully answerd. As for Instance›[40] I would Desire to know after what manner the motion of a man['ʟ]s hearth is perform'd, here though It cannot ‹[41]with any kind of humanity be› attempted to be examined[42] in a man yet it may[43] as satisfactoryly perhaps, be resolv'd by the dissection of a[44] Dog or other creature whilst alive. There be others also that will not[45] afford one the opportunity ‹ever› of this way of tryall and that is such bodys as Are[46] and ever have been out of men's reach and of which they have noe means but their eye of Receiving any Information and those are severall kinds of coelestiall bodys such as ‹the stars planets and› comets.[47] Now those as they will admit but of a very few questions, Soe even of thos[e] the most of them can be but imperfectly and conjecturally answerd and the Same history, experiments and observations may ‹be found to› serve as well to recommend one hypothesis as at[48] first sight It may ‹seem to› Doe an other quite contrary supposition ‹in this case therefore we ought to examine what kind of experiment or observation would afford›[49] a way how to Determine the controversy ‹what tryall would prove experimentum crucis as Lord Bacon calld it›[50] and In order hereunto many new quaerys will be necessary to be added to the former. and various ways must be thought of for prosecuting the Inquiry. and great Diligence and circumspection must be used in making experiments and taking notice of Observations ‹for it is ‹not less›[51] considerable as to the prosecution of this Designe to be able to judge when a new experiment or observation is requisite then

---

[37] Deleted: I.

[38] Deleted: perhaps.

[39] Deleted: ‹I want examine›.

[40] Corrected: which I could not Doe in the body it self I would examine. Deleted: Its.

[41] Deleted: (…).

[42] Deleted: this.

[43] Deleted: be.

[44] Deleted: live.

[45] Deleted: give.

[46] Corrected: h.

[47] Deleted: (…).

[48] Deleted: their.

[49] Deleted: (…).

[50] Cf. Bacon (2004), 318–21.

[51] Corrected: (…).

to be able to make it such. Great care also must be›⁵² in ranging and comparing those collections. The fourth thing therefore is to know after **'f. 105r'** what manner to ‹make range and register›⁵³ Experiments and observations. And for this end I cannot think of any better method then according ‹to› the order of the Quaery propounded. Supposing that they be made according to some good and very compraehensive method. And indeed in the making of them if rightly done a great Deal of the Difficulty of solving them will be avoyded, for if the quaerys Doe proceed from the most obvious proprietys and qualitys of a body to the most abstruse, and intricate texture constitution and virtues of it then the progress in answering them will be much more naturall and Easy, and A thorough consideration of the first will hugely facilitate the solution of the later. Even as we see in the progress of a mathematicall Ratiocination, where by ascending by degrees and steps we easily make our way to the last and most abstruse⁵⁴ Problem. whereas should that be leapd on and attempted at first 'tis hardly to be thought feasable. Now as this method has with great successe been practised in mathematicall Learning⁵⁵ Soe 'tis not to be doubted but were it followd in all other Inquirys It would much facilitate and assertaine the undertaking. For the progress of the mind or ‹Reason›⁵⁶ is the same in all kinds of Ratiocinations and though we doe it as it were⁵⁷ without our owne advertency yet whatsoever knowledg soe acqu[ire]d that is more abstruse, 'tis collected all by this way, or by note. And could we but use the same Synthetick way in all our Indeavours we might certainly make much greater Discoverys in Physicks on the Nature of bodys then have been hitherto made. But 'tis an imperfection of our Nature that we are quickly ‹weary› of this more tedious and sure way and are impatient to be leaping to the extreames without touching or thinking on the **'f. 107r'** interposed meanes. And thence doe not only for the most part miss of your aim but are whol[l]y discouraged and give over our inquirys for the future. And hence 'tis we may meet with soe many Discouraging opinions from almost all persons, who are very apt to reason and Ask⁵⁸ to what end tends all your Inquirys? can you think there is any thing to be known that all the world hitherto has not been able to find out? have you soe high an opinion of your⁵⁹ selves as to think your⁶⁰ abilitys greater then any man ‹in the world besides› has or every any have had.⁶¹ If not what meanes your Inquirys why are you not contented to be as wise as your forefathers. To which I must answer that I cannot

---

⁵² Deleted: (…).

⁵³ Corrected: Range.

⁵⁴ Deleted: (…).

⁵⁵ Deleted: (…).

⁵⁶ Corrected: Ratiocination.

⁵⁷ Deleted: (…).

⁵⁸ Corrected: of.

⁵⁹ Deleted: self.

⁶⁰ Deleted: self (...).

⁶¹ Deleted: beside.

think the minds of men at praesent to be more perfect or ‹compleate in their assent›[62] then they have ever bee[n] in all ages, ‹though›[63] I am apt to think that the Ratiocinatio[n] may ‹have› been improv'd in this Later age to a somewhat greater perfection, But yet I cannot doubt to be very positive in affirming that the reason of man if rightly imployd is capable of exalting him to a much higher pitch of knowledg ‹concerning›[64] Naturall agents and causes then has hitherto been attaind. And that the reason why our praedecessors have not ‹arrived at›[65] it is because they have not made use of this method. What ever they light on was rather A product of Chance and an excellency of their particular Natures that were the Inventors, then the Naturall Consequents of Art: And As ‹the Child› I have lately heard of[66] which has a strange naturall faculty in the ‹speedy› resolving and[67] answering any question propounded about arithmeticall[68] multiplication soe as to answer it sooner ‹by note› then any man is[69] able to cast it up with his pen; ‹as this child I say is›[70] much more to be *admired* then the most skilful Arithmetician in the world. Soe is the Skilfull artist to be much higher *valued*. For whereas the child could only resolve questions propounded in one Arithmeticall Operation namely multiplication the Artist is able to answer all[71] **'f. 106r'** The Fifth thing therefore[72] to be explaind is the method of applying and making use of Philosophicall history toward the Raising of a New and active Philosophy, towards the improving of all kinds of trades and mechanicall operations and towards the producing of multitudes of New and usefull Inventions. and in a word for the improving the ratiocination of Man to the highest pitch of Perfection It is capable of. But the ‹praecepts and rules of›[73] this would be too long a work for this present ‹exercise› I shall rather therefore a little to Divert your thoughts ‹by an example›[74] produce somewhat of my owne and others observations

---

[62] Deleted: consummate.

[63] Corrected: and yet.

[64] Corrected: of.

[65] Corrected: (…).

[66] Deleted: a child.

[67] Corrected: all.

[68] Deleted: calculation.

[69] Corrected: was.

[70] Corrected: he is.

[71] Deleted: **'f. 106r'** in any. And whereas the child does it by a naturall faculty of leaping as it were to the extreme or product and that I am apt to think uncertainly ‹enough› the artist is more slow but more secure, the child hitts on it he knows not how, the Artist is able to prove every part of it. the comparison I think may hold in Casuall inventors such as those of Printing gunpowder Optick glasses and the like and those that by following a method should be able to produce not only aequall but greater inventions. The method therefore of making use of the materialls collected ‹in a Philosophicall› history is the chief thing to be lookd after for without this all the whole collection will signify very lit[l]le, we shall but with our knowledge increase our (…) that we know not how to use this vast treasure.

[72] Circled word.

[73] Deleted, in different ink: this method of applyind and.

[74] Deleted: and (…).

about a very Notable ‹appearance›[75] of Nature which Happ[en]ing but seldome
and[76] its cause or originale being unknown has acquird the name of a wonder or
Prodigie ‹and therefore is[77] soe much the more proper to be taken notice of in the
beginning of our Philosophicall History the subject I meane is›[78] the Comet or
Blazing Star which has soe lately abrightned the greatest part of the world.[79] ‹To
Prosecute therefore the lately defined method›[80] ‹now›[81] In order[82] to the finding out
the true Nature of this unusuale body, it will be requisite to propound such quaerys
as seem necessary to be Answerd for that ‹purpose›[83] **'f. 107v'**[84] **'f. 106v'** may[85] be
some such as these ‹1› In what place the comet was ‹2› or how far distant from the
earth? ‹3›[86] Of what kind of substance or constitution ‹4› whence it has its light? ‹5›
Of what magnitude it is? ‹6› After what manner it is movd? ‹7› Of what duration it
is ‹8›[87] And whether this ‹comet will ever appear again or whether it will never.[88]›
comet may not after a certaine space of time appear again ‹9›[89] what that time it may
probably be and where it is likely to appear?

‹10› whether it be a momentary or a lasting body? ‹11› whether it be newly generated when it first appears and annihilated when it disappear[s] or whether it be
made visible and invisible only upon the Account of its nearnesse or great distance
from us. ‹12› whether it shine[s] by its own light or by[90] the Beames of the Sun?
‹and if from›[91] the beames of the Sun ‹13› whether they cause that light by[92]
Refraction as most of our moderne writers of Comets affirme, or by reflection which
seems the more naturall? ‹14› whether the blaze form it be moved like flame or fixed

---

[75] Corrected: Phenomenon.

[76] Deleted: now be in the remark of.

[77] Deleted: in a respect.

[78] Corrected: (…).

[79] This comet was visible from November 1664 to April 1665, it was observed by Hooke on 23
December 1664. Cf. Hooke (1678), 7–8, 19; Birch (1756–57), vol. I, 511.

[80] Corrected: (…).

[81] Corrected, in a different ink: (…).

[82] Deleted: therefore.

[83] Corrected: finding out the Nature.

[84] Deleted: of this method of applying and making Philosophicall History towards the raising of a
new and active philosophy towards the improvement of Trades and meckanicall operations, and
towards the producing new and usefull Inventions.

[85] Deleted: And those.

[86] Corrected: next.

[87] Corrected:?.

[88] Deleted: whether.

[89] Corrected: (…).

[90] Deleted: the Reflection of.

[91] Corrected: whether.

[92] Deleted: Reflection.

like the[93] Sun beames[94] between the cloudes in a vaporous and thick air. ‹15› whether it hath any particular pointing or direction more to one place of the world then an other and so what part it Doth most point at if such there be[95] '**f. 108r**'[96] ‹16› from whence it has its radiation or tayle opposite for the most part to the Sun? ‹17› And what may be the Reason that It does sometimes considerably deflect from it, ‹18› what may be the reason of its curvity ‹19 How often cometts have been observd? 20›[97] what its motions are. ‹21› by how many hypotheses all the Phaenomena may be Solvd. ‹22› what means there may be of Distinguishing which is the true ‹hypothesis›[98] and which the false. ‹23› whether It be likely to produce any Considerable effect upon the Earth. there might be severall other quaerys ‹under every of these heads› propounded concerning it but because I see not at praesent any likely way of Answering them I shall content my self only with these few, this being one of those bodys that is out of our reach and[99] hap[pen]ing neither but very seldom we cannot make our tryalls of that kind as various as are[100] requisite.

As to ‹the answering of› the first question In what place or how far Distant[101] As for the ‹first sort of› parallax. By ‹(comparing)› two observations I have which were made at a Considerable Distance[102] North and South of each other ‹the one at Sevill in Spain the other at Paris› can not find any sensible parallax but Rather the contrary for the observations being only figures or schemes of the place of the comet among the ambient starrs the Sevill observator seems to place him somewhat lower or more southward then the Parisian, and therefore though perhaps neither of them were accurate yet[103] 'tis very probable if there were any sensible parallax it must needs be very small. by comparing also some later Observations made at Portsmouth and

---

[93] Deleted: radiations of the.

[94] Deleted: through the.

[95] Deleted: X. It was the opinion of most writers before Ticho Brahe and Kepler that ‹Comets› were sublunary meteors drawn up into the higher region of the air and there set on fire and soe continued burning like the meteor were wholy consumed and according as the matter increased or wasted soe did the appearance of the comet, But this noble Dane and severall other about that time found by accurate observations made that its parallax was lesse than that of the moon and consequently that it was further Distant from the Earth, that it must be a body of an other magnitude and nature then most till that time had Imagined. and therefore that it ought to be otherwise thought of then what the Generality of mankind believd concerning it.

[96] Deleted: As ‹1› of what Distance. In ‹2› what place? or how far distant? after what ‹5› manner movd. Of what ‹3› magnitude of what ‹4› Constitution from whence it has its light.

[97] Deleted: (…).

[98] Deleted: (…).

[99] Deleted: bacause.

[100] Corrected: is

[101] Deleted: there seeme to be two wayes by which we ‹may› came to some certainty of it. The first is by the parallax, and the second is by the way or motion of it.

[102] Deleted: almost.

[103] Deleted: I found.

here at London[104] there seems not to be any sensible Parallax, and consequently[105] the comet must be much higher then the Air or atmosphaere unless we **'f. 109r'** suppose it Indefinitely Extended upwardly and to reach[106] some thousands of miles in height. But this is Insignificant. I pass by **'f. 110r'** yet the Refraction or Inflection of the air will not amount to ‹many seconds›[107] ‹both the objects being almost aequally raysd by refraction› whereas in the other way of tryall by the altitudes ‹of the comet from the horizon› taken with a quadrant ‹the comet alone is raysd by refraction the horizon remaining soe that›[108] the Refraction is oftentimes greater then the parallax and consequently what to ascribe to the one and what to the other is uncertain.[109] plaine by a scheme. Let A B C Repraesent the earth A E the sensible horizon[110] Let E repraesent the Star in the horizon and D the comet and let E A D be the ‹true› Parallax of about 30[111] Degrees Now by the Refraction of the air the rayes[112] both from the Star and comet are refracted or bended up nearly soe that D appears in G and E in T. I say the refraction of the air shall not make this ‹refracted› parallaxticall angle less then the true one by much more then 2 or 3 min[113] for both the star and comet suffering a refraction in the air will[114] be both Elevated ‹alike› but only ‹with this difference that› that which appears lowest namely the comet, will suffer the Greatest refraction. For since the rayes from D must have a greater inclination in the air then the rayes from E that touch the eye in A it follows it must suffer a ‹somewhat› greater refraction ‹though not considerably much more› now if the Paral[l]ax be calculated from an elevation ‹of the comett› taken when pretty near the horizon the Paral[l]ax and refraction will come toward[s] other[115]:

'f. 110v' This Relates to a Natural Philosophy I suppose of no worth
Lecture about the Improvement of Natural History
(3)
NP[116]

NB A Child that could multiply any Numbers is sooner by a natural Faculty then the best Arithmetician with his Demonstration is hence related

---

[104] Hooke compares different observations of this comet in his writing on comets published in 1678, cf. Hooke (1678), 18–22.

[105] Deleted: it must.

[106] Deleted: beyond the.

[107] Corrected, in different ink: more then a minute or two.

[108] In different ink.

[109] Deleted in different ink: To make this somewhat more.

[110] Deleted: let H be the comet in the meridian (...) to which let I signify a small Starr which touches it or appears very near it when in the meridian.

[111] Deleted: (…).

[112] Deleted: (…) of the.

[113] Deleted: almost.

[114] Corrected: be

[115] Deleted: I doe not wonder.

[116] Deleted: (4).

Guildhall Library MS 01757, ff. 111r–121v

'f. 111r' Quanquam sapientia omnis humana per angustis limitibus circumscribatur vt via vllus sit mortalium qui non in aliquid labatur quamcunque is demum philosophiae partem attigerit, fatendum summam est, inter omnes disciplinas et scientias, nvllam esse quae firmioribus nitatur fundamentis, nvllam esse quae certiora principia magisque perspicuas dimostrationes habeat, quam mathesis. Nam cum reliquae omnes circa eas versantur res, quae partim propter maximam svbtilitatem atque obscuritatem cognosci nequeunt, partim ob perpetvam mutabilitatem certis Scientiae legibus non tam facile adstringuntur, mathesis medium quasi sortita locum ea. considerat quae a sensibus atque intellectu[117] haud difficulter percipiuntur; mutationibus tamen ita carent vt facile scientiae legibus obtemperent. Et quamvis fortasse in Omnibus matheseos partibus, non tam clare tamque evidenter demonstretvr in Geometricis tamen atque adeo etiam in arithmeticis (quae ‹vere› pars[118] geometriae censenda est) hoc ipsum adeo manifesto[119] deprehenditur, vt nemo[120] qui veritatem ‹amet et sedulo› quaerat non facillime perspiciat. eo enim tendit scientia et disquisitio mathematica vt veritatem asseqventur, vnde Aristoteles Geometriam, veritatis speculatorem appellat, cum preterea Reliquae Scientiae Raro evidentiam pariunt, raro certitudimem, sed Opiniones et incertas conjecturas gignent. Vnde rerum vmbras et spectra tantum, non solida ‹et vivida›[121] Determinata Rervm ‹corpora›[122] amplectuntur. Vnde eas merito non Scientias sed conjecturas Plato voluit appellari.[123] Huic notitia huius[124] excellentis, a veritate tanquam fine vel scopo. Quo omnis tendit operatio siue,[125] disquisitio mathematica, ‹appellatio› desumitur.

'f. 112v' First Latin Lecture[126] about Algebra.[127]

'f.113r' Vti in[128] Scientijs, mathematicae sunt[129] praestantissimae et certissimae, Ita ex mathematicis ea. maxime praecellit, quae minime omnium scientijs alijs non aequa certis et Demonstrativis, implicatur, Hinc ‹cognitio› scientia mathematica pura et simplex quoad subjectum suum multo excellentior est eâ qua cum alia physica scilicet vel minus certa componitur et confunditur, in ea. quippe Ratio nos-

---

[117] Deleted: sine.

[118] Deleted: etiam (…).

[119] Corrected: manifestandi.

[120] Deleted: est.

[121] Corrected: et.

[122] Corrected: fabrica. Deleted: amplectetur.

[123] Cf. Biancani (1615), 27.

[124] Deleted: ex.

[125] Deleted: actio.

[126] Deleted: s

[127] End of the first algebraic lecture, Guildhall Library, London MS 1757.12.1, ff. 111r–112v.

[128] Deleted: omnibus.

[129] Deleted: omnium.

tra magis abstrahitur a materiâ, et conclusiones sunt magis vniversales. Hujus clas-
sis Sunt[130] geometria, quae proprietatum et affectionum quantitatis continuae est
Scientia, Atque Arithmetica quae ‹Scientia est› proprietatum et affectionum quantitatis discretae. ‹vna est proprietatum›[131] extensionis,[132] ‹Scientia est› prout ‹ista
extensio›[133] indiuisa et comparatim solummodo spectata. Altera vero proprietatum
extensionis quatenus ‹ea. extensio› in distinctas partes Divisa vel separatim
Consideratur. Geometriae igitur est Abditissimas diuersarum quantitatum continu-
arum, Arithmeticae vero quantitatis discretae siue numeri proprietates inuenire. Ad
quod vtrumque praestandum nulla datur via ‹methodvs›[134] magis demonstratiua,
magisque compendiosa quam Algebra. Ars tantae et tam mirandae ‹indolis›[135] vt
nullus humani ingenij foetus ei possit compararj Ars cujus beneficio, oculus, manus,
quin et calamus propemodum, praestare majora valent, quam ipsum cerebrum, ipsa
ratio, immo quam ipsa ‹mens vel› anima humana sine hac arte[136] exequi potest. Ars
inquam in qua conspicere licet ipsa rationis fondamenta nexumque; quibus nempe
medijs ad conclusiones et axiomata, ad ipsa etiam Rationis ipsius fundamina peru-
eniamus. In ea. quippe videre est quomodo ab objecto maxime obuio et sensibili ad
summum ratiocinationis fastigium provehamur, quibus fundamentis axiomata
superstruantur, quo denique pacto illa omnia a[137] maxime obuiis sensuum informa-
tionibus[138] deriuentur. Nequaquam[139] vero nobis congeniti sunt vel infusi habitus,
sed deductiones potius, per continuam ratiocinationis seriem, comparando scilicet,
componendo et separando, ‹Resolvendoque› perque generales modos examinandi et
applicandi maxime sensibiles corporum proprietates, institutae. Hujus artis beneficio evidenter omnino[140] oculis objicitur, quomodo inventio promoueatur in ‹mente›[141]
humana, quomodo ab evidentibus hujusmodi[142] obuijsque[143] principijs quale est
exempli gratia quae conueniunt in tertio conueniunt inter se[144] ad inueniendum mys-
teria abstrusissima progrediamur quomodo quando res aliqua proponitur invenienda
‘f. 113v’ ‹mens›[145] humana operatur, cogitet, et ad id inveniendum se, verset. Si quis

---

[130] Corrected: est.

[131] Corrected: Altera est scientia.

[132] Deleted: est.

[133] Corrected: (…).

[134] Corrected: praestantia.

[135] Corrected: excellentiae.

[136] Deleted: praest.

[137] Deleted: planissima et.

[138] Corrected: informationes.

[139] Deleted: Neque nobis.

[140] Deleted: ad.

[141] Corrected: cerebro. Deleted: et.

[142] Deleted: et.

[143] Deleted: simplicissimisque.

[144] Deleted: vel duo et duo sunt quatuor.

[145] Corrected: cerebrum vel ratio.

enim serio pensitet quando in hoc illudue inuentum inciderit, examinatae sigillatim suae circa rem istam ratiocinationis processu, depraendet, Rationem suam eâdem plane methodo et viâ fuisse operatam quâ ratio et manus ejus in geometricarum vel arithmeticarum disquisitionum prosecutione, mediante Algebrâ Operantur.

Algebra Itaque (quicquid demum siue Arabicae siue alterius originis,[146] vocabulum significet, nil refert, dummodo de rei significat[147] notione[148] conueniat) Algebra inquam siue methodicus rationis processus quo a maxime obuijs cognitisque subjecti alicujus proprietatibus, ad abditissimi misterij inuentionem progredimur. Haec inquam Algebra non est duntaxat, vnica hujus[149] inquisitionis,[150] qua scilicet ad inueniendam quantitatis proprietatem spectat, cancellis coercenda, sed multo longius ad complurium aliarum rerum proprietates inquirendas inueniendasque se extendit. ⟨Quod ipse post hac fortassis si detur occasio fusius possim explicare⟩ Ea tamen est ⟨plurimarum scientiarum⟩[151] imperfectio, vt non fuerit haec nostra ⟨Algebra scilicet⟩ in vllis alijs ⟨disquisitionibus⟩ quam,[152] circa quantitatis proprietatis,[153] in certam formam digesta, et in pluram facilemque methodum[154] redacta.

Haec ⟨huiusci Scientiae⟩[155] pars non modo faeliciter caepta, sed et majorem in modum perfecta et egregie adornata ⟨est⟩[156]; licet reuera in quibusdam (vt deinceps fusius ostendam) nec illis ita paucis ad huc deficiat.

Hanc ⟨itaque⟩ Ratiocinationis artem ⟨circa quantitatis scilicet Proprietates⟩ siue algebram primo conabor explanare vt hoc expedito negotio, promiori et faciliori viâ ad eam prouehendam pergere liceat, vbi et ipsius ad resoluendas quasuis alias questiones vsum et applicationem ostendam. Praecipuum facilitatis mathematicae scientiae per algebram inueniendae et examinandae fundamentum cui in primis innititur, in eo consistit quod parue coercere spatio integram ratiocinationis seriem potest, ita quidem vt uno quasi obtutu **'f. 114r'** et tantum non in infranti ⟨simul et semel⟩ facultas detur, totam Ratiocinij nexum oculis vsurpare, et hac ratione quamlibet ipsius partem pro libitu, facili negotio, et maximâ certitudine examinare, comparare, mutare, transponere ordinare.

Quocirca omnium algebrarum ea. certe praestantissima est et perfectissimae ipsius ideae proxima, quae characteribus ⟨siue simbolis⟩ vtitur omnium paucissimis,

---

[146] Cf. Wallis (1685), sig. a2r, a3v, b1r, 1–3. Even if Wallis considered the work of François Viète the first great improvement of algebra since its Arabic foundation, according to the French mathematician, modern algebra or analytic art originates from Diophantus's zetetic, cf. Viète (1646), 1, 10.

[147] Deleted: (…).

[148] Deleted: (…).

[149] Deleted: des.

[150] Corrected: disquisitionis.

[151] Corrected: multarum utilium et valde necessarium scientiae partium.

[152] Deleted: in hac.

[153] Deleted: disquisitionibus.

[154] Deleted: dig.

[155] Corrected: itaque.

[156] Corrected: fuit

quaeque in operatione omnium minime prima fundamenta siue principia ‹siue symbola› quibus superstruitur ratiocinatio vel disquisitio, confundit[157] quaeque in ratiocinij serie omnium distinctissimè et[158] ‹euidentius›[159] apparet. Quare in ‹symbolorum›[160] tum significandis quantitatibus siue materijs, in quas operamur, tum exprimendo operationis vel processus modo, inseruientium selecta, isti equidem sunt praestantissimi proindeque omnibus alijs praeferendi, qui sunt omnium simplicissimi,[161] et maxime obuij.[162]

Et methodum quod attinet ipsos disponendi vel in disquisitione procedendi ea.[163] Dubio procul omnibus alijs anteponenda in qua sequentes Ab immediate praecedentibus deductionibus vel saltem in qua unaqaeque pars siue propositio siue deductio dependens vel a principio siue deductione quadam praegressa vel ab axiomate quodam prius demonstrato vel per se evidenti ita notatur, vt uno eodemque quasi obtutu, deductio ipsa et ipsius deductionis ratio evidenter patescant. Hac quippe methodo mens humana, investigatione rerum indulgens minus turbatur, et memoria retinendi, depromendique deductiones praegressas anxietate minus fatigatur.

Vietae igitur algebra speciosa,[164] omnibus algebrae cossicae[165] siue numerosae methodis est praeferenda, quia scilicet in operationibus numerorum, prima[166] principia ‹vel primae quantitates›, siue characteres et symbola primo posita et simplicissima, breuis pereunt et obliterantur. In operationibus vero speciosis vbi ‹symbola›[167] primo posita et constituta non variantur, et species et operationis modus seruentur incolumes.

Methodus vero Conterranej nostri Harriotti tantundem est praeferenda vietanae, ac vietana est antiquae.[168] Quocirca etiam **'f. 114v'** Oughtredi methodus in qua breuia aliquot et significatiua introducit ‹symbola›,[169] vti in demonstratione Decimi[170] libri Elementorum Euclidis[171] methodo Harriottana nonnihil praestat et vero methodus[172] ipsius quomodo scilicet in inuentione problematum sit

---

[157] Deleted: vel obscurat.

[158] Deleted: et planissime.

[159] Corrected: planissime.

[160] Corrected: characterum.

[161] Deleted: planissimi.

[162] Deleted: compendiosissimique.

[163] Deleted: sine.

[164] Cf. Viète (1591). This edition is present in Hooke's library. Cf. *Bibliothecha hookiana*, 5.

[165] The term *algebra cossica* refers to the algebraic tradition developed from the works of Christoff Rudolff and Michael Stiefel. Cf. Rudolff (1533); Stifel (1544); Wallis (1685), 3.

[166] Deleted: capita et.

[167] Corrected: characteres.

[168] Harriot (1631). Cf. *Bibliothecha hookiana*, 5.

[169] Corrected: characteres.

[170] Corrected: primi.

[171] Cf. Oughtred (1652a), 1–2.

[172] Deleted: (…).

procedendum, provt in claue[173] ejus traditur quâque in alijs operibus vtitur, methodo Harriotti quamvis ea. fuerit posterior succunbit.[174] In ea. quippe methodus ipsius vtpote lineas siue propositas quantitates per characterem duplicem designans et vocis complures introducens et immiscens, tum etiam vnamquamque deductionem siue propositionem, suâ, sigillatim lineâ consignare omittens, perplexitatem gignit.[175]

Cartesius vero quamuis reuera maxima algebrae ipsius pars eadem sit cum harriottanâ, quae diu ante Cartesianam lucem viderat, quoad methodum tamen suas digerendi demonstrationes Harriottanam excellit.[176] At Herigonij modus mihi quidem hac inventione[177] siue additamento videtur vtrique illorum praecellere. Distinctam quippe marginem seruat in quo cujusque deductionis rationem modumve quo producta erat consignat i.e. quo operationis genere, a quâ definitione postulato, axiomate, propositione, corollario.[178] Et (ob hanc quippe et alias causas eximia) methodus[179] praestantissimj illius mathematici ‹Nostratis›[180] Doctoris Pellij, ceteris in quas quidem ego incidi omnibus, tum ob characterum quos adhibet vim significantem, tum ob methodj ipsius in operando et deducendo euidentiam et ordinem est anteferenda.[181]

Primum itaque quod spectandum in algebra est vt characteres ‹symbola› siue signa sint omnium planissima, simplicissima, breuissima et maxime significatiua, quibus scilicet quantitates et operationes et producta et quicquid tandem notatu dignum sit possint omnium[182] planissime breuissime et efficacissime significarj. Et primo quidem quoad quantitates primas siue principia quae inceptis nostris inseruiunt cum vnamquamque earum quâ vtimur tamquam[183] vnum ‹quid› et distinctum ab alijs omnibus consideremus significare id praestat per characterem aliquem exiguum et simplicem, puta per characterem simplicem vnius alicujus literae '**f. 115r**' alphabeti cujusdam. Qui modus hactenus inter Doctos obtinuit ob typorum, typographijs

---

[173] Cf. Oughtred (1631).

[174] Even John Collins, the mathematician, in an undated answer to Wallis' letter of February 1667, attested Hooke's reservations on the *Clavis*. Referring to the new Latin edition of Oughtred's work, Collins states: "the aim of those objections was not to disparage the author, but to incline you to supply the defect of him, that his book, together with yours, might be of the more durable esteem, and not undervalued, (as that author now is by Mr. Hooke and Dr. Croone,) as wanting the most material parts of algebra". Rigaud (1841), 480.

[175] Cf. Oughtred (1631), 3.

[176] Cf. Wallis (1685), sig. a4r.

[177] Deleted: (…).

[178] Cf. Hérigone (1634), vol. II, 27, 40, 43–4, 55–67, 72–3, 171–296.

[179] Deleted: quam uti.

[180] Corrected: Domini.

[181] John Pell's ideas on algebra are expressed in his revision of the work of one of his Swiss students, Johann Heinrich Rahn. Pell's work was completed on 18 May 1665, long before the publication, and circulated among Royal Society's virtuosi. Cf. Rahn (1668), Sig. a 2 r; Wallis (1685), 218–24.

[182] Deleted: distinctissime.

[183] Deleted: (…).

inservientium commoditatem. Vel etiam per quoslibet alios characteres fictos siue pro re nata inuentos siue primo constitutos, Quorum singuli suum habeat significatum ‹vel› ad placitum, vel positiue distinctum et determinatum; vt euidentius id significent cuj designando adhibentur. Id quod equidem alphabeti etiam literae possunt praestare. Sed quo minores[184] planioresque fuerit isti characteres, eo certe meliores et ‹ad hanc vsum› magis ‹accommodati›.[185] Oughtredj igitur commoda via est, qua omnes consona majuscula B C D F G etc. ‹vt videri est in prima schematis linea et› significant quantitates aliquas cognitas et definitas, characteres vero majusculi vocalium Scilicet A E I O U quantitates incognitas siue quaesitas. Praeterea etiam hac methodo praestat quod si duae sint quantitates incognitae siue quaesitae quarum una sit altera major, tum A semper majorem E vero minorem ‹Designat› Consonantibus etiam nonnullis distinctam propriamue et vniversalem significandi vim ‹assignat›.[186] Sic Z semper significat duarum istarum quantitatem incognitarum summam. X vero differentiam. Z cum commate summam quadratorum ipsarum. et X cum commate quadratorum differentiam.[187] Sic etiam Harriottj aptus modus est vbi minores vocalium characteres a, e, i, o u, y ‹vt dispositi sunt in 3 linea›. Significant quantitates incognitas, minores vero consonarum characteres puta b, c, d, etc. quantitates cognitas siue in operationibus coefficietes.[188] Via etiam Cartesij non est multum istis quas reconsuimus ab similis, qui primos omnes c[h]aracteres minores literarum alphabeti[189] Italicj capit pro quantitatibus cognitis, vti ‹in eadem 3 linea› a b c d etc. et characteres minores literarum posteriorum alphabeti, puta z, y, x, u, t, etc. pro quantitatibus incognitis.[190] Multo etiam faciliorem reddit operationem, memoriamque in singularum notarum vel symbolorum[191] significatione retinenda et prompte reddenda leuat, si vel lineae superficies vel corpora quibus significandis adhibentur, charactere notentur eodem.[192] Vel si lineae superficies vel corpora habeant nomina propria, tum memoria multam juuatur, si prima ipsius nominis litera ponatur ad rem significandam sed cum puncto aliquo siue accentu vel nota, quae indicat eam habere '**f. 115v**' significatum peculiarem secus enim confusionem potius inferret quam simplicitatem et distinctionem.[193] Puta si notandis trianguli rectanguli lateribus. B significat Basin P perpendicularem et H hypothenusam. Vel si notandis in Parabola linejs O significaret ordinatum T Parametrum I Interceptam partem Diametrj, t tangentem et similiter.

---

[184] Deleted: brevioresque et significationis.

[185] Deleted: adaptae.

[186] Corrected: habent. Cf. Oughtred (1631), 50.

[187] Cf. Oughtred (1631), 30–1.

[188] Harriot (1631), 4–5, 46.

[189] Deleted: Latinj.

[190] Descartes (1964–74), vol.VI, 371, 374.

[191] Deleted: notatione.

[192] Deleted: (…).

[193] Deleted: ‹quae methodo vtitur clarissimus Wallitius›.

Quin etiam in problematum Geometricorum examine, multum juvat memoriam, et multo minus turbat ratiocinationem; si extrema linearum schematis diuersis illis characteribus notentur qui nunc passim ‹brachy›graphia[194] inseruiunt adeoque vnaqueque lineae notari potest duabus literis, et tamen non nisi vno charactere paruo et compendioso id quod memoriam et imaginationem plurimum leuat. Ponamus ex. gr. ‹symbola in 5ᵃ linea schematis›[195] esse charateres literarum alphabeti ‹suprapositi›[196] in 4 linea›[197] et problema ‹Quodcunque› proponatur solvendam,[198] Dico Symbola aeque fore compendiosa tamque arcte et scite ad se inuicem in collocatum, si vnaquaeque linea ‹sc[h]ematis Problemati inservientis› charactere notetur literarum quae terminant linearum extrema, quam si vno duntaxat charactere singulo juxta modum ‹cartesianum siue› vulgarem ‹signetur›[199] vt si loco (ab) ponatur ⌐loco b ponatur ϛ, loco x ponatur ↳ ‹et ad quantitatem incognitam denotandam adjiciatur punctum siue accentus supra symbolum›.[200] eritque aequationis operatio multo facilior, cum multo minus fatigetur memoria maxime si geometrice procedat ratiociniatio ‹vti experienti constavit›. At in questionibus arithmeticis vbi[201] operatio procedit eodem modo quo in arithmetica vulgari, isto in casu perinde est, literis né singulis, an duplici charactere res peragatur. Sed si methodo vtamur Cartesianâ quae sane rationi valde est consentanea et egregia, puta si omnis generis quantitates qualescunque siue continuae siue discretae fuerint, reducamus ad lineas ‹rectas› quoniam quocunque demum fuerit inter quaeuis duo corpora, superficies numerosue siue[202] inter quasuis duas quantitates proportio eadem esse 'f. 116r' eadem esse potest inter lineas.[203] Eo inquam in casu multum praestare autamen characteribus potius quam ‹literis›[204] vti poteruntque litera varijs operationis modis insevire vt ‹post hac›[205] tradam.

Proximum in Algebra observandum est respectus quam quantitates habent ad se inuicem. Atque hi respectus distinctis et compendiosis notandi sunt characteribus,

---

[194] Corrected: Typographia.

[195] Corrected: hos.

[196] Deleted: (…).

[197] Corrected: a, b, c, d, f, g, h, k, l, m, n, p, q, etc.

[198] Deleted: hujusmodj (...) Data recta AB utcunque sectâ in C erectaque ex ejus termino B super ipsa perpendiculari indefinitâ BD ex altero ejus termino A rectam lineam ductu AD huic occurrentem ita vt ipsa aequalis sit rectis DB, BC simul sumptis. Supponamus juxta modum Cartesij (qui omnium quidem adhuc brevissimus est). Supponamus inquam AB vocari (...) a et CB vocari b et BD denotari per x.

[199] Corrected: notetur.

[200] Corrected: ‹una› cum puncto siue accentu superius ad quantitatem incognitam denotandam tumque.

[201] Deleted: ratiocinatio siue.

[202] Deleted: inter.

[203] Cf. Descartes (1964–74), vol. VI, 371.

[204] Corrected: linejs.

[205] Corrected: non.

qui planissime denotant ea., quibus significandis adhibentur, quoque omnium minime species quibus subserviunt confundant.

Respectus ille triplex est, vel comparativus vel conjunctivus vel cooperativus, Comparativus vel Definitus vel indefinitus: Indefinitus est quando vna quantitas est indeterminate major vel minor quam alia. Atque ad hos Respectus exprimendum Harriottus vtitur hisce notis | > < quarum prior significat majorem posterior minorem. e.g. a > b significat a esse majorem b et a < b significat a esse minorem quam b. Secundo respectus ‹quantitatum› ad se inuicem est Definitus et juxta eum quantitates sunt vel aequales vel proportionatae prior notatur ab harriotto duabus linejs parallelis inter duas quantitates aequales positis hoc modo A = B significat A esse aequalem B.[206] posterior notatur ab Oughtredo ‹4ᵒʳ punctis quadrate dispositis›[207] nota hâc post eas quantitates positâ, quae isto respectu afficiuntur, cuj nota, proportio siue respectus postponitur. Exempli causa. ‹in 7ᵃ linea›[208] a.b::2.3 significat a habere eandem propo[r]tionem ad b quam 2 habet ad 3. quae proportio etiam vsitato notatur simbolis vt Oughtredus nostras communiter vtitur ‹literis› R S T V etc.[209] Cartesius vero omnibus indifferenter. Exprimunt alij nonnulli proportionem quantitatum, more fractionum, hoc modo $\frac{a}{b}$, $\frac{c}{d}$ [210] qui egregius sane et distinctus foret modus plurimumque lucis inferret naturae[211] proportionum dummodo nota proportionis interponeretur hoc modo $\frac{a}{b} :: \frac{c}{d}$ quo significatur a se habere ad b vt c ad d vel etc. id est vt quaeuis quantitas superior ad quamlibet inferiorem ita altera superior ad alteram inferiorem, siue vt quaevis quantitas inferior ad quamlibet superiorem ita altera inferior ad alteram superiorem. **'f. 116v'** Alius quantitatum ad se invicem respectus est, quo illa vel quantitatem aliam praecedentem augent vel minuunt. Ea quantitas seipsum vel aequalem sibi quantitatem addit quantitati praegressae quae signum affirmativum vel copulativum + praefixum sibi habet vt a + b significat b esse additum ad a. siue a esse auctum per b, vel a et b. et ea. quantitas minuit quantitatem praecedentem quantitate sibi ipsi aequali, quae signum negativum siue diminutivum – sibi habet praefixum. vt[212] hoc symbolum a – b significat quantitatem ‹b› demptam esse ex a. id est a diminutam esse b. quantum igitur b plus est quam a

---

[206] Harriot (1631), 10.

[207] Corrected: hoc modum::.

[208] Endorsed in lefthand.

[209] Oughtred (1631), 7; Id (1652a), 1.

[210] Circled.

[211] Deleted: aequationum.

[212] Deleted: hoc.

tanto a notatur hoc symbolo minus esse quam nihil.[213] Alia signa Algebraica in proximum diem reservo.[214]

'**f. 117r**' Ostendi lectione novissimâ, Algebrae insignem Excellentiam; eam scilicet usui esse tam facilitandis Demonstrationibus, quam Problematibus inueniendis
resolvendisque eamque aeque conducere juvandae memoriae quam excolendae
intelligentiae et rationi. Conatur etiam ostendere qua in re,[215] ista ejus Excellentia
consistat, ex quibusque causis et principiis oriatur. Tradebam insuper quaenam sit
Idea Algebrae perfectissima, quaeque ex omnibus methodis, hactenus receptis,
quam proxime ad ‹eam› accedat, quaque in re[216] singulae earum deficiant; subindicatis etiam mediis, quibus istae methodi dirigi atque emendari ‹possint›[217] quibus
expeditis, ad particulariorem pergebam modorum illorum explicatione, qui solennes
hactenus fuêre, notandi disponendique quantitates in his inquisitionibus adhibendas; ostendens, quibus notis et juxta quem modum illae sint exprimendae. Atque
hanc ‹in› rem[218] explicabam symbola sive characteres, omnium optimos exprimendis lineis in geometrici alicuius Problematis resolutione usurpatis esse duas literas,
quibus linea quaeque notatur in Schemate sive figura, de qua quaeritur, easque '**f.
117v**' non expressas ‹Uncialibus[219] illis characteribus›[220] literarum Alphabeti, ut
moris est Oughtredo, per omnes eius Tractatus,[221] sed notis sive characteribus
‹illis›[222] quibus utimur in Brachygraphia.[223] Hac enim ratione symbola multo facilius
haererent memoriae comprehensa quippe parvo spatio, proindeque simul et semel
in aspectum incurrentia[224]: unde etiam longe faciliorem reddunt totam operationem,
ut cuilibet experturo patebit. Etenim ob characteris parvitatem, disponi, digeri et
compingi ea. possunt in locum ‹aequè›[225] parvum, si non minorem,[226] ac methodo
Cartesianâ, quâ una singula adhibetur literula, ad unam singulam lineam sive

[213] Viète, Harriot and Descartes regularly adopted the cossic symbols + and – for addition and subtraction instead of plus and minus used before Rudolff and Stiefel. Oughtred, on the contrary,
maintained both ways of indication, the symbolic and the verbal. Cf. Viète (1591), 5; Harriot
(1631), 7; Descartes (1964–74), vol. VI, 371; Oughtred (1631), 3.

[214] End of the second algebraic lecture, Guildhall Library, London MS 1757.12 ff. 113r–112v. The
third lecture follows, Guildhall Library, London MS 1757.12, ff. 117r–121v. This latter is not in
the hand of Hooke.

[215] Deleted: et.

[216] Deleted: sin unaqu.

[217] Corrected: queant.

[218] Deleted: ostend.

[219] Deleted: per.

[220] Corrected: aliis characteribus.

[221] Cf. Oughtred (1631), 3.

[222] Corrected: earum.

[223] Deleted: ista

[224] Corrected: occurrentia.

[225] Corrected: adeo exiguum.

[226] Deleted: qu.

quantitatem denotandam: molestiaque, quae creatur memoriae,[227] dum retinere debet, quaenam linea sive lineae pars[228] significatur per ipsam, penitus devitatur; quae in nonnullis Problematibus et disquisitionibus, complicatae si fuerint, satis gravis est ut esploranti cuivis fiet manifestum.

Post haec, in medium afferebam diversos illos characteres, exprimendis omnium simplicissimis operationibus, quantitatumque ad invicem relationibus, usitatos, sive id fuerit ad exprimendam earum Affirmationem **'f. 118r'** ‹vel›[229] Negationem, additionem vel subductionem, multiplicatinem vel divisionem, quadrationem vel extractionem; nec non earum proportionalitatem, sive continuam sive discretam, sive directam sive reciprocam: ut et majoritatem, minoritatem et aequalitatem earum,[230] multosque istius generis plures, quorum aliqui[231] usum ‹obtinuerint›[232] Catholicum in operationibus Algebrae, proindeque deesse nequeunt: quanquam multi esse[233] possint alii, qui usum habent magis peculiarem,[234] in quibusdam[235] duntaxat quaestionum generibus, quos Characteres peritus quivis Algebraicus invenire facile potest; et pro re nata usurpare: Quales sunt quos introducit[236] Oughtredus ad Demonstrationem decimi libri Euclidis, Herigoniusque[237] invenit ad Algebram suam ‹deducendam›[238]; quorum pars maxima omitti[239] in diversis aliis problematibus potest,[240] talibusque duntaxat, quae de isto solum sunt generes, congruunt.[241] Haec notarum genera excogitari et constitui facile possunt, quotiescunque fert occasio, pro natura Subjecti, circa quod ‹praecipue› instituitur Disquisitio, ad denotandum diversas proprietates, operationes, relationes, ‹vel› notiones, peculiari isti subjecto proprias. In quorum **'f. 118v'** Omnium excogitatione, observanda est haec Regula, ut quam fieri potest sint brevisissimi, nec non recordatu[242] et[243] intellectu faciles, hoc est, omnium optimi sunt Characteres illi, in quibus quod maxime significat characteribus omnium brevissimis simplicissimisque planissime exprimitur. Sic (e.g.) Oughtredi Character r, positum pro *Rationale*, et ∩ positum pro *superficies*

---

[227] Deleted: q.

[228] Deleted: per ipsam.

[229] Deleted: sive.

[230] Deleted: multique.

[231] Deleted: multa usus est.

[232] Corrected: obtinere possunt.

[233] Deleted: queant.

[234] Deleted: non.

[235] Deleted: solum.

[236] Deleted: Ought.

[237] Corrected: Herigonque.

[238] Corrected: provehendam.

[239] Deleted: postest.

[240] Deleted: sunt.

[241] Oughtred (1652a), 1–2; Hérigone (1634), vol. II, 5–10.

[242] Deleted: faciles.

[243] Deleted: obvii.

curva, valde convenit et complures characterum Herigoni sunt ejuscemodi, quos brevitatis gratiâ omitto.[244] Quandoque etiam in Quaestionis Algebraicae processu abbreviando, duae, tres, quatuorque, pluresve quantitates complicatae significari supponuntur, unâ simplici literâ, uti deinceps,[245] uberius exponam in Harrioti et Cartesianae methodi explicatione.

Sunt ‹complures›[246] alii[247] Characteres amplissimi usûs in Algebra, qualis est hic ‖ qui differentiam significat duarum istarum quantitatum, quas interponitur, sive major sit prior sive posterior[248]: Sic etiam $\sqrt{}$,[249] praepositum ulli quantitati, significat ‹latus› quadratum[250] vel radicem quadratam illius quantitatis, et $^2\sqrt{}$ radicem cubicam, et $^3\sqrt{}$ radicem quadrato-quadratam etc. Et si plures sint quantitates unâ, quarum aggregati vel differentiae radii significanda sit, tum post rotam hanc $\sqrt{}$ **'f. 119r'** simplex ducenda est linea super omnes istas quantitates, quae ad istud signum sunt referendae.[251]

Multiplicatio etiam ab Oughtredo denotatur, signo crucis Andreanae X, quantitatibus ‹sic›[252] multiplicandis interposita[253]; quae ‹utique› in ‹ipsius› methodo,[254] in qua duabus ‹utitur› literis[255] ad denotandam lineam vel quantitatem,[256] est necessaria ‹ut visum est in 3 Linea›; at ne[257] modo Cartesiano, in quo nonnisi una adhibetur litera vel nota pro quantitate simplici, non item.[258] Nec ullâ opus est Oughtredi notâ divisionis in methodo hac Cartesiana, nec in methodo a me proposita per symbola Tachygraphica. Sed[259] nonnulli alii dantur Characteres, non utiles modo, sed plane necessarii, ad faciliorem et accuratiorem quaestionis alicujus Algebraicae deductionem; quas exponere conabor, postquam prius ostendere, quae ‹nam› sint simpliciores planioresque Algebrae operationes in methodo recepta; iis quippe intellectis, multo erit facilius, reliquas expedire.

---

[244] Oughtred (1652a), 1; Id. (1652b), sig. a2r.

[245] Deleted: plenius.

[246] Corrected: et.

[247] Deleted: aliquot.

[248] A distinctive symbol for difference between quantities of which it is not stated the greater and the less, is employed by Viète. Hooke probably inverted Viète's symbol because of its possible confusion with the equality symbol introduced by Recorde (1557), but not adopted by Viète. Cf. Viète (1646), 5.

[249] Deleted: positum.

[250] Deleted: latus.

[251] Unlike Harriot and Descartes, Hooke denotes the index by means of numbers and places it out of root. Cf. Harriot (1631), 99–102; Descartes (1964–74), vol. VI. 371.

[252] Corrected: ita.

[253] Deleted: quod.

[254] Deleted: uti.

[255] Deleted: utitur.

[256] Deleted: opus.

[257] Deleted: method.

[258] Cf. Oughtred (1631), 10; Descartes (1964–74) vol.VI, 371.

[259] Deleted: complures.

Quod igitur attinet Additionem; Quantitas significatur[260] alii quantitati ‹addenda›[261] si signum, quo quantitas addenda notatur, ipsis interponitur, sive signum illud sit ‹nota› affirmativa ‹ + › sive negativa ‹-›, hoc modo etc. A + B . A − B. Nam additio quantitatis positivae, alteram reverâ auget 'f. 119v' istâ quantitate, uti additio quantitatis negativae, hoc est, carentiae sive defectus istius quantitatis, reverâ minuit.

Deinde, Quoad Subtractionem; Quantitas aliqua significatur ab alia quantitate[262] abstracta, si abstrahenda quantitas postponatur alteri quantitati cum signo contrario ei quod obtinebat ante hanc abstractionem, hoc modo etc. ‹si ab A. sit abstrahenda + B. manebit A − B›. Etenim abstractio positivae quantitatis ab alia, relinquit eam minorem, quam erat prius, ista quantitate: et subductio Defectus quantitatis ejusmodi, reverâ eam[263] tanta quantitate auget.

Ubi, in transitu notandum, passim primam seriei quantitatum, si sit affirmativa, nullo signo notari, sed affirmativum signum subintelligi; at quantitatem negativam semper[264] indicari[265] praefixâ notâ.

Tertiò; Quantitas aliqua significatur multiplicata per aliam quantitatem, quando duae hae quantitates conjunguntur absque ulla nota interposita, uti etc. ‹Si a multiplicanda sit per b conjunguntur a et b simul ut ab› Sed quoad speciem quantitatum multiplicandarum, quae denotantur per signa praefixa, sive negativa sive affirmativa, cape brevem hanc Regulam: Quodsi utraeque quantitates multiplicandae notantur eodem signo, sive negativo sive affirmativo tum productum notatur signo affirmativo uti etc. ‹si + a per +b fit +ab et si −a per −b fit +ab›. Etenim affirmatio affirmationis, 'f. 120r' et negatio negationis,[266] ambo sunt modi affirmandi: Sed si notantur signis differentibus, tum signum, producto praefixum, est negativum, uti etc. ‹Si + a per −b. vel −a per +b fit −ab› Affirmatio quippe negationis, et negatio affirmationis, ambo sunt modi negandi.

Quarto, Quantitas aliqua significatur divisa vel dividenda per aliam, quando dividenda quantitas supraponitur, et quantitas dividens, infra eam, lineâ inter ipsas[267] trajectâ: quae Cartesii est methodus, quando istae quantitates in specie[268] dividere se invicem actu non possunt, uti ‹Si a dividenda sit per b scribit a/b› etc.[269] Sed Oughtredus aliâ hac utitur methodo, ad Divisionem significandam, nempe etc.[270]

---

[260] Deleted: addita.

[261] Corrected: addita.

[262] Deleted: sub.

[263] Deleted: auget.

[264] Deleted: not.

[265] Deleted: no.

[266] Deleted: sunt.

[267] Deleted: traductâ.

[268] Deleted: non pos.

[269] Descartes (1964–74), vol. VI, 371.

[270] Cf. Oughtred (1631), 11.

Quae non est aequè commoda, ac prior, quia aliquanto magis[271] implicat speciem, methodumque propositiones in quaestione aliqua Algebraica digerendi[272] turbat; uti deinceps demonstrabitur.

Sed si Dividendus in specie capax sit actualis divisionis per Divisorem in specie, tum productum est; Dividendus, cum Charactere Divisoris inde expuncto, uti etc. ‹si ab dividenda sit per b deletur b et scribitur a› ex quo casu si quantitates operantes utraeque notantur eodem signo, hoc est, si ambae sunt affirmativae; vel ambae negativae quantitates, tunc Productum erit affirmativum, uti etc. Sed si signa earum sunt dissimilari **'f. 120v'** hoc est, unum affirmativum, alterum negativum, tum etc. quota quantitas erit negativa.

Quinta operatio est Extractio Radicis ‹ex aliqua›[273] quantitate, hoc est, Investigatio alicujus quantitatis quae multiplicata in seipsam, produceret quantitatem aequalem quantitati, cujus radix desideratur, sive Inventio medii Proportionalis, inter quantitatem istam et Unitatem in ‹Arithmetica vulgari› Iam vero, quia hoc semper fieri nequit in specie, ideo[274] factum duntaxat supponitur, et quasi factum esset, notatur figurâ hac, $\sqrt{ab}$: qui character exprimit radicem sive latus *ab*, sive quod radix de *ab* desideratur vel supponitur quaerenda. Sed si inveniri potest quantitas in specie, quae in seipsam multiplicata producere potest istud quadratam, ‹vel›[275] cubicam, vel Biquadraticam quantitatem, tunc ista quantitas scribitur, non verò character antedictus; uti si radix de *aa* desideratur, tum *a* scribitur pro ea.,[276] quia a[277] in *a* producit aa. Si species multiplicationis vel Divisionis numeros habeat ipsis praefixos procedendum est cum ipsis, ut in Arithmetica communi; uti et in Additione ac subductione, nec non in Extractione Radicis; cuius operatio vulgaris scienda est, priusquam negotium Algebraicum intelligi perfectè queat.

Expeditis hunc in modum simplicium quantitatum operationibus, progrediemur ad methodum procedendi in quantitatibus compositis: quarum operandi modus licet primâ fonte videri possit nonnihil implicatior, idem est vel parum discrepans ab eo, quo simplicium quantitatum perficiuntur operationes; (...) patebit melius quam ex multitudine praeceptorum.

'f. 121r'

(1) r. pro Radice

(2) ∩ pro Curva

(3) ‖ pro differentia

(4) $\sqrt{.}$ $^2\sqrt{.}$ $^3\sqrt{.}$ vt $\sqrt{ab}$. $^2\sqrt{abc}$. $^3\sqrt{ab}$.

---

[271] Deleted: amplius.

[272] Deleted: per.

[273] Corrected: alicuius.

[274] Deleted: sic.

[275] Corrected: sive.

[276] Deleted: be.

[277] Deleted: deinde.

(5) ^{278} $\sqrt{ab+bc}.\sqrt[2]{abc-ddd}$

(6) x. Ut AB x BC.

(7) +vt a + b. vel aa + bc

(8) −vt a − b vel aa − bc.

(9) $\dfrac{a}{\dfrac{+b}{a-b}} \dfrac{a}{\dfrac{-b}{a+b}}.\dfrac{a}{\dfrac{-b+c}{a+b+c}}$

(10) a per b fit ab.

(11) +a per + b facit + ab. et − a per − b facit + ab.

(12) +a per − b facit − ab vel − a per + b facit − ab.

(13) Si a dividenda per b scribitur a/b

(14) Si ab dividenda per b scribitur a.

(15) $\sqrt{ab}$. $\sqrt{aa}$. scribitur a . et $\sqrt{aabb}$ denotatur per ab

'f. 121v'

1. r radicale ∩ curva ‖ differentia

2. $\sqrt{}$ latus. $^2\sqrt{}$ latus cubicum. $\sqrt{ab}$. $^2\sqrt{abc}$.

3. $\sqrt{ab+bc}$.$^2\sqrt{abc-ddd}$. AB x CD

4. $\dfrac{+a}{\dfrac{+b}{a+b}} \dfrac{+a}{\dfrac{-b}{a-b}} \dfrac{-a}{\dfrac{+b}{-a+b}}$

5. $\dfrac{+a}{\dfrac{+b}{a-b}} \dfrac{+a}{\dfrac{-b}{a+b}} \dfrac{-a}{\dfrac{-a}{-a+b}}$

6. $\dfrac{+a}{\dfrac{+b}{+ab}} \dfrac{-a}{\dfrac{-b}{+ab}} \dfrac{+a}{\dfrac{-b}{-ab}} \dfrac{-a}{\dfrac{+b}{-ab}}$

7. $\dfrac{a}{b} \dfrac{aa}{bc} \dfrac{ab}{b} = a \dfrac{abc}{b} = ac$

8. $\dfrac{+ab}{+b} = +a \dfrac{-ab}{-b} = +a \dfrac{+ab}{-b} = -a \dfrac{-ab}{+b} = -a$

9. $\sqrt{ab}$. $\sqrt{aa} = a$ $\sqrt{aaaa} = aa$. $\sqrt{aabb} = ab$.

Second Latin
Lecture of
Algebra^{279}
(2)

---

^{278} Deleted: $\sqrt{ab}$ + $\sqrt{(\ldots)}$.

^{279} Deleted: (3).

Royal Society Classified Papers, vol. XX, ff. 65r–66v

'f. 65r' Mr Hooke first Algebraic Lecture[280] june 101,665[281]

As of all sciences, mathematics are the best and most certain, so of Mathematicall that is most excellent, which is least incumbred and rouind with other knowledg, and there Pure and Simple mathematical knowledge is, for its speculation much more excellent, than that, which is implicated and compounded with other physicall and sensible knowledge, for in this the reason is more abstracted from matter, and the conclusions more universall. Of this kind is Geometry, which is the knowledge of the proprieties and affections of continued quantity; and Aritmetick, which is the knowledge of the proprieties and affections of discontinued quantity: the one, of an extension, as it is considered undivided and only comparatively; the other, of extension, as it is considered divided or separated into distinct parts. The busines therefore of Geometry is, to find out the most abstruse proprieties of severall continued quantities, the busines of Aritmetick, to find out the most abstruse[282] proprieties of discontinuous quantity or number. For the performing of both which, there cannot be a more excellent, more demonstrative and more compendious way, than by Algebra. An Art of so excellent a nature and sure invention that there is no one product of human witt or ratiocination, comparable to it: an art, by which the eye, the hand, nay the pen almost is made to doe greater things, than can be done by the braine, reason or the very soule of man[283]: An art, wherein we may see the very grounds and proceedings of ratiocination, by what means we come to conclusions and axioms, and even to the very foundations of reason itself. For in this we see, as it were, how from the most obvious and sensible object we are carried to the highest pitch of ratiocination, upon what ground axioms are built, and how they are all derived from the most plain and obvious information of the senses: and that they are not innate in us or infused, but acquired habits, deductions drawn by a continued series of ratiocination by comparing, compounding and separating, and generall wayes of examining and applying the most sensible proprieties of bodies. Here we see evidently spread before[284] our eyes, how invention is prosecuted and carried on in the Braine,[285] how from such plaine and obvious and slight truths, as that 2 and 2 make 4[286] we proceed to find out the most abstruse mysteries; how, when a matter is

---

[280] Deleted: Re.

[281] This text is not in the hand of Hooke.

[282] Deleted: quantities.

[283] The corresponding Latin text reads: "‹mens vel› anima humana"; Guildhall Library, London MS 1757.12, f. 113r.

[284] Deleted: us.

[285] In the Latin version the original "cerebro" had been subsequently replaced by "mente"; Guildhall Library, London MS 1757.12, f. 113r.

[286] The corresponding Latin text reads "exempli gratia quae conuentiunt in tertio conueniunt inter se," and it is followed by the deleted example "vel duo et duo sunt quatuor"; Guildhall Library, London MS 1757.12, f. 113r.

propounded to be found out, the brain or reason[287] of man works and contrives and turns itself as it were to find it out. For if a man shal[l] seriously examine how he came by any kind of invention, he hath lighted upon, if he doe[s] more particularly examine the process of his ratiocination in that particular, he may find, that his reason hath wrought by the same method, and after the very same way, that his reason and hand works in the prosecution of Geometricall or Arithmeticall inquires by Algebra. Algebra therefore,[288] or ⟨a⟩ methodicall **'f. 65v'** proces[s] of ratiocination, whereby we proceed from the most obvious and known proprieties of a subject, to the invention of the most abstruse inquiry or mystery: This Algebra is not merely to be bound up to this one only subject of inquiry, and[289] to the invention of the proprieties of quantities, but can be made use of and does extend itself much further for the seeking and finding out the proprieties of divers other things[290]: though in truth such is the ⟨im⟩perfection of many usefull and indeed necessary parts of knowledge,[291] that it[292] hath not been drawn up to a certain forme and reduced into a plain and easy method in any other inquiry, than in this, after the proprieties of quantity. This therefore is a thing already begun and in great part perfected, at least very far promoted, and very well adorned, though in truth is yet in many particulars deficient, as I shall hereafter more largely manifest. This art of Ratiocination[293] or Algebra I shall endeavor first to explain, whereupon I shall be able the better and more easily to proceed in the promoting of it, and making it usefull and applicable for the invention and finding out of any other inquiry.

The main ground of the facility of inventying and examinying mathematical knowledg by Algebra, upon which it is chiefly built, is the comprising in a small space a whole series of ratiocination[s] so, as to a small cast of an eye as it were, and in an instant almost, one is enabled to examin, and compare and change, and transpose and order any part of it, as he pleases, with very litl[l]e trouble and the greatest certainty. And therefore of all kinds of Algebra that is the best and the nearest to the most perfect idea of it, that hath fewest characters,[294] and that in the operation does

---

[287] In the corresponding Latin text "mente" subsequently replaced the original "cerebrum vel ratio"; Guildhall Library, London MS 1757.12, f. 113v.

[288] The following part, in round brackets, of the original Latin version had been here omitted: "quicquid demum siue Arabaicae siue alterius originis, vocabulum significant, nil refert, dummodo de rei significant notione conveniat"; Guildhall Library, London MS 1757.12, f. 113v.

[289] Deleted: of.

[290] In the corresponding Latin version "Quod ipse fortasse si detur occasio fusius possim explicare" had been inserted; Guildhall Library, London MS 1757.12, f. 113v.

[291] The corresponding Latin text "multarum utilium et valde necessarium scientiae partium" had been corrected in "plurimarum scientiarum"; Guildhall Library, London MS 1757.12, f. 113v.

[292] The English "it" translates the Latin "haec nostra ⟨Algebra scilicet⟩"; Guildhall Library, London MS 1757.12, f. 113v.

[293] Deleted: of. In the corresponding Latin text "circa quantitatis scilicet Proprietates" had been inserted; Guildhall Library, London MS 1757.12, f. 113v.

[294] After the corresponding "characteribus," "symbolorum" had been inserted in the Latin text; Guildhall Library, London MS 1757.12, f. 114r.

least ‹of all› confound and bury or obscure the first grounds or principles,[295] upon which the ratiocination or inquiry is built, and that, which in the series of the ratiocination[s] does most distinctly and plainly appear, and therefore in the cho[o]sing the characters[296] both for signifying the quantities or matters wrought upon, and also for signifying the manner of operation or the method of proceeding, those certainly are the best and most of all to be preferd, which are the most simple, plain, obvious and short: And for the method of planning them or proceeding in the inquisition, that is certainly most to be preferd, wherein the succeeding does depend upon the immediately preceding, or at least, wherein, each part or proposition or deduction depending either upon some principle or deduction preceding, or upon some formerly demonstrated or evident axiom, is so markd, that by the same cast of the Eye as it were, the deduction itself and the reason of that deduction is evidently to be seen. For thereby[297] the reason and understanding of a man upon an Enquiry is lesse disturbd, and the memory les[s] burthend with a needles[s] anxiety in the returning and producing of those deductions, that have preceded. **'f. 66r'** Vieta therefore in his method of proceeding by Species, is much to be preferrd before all the ways of Cossick Algebra or numb[e]ring Algebra, that was known before him[298] ‹because in the working of numbers the first heads[299] and principles are quickly lost and buried, ‹whereas› in the operation by species both the species and the manner of the operation is preserved intire[300]›.[301] And the method of our English Harriot is as much to be preferrd before his, as his was to be preferrd before that of the Antients. Therefore the method of Oughtred, wherein he introduces severall short and significant characters[302] as in his Demonstration of the 10 Book of Euclid is somewhat better than Harriot[']s: but in his methods of proceding in the Invention of problems in his Clavis (though he succeeded Harriot) is yet worse; for therein his method both of marking the lines or proposed quantitys by double characters, and his introducing and intermixing severall words, and his not putting down each

---

[295] After the corresponding "fundamenta siue principia," "sive symbola" had been inserted in the Latin text; Guildhall Library, London MS 1757.12, f. 114r.

[296] In the Latin version the corresponding "characterum" had been replaced by "symbolorum"; Guildhall Library, London MS 1757.12, f. 114r.

[297] The English "For thereby" translates the Latin "Haec quippe methodo"; Guildhall Library, London MS 1757.12, f. 114r.

[298] The phrase "that was known before him" has not correspondence in the Latin version; Guildhall Library, London MS 1757.12, f. 114r.

[299] The corresponding term "capita" had been deleted in the Latin text, where "prima principia" is followed by "‹vel primae› quantitates, sive characters et symbola primo posita et simplicissima," here omitted; Guildhall Library, London MS 1757.12, f. 114r.

[300] The English text "‹whereas› in the operation by species both the species and the manner of the operation is preserved intire" translates the Latin "In operationibus vero speciosis vbi ‹symbola› primo posita et constituta non variantur, et species et operationis modus seruentur incolumes"; Guildhall Library, London MS 1757.12, f. 114r.

[301] Endorsed in lefthand.

[302] In the Latin version the corresponding "characterum" had been replaced by "symbolorum"; Guildhall Library, London MS 1757.12, f. 114v.

deduction[303] or proposition in a distinct Line is perplex and difficult. But Des Cartes, though in truth the greatest part of his Algebra be the very same with that of our Harriot, which was long before, yet for his method of ranging his demonstration he does excell Harriot. But Herigon['']s way to me seems much better than either. For he allows a distinct margin, wherein to set down the reason of each deduction or how it was produced, i.e. by what operation and from what definition, postulation, axiom or proposition, or Corollary. And (which for this and severall other particulars is very excellent) a method I have seen of Dr Pell,[304] is to be preferrd before all the rest, that I have yet met with, both for the significancy of the Characters, and the plainnes and regularity of his method of proceding and deducing.

The first thing therefore to be lookd after in Algebra is a most plain, simple, short and most significant character,[305] whereby both the quantity and operation and effects[306] may be most distinctly, plainly, briefly and significantly exprest. First, for the first quantities or principles we begin with, because we consider each, we have occasion to make use of, as one distinct thing, it is best to signify it by some small and single character, as by the simple character of some one letter of some Alphabet, which is the way that hath hitherto obtained precedency in the world,[307] because of the conveniency of the types for printing, or else by any number of invented characters,[308] that shall each of them have their significancy[309] so as to signify more evidently what they stand for, which the letters indeed of the Alphabet may also doe. But the smaller and shorter, ‹and›[310] the more significant or plain[311] the character be, the better.[312] Mr Oughtred['']s way therefore is good, where all consonants[313] signify

---

[303] Deleted: and.

[304] In the corresponding Latin texts "Doctoris Pellij" is preceded by "praestantissimj illius mathematici ‹Nostratis›," here omitted; Guildhall Library, London MS 1757.12, f. 114v.

[305] In the Latin version, the corresponding "characteres" is followed by"‹simbola› sive signa," here omitted; Guildhall Library, London MS 1757.12, f. 114v.

[306] The following part of the corresponding Latin text had been here omitted: "et quicquid tandem notatu dignum sit"; Guildhall Library, London MS 1757.12, f. 114v.

[307] The English "in the world" translates the Latin "inter Doctos"; Guildhall Library, London MS 1757.12, f. 115r.

[308] The following part of the corresponding Latin text had been here omitted: "sive pro re nata inventos sive constitutos"; Guildhall Library, London MS 1757.12, f. 115r.

[309] The corresponding Latin text, "habeat significatum," is followed by "‹vel› ad placitum, vel positive distinctum et determinatum," here omitted; Guildhall Library, London MS 1757.12, f. 115r.

[310] Corrected: or.

[311] The corresponding Latin text, "brevioresque et significationis," had been deleted; Guildhall Library, London MS 1757.12, f. 115r.

[312] The following part, here omitted, had been inserted in the Latin version: "et ‹ad hunc vsusm› magis ‹accomodati›"; Guildhall Library, London MS 1757.12, f. 115r.

[313] The Latin "majuscule B C D F G etc.," followed by the inserted reference to an unpreserved figure "ut videri est in prima schematis linea et," had been here omitted; Guildhall Library, London MS 1757.12, f. 115r.

the known quantitys,[314] and the vowels, the unknown,[315] and if there ‹be›[316] two unknown quantities, that are one bigger than the other, then A alwaies signifies the greater, and E the lesse. Some of the Consonants also have[317] distinct[318] or appropriat[e] and universall significancy, as Z alwaies signifies the summ of those two unknown quantities, and X the difference, Z with a comma the summ of their squares, and X with a comma the difference with their squares. Thus Harriot[']s way is good, where the small characters of the vowels a e i o u y[319] signify the unknown quantitys, and the small characters of the consonants b c d etc. signify the known quantitys or coefficients in the operations. Des Cartes['] way also is much of the like nature, who takes all the first small characters of the letters of the Alphabet[320] for the known quantitys, as[321] a b c d e f etc. and[322] z y x v t etc. the small characters of the latter letters of the Alphabet for the unknown quantitys. It does much facilitate also the operation and help the memory in the retaining and readily producing the signification of each mark[323] 'f. 66v' if either the lines, superficies or quantity,[324] they are put to signify, be marked with the same character, or, if the lines, or surfaces or bodies have any appropriat[e] names, then it helps the memory much, if the first letter of that Name be set to signify the thing, but with some point or mark, to shew that it hath a peculiar signification, otherwise it may tend to confusion rather than plainnes and distinction,[325] as if for marking the sides of a right angled triangle B signified the Base, p the perpendicular, and H the Hypothenusa; or if for making the lines in a parabola O signified the ordinate, P the parameter, I the incepted part of the

---

[314] The corresponding Latin text reads "quantitates aliquas cognitas et definitias"; Guildhall Library, London MS 1757.12, f. 115r.

[315] The corresponding Latin text reads "characters vero mjusculi vocalium Scilicet A E I O U quantitates incongitas siue quaesitas"; Guildhall Library, London MS 1757.12, f. 115r.

[316] Corrected: therefore. The following text in the corresponding Latin version had been here omitted: "Praeterea etiam hac methodo praestat quod"; Guildhall Library, London MS 1757.12, f. 115r.

[317] The corresponding Latin verb, "habent," had been corrected in "assignat"; Guildhall Library, London MS 1757.12, f. 115r.

[318] Deleted: and.

[319] In the corresponding Latin text the reference "ut dispositi sunt in 3 linea" had been inserted; Guildhall Library, London MS 1757.12, f. 115r.

[320] The corresponding Latin text reads "alphabeti Italicj," correcting "Latinj"; Guildhall Library, London MS 1757.12, f. 115r.

[321] In the corresponding Latin text the reference "in eadem 3 linea" had been inserted; Guildhall Library, London MS 1757.12, f. 115r.

[322] Deleted: x.

[323] The corresponding Latin term was followed by "vel notatione," corrected in "symbolorum," here omitted; Guildhall Library, London MS 1757.12, f. 115r

[324] The English "quantity" translates the Latin "corpora"; Guildhall Library, London MS 1757.12, f. 115r.

[325] In the corresponding Latin version a reference to Wallis' work had been inserted and then deleted: "quae method vtitur clarissimus Wallitius"; Guildhall Library, London MS 1757.12, f. 115v.

diameter, and T the Tangent.[326] Or, indeed, in the examining of Geometricall Problems it does much help the memory and much les[s] disturb the ratiocination, if the Ends of the Lines of the Scheme be markd with the severall Characters used now commonly for writing shorthand, and so each line may be markt with 2 letters, and yet but with one small and short character, which much helps the memory and imagination, or suppose these[327] the characters of the Letters of the Alphabet b, c, d, f, g, h, k, l, m, n, p, q, etc. ⟋ ⟨ ⟩ ⌒ ♪ ⟩ ⌣ ⌒ ⟍ ⟋ ⁓ ♪ etc.›.[328] And a problem propounded to be solvd is this: the line BC being given out in D, and a perpendicular being drawn from C, it is desired to draw a Line from the other end B to cutt the perpendicular so in F, that FC and CD be equall to BF. Suppose according to Des Cartes['] way BD be calld a and CD be calld b, and FC be markt x.[329] I say, the symbols will be as short and stand as close and handsomly together, if every line[330] be markt with the c[h]aracter of the letters,[331] that terminate the ends of the Line, as if it were only markt with one single character in the common way[332]; as if instead of a[333] be put ⟩ instead of b be put ⟨, instead of x be put ⟨[334] with a point over it, to denote the unknown quantity[335]; and then the working of the Equation will be much easier, there being lesse burthening of the memory, especially when the ratiocination proceeds geometrically[336]: but in arithemeticall questions, where the

---

[326] In the Latin version the triangle's perpendicular, parabola's parameter and tangent had been denoted, respectively, by P, T and t. Here the corresponding symbols employed are p for the perpendicular, P for the parameter, and T for the Tangent; Guildhall Library, London MS 1757.12, f. 115v.

[327] In the correspondent Latin version a reference to "symbola in 5ª linea schematis" had been inserted; Guildhall Library, London MS 1757.12, f. 115v.

[328] Symbols inserted in correspondence of the letters. These symbols do not appear in the corresponding Latin version, where the letters had been replaced by the reference to symbols "suprapositi in 4 linea"; Guildhall Library, London MS 1757.12, f. 115v.

[329] The whole description of the problem had been deleted in the Latin version; Guildhall Library, London MS 1757.12, f. 115v.

[330] In the corresponding Latin text a reference to the "sc[h]ematis Problematis inservientis" had been inserted; Guildhall Library, London MS 1757.12, f. 115v.

[331] Deleted: of.

[332] In the corresponding Latin text, between "modum" and "vulgarem," "cartesianum sive" had been inserted; Guildhall Library, London MS 1757.12, f. 115v.

[333] Instead of "a" the corresponding Latin texts reads "(ab)"; Guildhall Library, London MS 1757.12, f. 115v.

[334] The symbols here employed are different from those employed in the Latin version; Guildhall Library, London MS 1757.12, f. 115v.

[335] The correspending Latin text ("‹una› cum puncto siue accentu superius ad quantitatem incognitam denotandum tumque") had been replaced by "et ad quantitatem incognitam denotanduam adijciatur punctum siue accentus supra symboloum"; Guildhall Library, London MS 1757.12, f. 115v.

[336] In the corresponding Latin text "uti experienti constatvit" had been inserted; Guildhall Library, London MS 1757.12, f. 115v.

ratiocination or[337] operation proceeds after the manner of numbers,[338] in that case it is indifferent, whether by single letters or by double characters. But, if we proceede by Descartes['] method, which is indeed very rationall and good, that is, to reduce all kinds of quantities whatsoever, whether continued or discret, to lines,[339] because whatsoever proportion there is ‹between›[340] any 2 bodies, surfaces or numbers,[341] the same may there be of lines: In that case I think it much better to make use of characters, than letters: and the letters may serve ‹to denote› the various ways of operation.[342]

The next to be taken notice of in Algebra, is the relation this quantities may have to one another. And these are to be noted with distinct and short characters, such as may most plainly denote, what they are set to signify, and may least confound the species, to which they are subservient. The relation is either comparative, or conjunctive, or [343] cooperative: the Comparative is either defind or undefind. Undefind is, when one quantity is uncertainly bigger, or uncertainly lesse than another, and here the marks are [344] › for bigger, and < for les[s], as a › b signifies a is bigger than b etc..[345] When the relation of quantities to one another is defind, they are either equal or proportionat[e]; the first is markt by Harriot with =, as A = b, ‹i.e.› A is equal to be, the 2^d is markt by Oughtred with [346]::, as a·b::2·3 i.e.[347] a has the same proportion to b that 2–3 which proportion also is usually noted with symbols, as Oughtred makes use ‹commonly›[348] of [349] R, S, T, V etc. Des Cartes indifferently of

---

[337] In the Latin version the corresponding "ratiocinatio siue" had been deleted; Guildhall Library, London MS 1757.12, f. 115v.

[338] The English text "after the manner of numbers" translates the Latin "modo quo in arithmetica vulgari"; Guildhall Library, London MS 1757.12, f. 115v.

[339] In the corresponding Latin text "rectas" had been inserted; Guildhall Library, London MS 1757.12, f. 115v.

[340] Corrected: to.

[341] The corresponding Latin text after "numerosve" reads "sive inter quasuis duas quantitates," here omitted; Guildhall Library, London MS 1757.12, f. 115v.

[342] The last part of the corresponding Latin text ("vt ‹post hac› tradam") had been here omitted; Guildhall Library, London MS 1757.12, f. 116r.

[343] Deleted: sub.

[344] The corresponding Latin text includes a reference to Harriot, here omitted. Instead of "here the marks are," the Latin text reads "ad hos Respectus exprimendum Harriottus vtitur hisce notis"; Guildhall Library, London MS 1757.12, f. 116r.

[345] The corresponding Latin texts "et a < b significant a esse minorem quam b" had been here omitted; Guildhall Library, London MS 1757.12, f. 116r.

[346] In the corresponding Latin text the symbol:: had been replaced by the following text: "‹4or punctis quadrate dispositis› nota hâc post eas quantitates positâ, quae isto respect afficiuntur, cuj nota, propriortio siue respectus postponitur"; Guildhall Library, London MS 1757.12, f. 116r.

[347] The text endorsed in lefthand in the Latin version ("‹in 7a linea›") had been here omitted; Guildhall Library, London MS 1757.12, f. 116r.

[348] Corrected: sometimes.

[349] In the corresponding Latin text the term ‹literis› had been inserted; Guildhall Library, London MS 1757.12, f. 116r.

⟨any⟩ others expresses the proportion of quantities after the manner of fractions[350] $\dfrac{a}{b}\ \dfrac{c}{d}$, which were a very good and distinct way, and would [shed] a great light into the nature of proportions, if only the mark[s] of proportion were interposed as $\dfrac{a}{b} :: \dfrac{c}{d}$ i.e.[351] a is to[352] b as c to d, or b to a as d to c, or b to c as d to a or a to d as c to b.[353] A 2^d respect of quantities one to another is, as they do augment another ⟨preceding quantity, or doe detract from it, that quantity do[e]s add itself or an equal quantity with itself to the preceding quantity, that has the affermatif or copulative signe + prefixt to it so a + b i.e. b is added to a.[354] And that quantity does diminish from the preceding quantity, a quantity equal to itself, that has the negative, or diminishing signe – prefixt to it, as a – b signifies the quantity of b to be taken out of a that is, a lessend by b as much therefore as b is more than a by so much is a by this symbol denoted worse than nothing[355]⟩.[356]

Royal Society Classified Papers, vol. XX, ff. 171r–174r

'**f. 171r**' July 3^d 1689. read the 24[357]

I discoursed the Last Day concerning an Experiment of the Penetration of two liquors one into an other, which though it be properly a physicall Experiment yet tis of that Nature that It is not Sensible of it Self without the help of Art. The like may be sayd of multitudes of other physicall ⟨powers⟩ operations or effects, which though they may be oftimes very vegete and Active in themselves yet the Reason of their insensible manner of operation we have noe meane of comming to the knowledge of them till by Some Lucky hit or Accidentall observation we find a medium to Discover that Secret way of working and Soe to make it affect Some of our Senses. Thus the Virtue of the Loadstone in attracting of Iron becomes Sensible by

---

[350] The corresponding Latin text reads "Cartesius vero omnibus indifferenter. Exprimunt alij nonnulli proportionem quantitatum, more fractionum"; Guildhall Library, London MS 1757.12, f. 116r.

[351] Deleted: as.

[352] Deleted: bc.

[353] The corresponding Latin text reads "quo significatur a se habere ad b vt c ad d vel etc.," followed by the verbal explication of the proportion, here omitted: "id est vt quaeuis quantitas superior ad quamlibet inferiorem ita altera superior ad alteram inferiorem, sive ut quauis quantitas inferior ad quamlibet superiorem ita altera inferior ad alteram superiorem"; Guildhall Library, London MS 1757.12, f. 116r.

[354] The following text of the corresponding Latin version had been here omitted: "siue a esse auctum per b, vel a et b"; Guildhall Library, London MS 1757.12, f. 116v.

[355] The Latin text of the lecture continues: "Alia signa Algebraica in proximum diem reservo"; Guildhall Library, London MS 1757.12, f. 116v.

[356] Endorsed in lefthand.

[357] In different ink.

seing a piece of Iron lifted up to it, and there kept Suspended by it; the Polar Virtue also of it or the making a needle of iron or steel conveniently suspended on the point of a Needle ‹after the end of the said needle hath been toucth or Rubbed on the polar part of the Stone› to turn and place it self in this or that position with Respect to this or that part of the Stone, and when that Loadstone is Removd to Respect the North and south parts of the Earth, this[358] Virtue of the Stone is not Discoverable to any sense, for we can neither by the seeing hearing Smelling tasting feeling perceive any such qualification in the substance of it, which is little Differing from some other kinds of Stones which have noe such qualification, and therefore it Lay[s] hid from the beginning of the world till to 2 or 3 ages since and was not known though the other vertue[359] or power of it were known for as Long a time as we have any Records of Naturall History. Again tis not to be doubted but that the pressure of the air hath[360] had its effects as long as the Earth hath been in being, and yet we find that Galileo was the first that found a way to make it Sensible by evacuating a cylindrick body first by a plug ‹drawn out by a weight›[361] and afterwards by quicksilver used in Stead of both the weight and plug, and this hath been since improved to Discover other qualifications of the air which as they were by the meanes then known altogether incapable of affecting the Sense Soe they became altogether Latent and unknown and Soe unthought of and not sought after.[362] Soe a further improvement of this medium or method of making this qualification more sensible might yet Discover other particulars which might be as Considerable as the Last improvement of it by the Barometer ‹was› for Discovering the changes and inclinations of the weather. For this purpose I invented and have here Discovered ways to make it as Sensible as it can be be Desired,[363] and if the magnifying of each primary inch of change into a foot were not Sufficient to Discover such latent proprietys motions or operations I have shewd a way how to make it yet more sensible by magnifying each inch to a yard ‹to›[364] to 10 nay ‹to› 100 ft if there be need of it and in truth to what magnitude can be desired and yet at the Same time to be[365] as easy in its way of motion and as sensible to the eye, that the motions that Shall happen in each inch of the 100 feet shall be as plaine and perspicuous as the same alterations[366] can be that are made in the primary single inch. This though possibly some may say is nothing yet I conceived by all unprejudiced men 't will be Looked upon as ‹the› highest improvement ‹that can be desired› of the way of Examining weighing measuring or corrupting of this qualification of the atmosphere namely of the variety of the

---

[358] Deleted: motion is.

[359] Corrected: virty.

[360] Corrected: have.

[361] Cf. Galileo (1890–1909), vol. VIII, 62–3.

[362] Deleted: A.

[363] Cf. Hooke (1726), 169–73.

[364] Deleted: a fathom or.

[365] Deleted: Soe.

[366] Deleted: would be.

pressure thereof from Severall causes which possibly may be yet Latent but are more Likely to be detected when the effects and operations are more fully known by Experiments And Observations made as they ought to be with a prospect of their use. Such Instruments and such Observations as this ‹I humbly conceive› might well Suite with the Reputation of this Society who are Expected by the Ingenious men of the World to Doe somewhat more than what is ordinary to ‹be› performd by any[367] private man.[368] Nor can it indeed be expected, that any one should Spend his otherwise more beneficiall time, in perfecting of such things as have not the Immediat[e] prospect of Benefitt, though they promise never soe fairly the Increase of **'f. 171v'** knowledge. We therefore find that the Greatest part of Learned men Respecting the Reward; Soon list themselves into ‹the› Societys of Divines, Lawyers or physicians, where their way to Canaan is already chalkd out. And if Some Straglers chance to be left behind the Caravan, they aim at Diverting their private trade to Some parts where they think there may be somewhat more then common advantages reaped.

We generally[369] Observe therefore that of the few that Remain for Experimentall philosophy, and to inquire into the knowled[ge] of Nature or Art, the Greatest part have been[370] Seekers after the philosophers Stone or the Perpetuall motion which every one ‹at first› thinks he has a[371] Prospect of, tho[ugh][372] ‹it prove[s] but a pisga sight and that› he never lives long enough to arrive at it. tis true ‹such seekers doe oft[en]›[373] discover many pleasant prospects in their way and find many curious experiments, but they are only Regarded as *in transitu* and not further sought into for other uses.[374] ‹we confesse› indeed we doe owe some very considerable Discoverys to these accidentall hits where they chance to be prosecuted to further improvements witnesse that Composition of Gunpowder, and Some other chymicall Inventions, but Alas how small ‹would[375]› the number of them[376] seem, if multitude of other Discoverys that have been light[ed] upon, had been followed as much as this I have named has had the ‹ill› Luck to be. And though new discoverys are generally at First sight and inquirers after them are generally Stigmatized with the names of Projectors and Scepticks because they are not contended to tread in the com[m]on tracts and to go round in the same way like the horse in the mill, yet some of their greatest enemys have been ashamed not to acknowledge that the world is beholding to them for many usefull Discoverys which none of the other multitudes of Endowed persons had ever made before them; As I could instance in some of the Adversarys

---

[367] Corrected: many.

[368] Corrected: men.

[369] Deleted: find.

[370] Deleted: (…).

[371] Deleted: (…).

[372] Corrected: the.

[373] Deleted: they.

[374] Deleted: tis true.

[375] Deleted: be.

[376] Deleted: it.

of the Deservedly famous Galileo. This man I the Rather instance in because he has given diverse Specimens of the benefit that may be made of very common and obvious experiments where industry and art is made use of, for the improving and perfecting of them to[377] Scientificall knowledge, how common and obvious was the experiment of the Vibrative motion of a pendulous body, and how little Regarded by all mankind before he took it into consideration, and yet we find what usefull Discoverys he made upon the contemplation of it, Soe that since that time clocks have been found to Divide and measure out aequall Spaces of time more exactly then the Sun it Self.[378] now what he hath Instanced in upon one[379] Common Experiment the Same may be Done[380] by 100 instances upon 100 other, if Art and Industry and Incouragement be not wanting, I think I have not been wanting to shew somewhat that may confirm the truth of what I now assert for I did first Discover the true cause of the motions of the Coelestiall bodys, by the contemplation of the circular[381] motion of a pendulum,[382] and the Ovall figure of the earth and Sea from the Contemplation of the effects of Vertiginous motion,[383] And thought Des Cartes Mr Hobbs and a great many others would make the gravity of bodys, to be causd thereby, yet I first shewd and proved that the quite contrary effect was caused by that motion.[384] I could instance in many other such consequences Drawn from the Contemplation of Seemingly very triviall and Obvious Experiments and Observations. But these may at praesent suffice to shew that many of those experiments which at first sight may be slighted may yet contain information enough to found the Discovery of Some very usefull part of[385] knowledge. Which if it be improved to the utmost may prove as considerable as the Rectifying of clocks and watches. Now this Discovery might easily be proved by pendulum clocks tryed in Differing parts of the world. As also in a great measure by the Barometer if rightly observed. Moreover.

The Barometer if made use of at Sea might certainly be improved to great benefit for Navigation, for[386] forwarning the Mariner of Stormes or calmes and of other alterations in the weather and Seasons besides that an account thereof truly kept would give great light for ‹making› a true theory of the height forme and gravitation of the Atmosphaere Now as it is[387] at praesent generally made use of at Land, for that purpose Soe it might with as much ease certainty and nicenesse be made use of

---

[377] Deleted: (…).

[378] Cf. Galileo (1890–1909), vol. VIII, 140.

[379] Deleted: Exp.

[380] Deleted: by the.

[381] Deleted: or Elliptical.

[382] Cf. Birch (1756–57), vol. I, 505–7; vol. II, 90–2.

[383] Cf. Hooke (1705), 346, 349, 351–3, 356, 363.

[384] Cf. Ibid., 183.

[385] Deleted: Kno.

[386] Deleted: prod.

[387] Deleted: made use of.

at Sea. if art and Industry[388] were not wanting. Which would not neither be long[389] Soe if Incouragements were not wanting to excite them and Malicious indeavours did not ‹eagerly› promote Discouragements.

'**f. 172r**' I have formerly propounded here another[390] problem about the alteration of the[391] vertiginous axis of the ‹Earth›[392] I doe not wonder to find some indeavouring to contradict it ‹as I have found others doe all my other proposalls› but I should be glad to see it either proved or Disproved, really and Sub[s]tantially as it ought. For I aim at nothing but to know the truth and have noe more concerne to have it be as I have suspected it then to have it fixt and Immovable. However I must needs Say that Nothing that I have yet heard alledged to the contrary, has any thing of argument to me[393] for one observation certainely made that shall prove it to have *varyed* but 1 min, will be more Significant then 1000 such arguments for the Stability thereof; There are not wanting Authors enough who before the Descovery of the variation of the Variations had asserted that Invariable but how Little Did all their Authorijs Signify after that Observations had ascertained the Contrary. the like may be instanced about the opinion of the Antipodes, and even against the Doctrine of Gravity and the motions of the Heavens. And yet the truth doth at Last praevaile against all gain sayers. I have shewd what ways and meanes there are of bringing these Doubts to a certainty which whosoever has a concern for the knowing of the truth ‹and›[394] is free from praepossession and praejudice ‹or interest to the contrary› may try, it seemeth to me a necessary consequence from my theory of things, but if it prove otherwise the theory will need some alteration. Homo sum nihil humani[395] a me alienum puto yet ‹since I have found that› severall other consequences have proved true I shall therefore not[396] Despair of this till Experiment has Evinced the Contrary which may possibly be worth the tryall. But It must be done with somewhat more then ordinary care or it will prove ineffectuall.

However I[397] am of the opinion that nothing can be more worthy the contemplation of this Society then things of this Nature, that is I meane of Stating and Determining the Limits boundary and Standards of knowledge. as to Determine the praecise Latitude of this place for this time, and the praecise meridian Line and this not to minutes only but even to Seconds. to determine the praecise Length of a pendulum that vibrates seconds of time by the fixt Starrs. or the 86,400 part of the time of the Revolution of a Starr. and the like for these would be as touchstones or

---

[388] Deleted: ‹as suitable (…)›.

[389] Deleted: ‹soe› wanting it.

[390] Deleted: qu.

[391] Deleted: mag.

[392] Cf. Hooke (1705), 345, 353.

[393] Deleted: ‹and they seem to savour more of the spirit of contradicting then evincing the truth›.

[394] Corrected: who.

[395] Corrected: humanum.

[396] Deleted: who.

[397] Deleted: must.

Standards for future as well as ‹own› for the Praesent age. And they cannot be soe
well performed by any single person. (But ‹else being my private thoughts only I
submit them›[398] to the judgement of the Society who are ‹much› better able to make
a determination ‹what is most likely to be for the best›[399]) The Like I could wish also
as to the Instruments[400]: that the best of all kinds of such as are for philosophicall
Use might here be to be found, Such as the best Telescopes the best of Microscopes
the best ballances the exactest weights and the exactest measures of Length Capacity
and time for while nothing but what is now common is here to be met with. Ingenious
Visitants Loose their Expectation and Soe their Esteem of this Most Excellent
foundation.[401]

whereas if whatever were here to be seen did as much exceed what is[402] to be
found elswhere as the fame and Reputation of this Society doth Exceed that of any
other, it would be not only a means of Increasing that Esteem, but[403] it would produce the consequences of it Namely the power of procuring from all Learned[404] as
well as from the unlearned whatsoever were desirable and should be found necessary for the promotion of Naturall Knowledge[405] in the world or in the parts thereof
that are already discovered. And which is yet much more considerable it would put
the Society it Self in a capacity of Discovering[406] new worlds or New parts of it yet
unknown, and possibly ‹now› as Little Dreamt of as America was before the
Suggestions of Columbus, for who has Stated the maximum or the minimum of this
Inquiry? 'tis Sufficiently plain how great an accession of Sensible evidence; the
praesent improvements of telescopes and microscopes have added to what was
known in ‹any›[407] praeceding age, and how much would that be Increasd if others
were found to exceed these as much as these doe the naked eye? which ‹Expectation›
though it may seeme to Depend upon a very unlikely Supposall, yet I conceive we
‹may› have reasons enough to Remove Despair Since I conceive it much more easy
to Improve a thing already Invented then[408] twas to invent the Instruments now made
use of. as the Discovery of the Rest of the parts of the earth have been more easy
then that ‹first› of Columbus[409] in the west Indies. The only[410] cause of fear Seems
to be the praesent Genious of such ‹patrons or› potentates as were able to putt

---

[398] Corrected: to submitt any thoughts.
[399] Corrected: (…).
[400] Deleted: of the.
[401] Deleted: (But (…) and
[402] Deleted: (…).
[403] Deleted: wo.
[404] Deleted: (…).
[405] Deleted: (…).
[406] Deleted: new wo.
[407] Corrected: the.
[408] Deleted: (…).
[409] Deleted: his Discovery.
[410] Deleted: fear.

forward Such an Expedition, which seems to be inclind to[411] undertakings of a contrary Nature, and to Destroy and Obliterate what is already known ⟨rather⟩ then to put a helping hand to any thing that may promo[te] it **'f. 172v'** Read before the Royal Society July 10, 1689. Praesent Sir John Hoskins. Mr Hill. Mr Henshaw. Mr Aubrey. Dr Sloane. Dr Mullen Dr Mills. Mr Saint George Ash. Dr Harwood. Mr Aubrey. Mr Hall[e]y. Mr Weeks. Mr Hunt. Sir John Hoskins said the Society would willingly be at the charge of any thing I should propound. and that if I did not doe all I was able I should be the Occasion of the Society[ˈ]s Ruine I answerd I would doe as much as any to uphold and more then I was obleigd to etc.

I did the Last day with Submission propound some things which I conceived were very desirable to be effected and such as I judged[412] did Lye within the Limitts of possibility. For that I conceive with my Lord Verulam that all those things may well be supposed possible and performable which may be accomplished by some one person, though not by every one, or which may be done by the united Labours of many though not by the single Labour of any one by himself or which may be effected in Severall ⟨years or in some⟩ ages though possibly not in one ⟨or two⟩.[413] or what may be finisht by the care and Charge of the Publique though not by the Abilitys and Industry of Private Persons. what ever therefore is necessary or at Least very desirable and of good use if Obteind when it Lyes within the Limits of possibility ought to be attempted in those ways or at Least where there Seems the most probability. Though it often times happens ⟨too⟩ that many things are Discovered by persons and ⟨by⟩ ways where there seems[414] the Least probability but they are accidentall and not to be relyed upon. However all the ways of Possibility may Lye some times within the Reach of some one, who by his own abilitys and Interest may be able to ⟨Excite and⟩ Sett all the other possible agents at work where the nature of the thing will conforme to the Limitts of time ⟨but where things cannot be hastned⟩ As in observations of Comets or meteors ⟨or such like⟩ which are the production of Nature and fall not within the Command of Art. to praescribe or Limitt their appearances or Actings ⟨there desine and art must cease⟩. But Art affords scope enough for Usefull and Necessary Disquisitions ⟨in such part of physick where it may be applyed⟩ And those other phaenomena of Nature which are but Rare are generally Lesse considerable if we respect the increase of Such a Scientificall ⟨naturall⟩ knowledge as tends to practise. There are *qu[a]esita* enough to employ all that are willing to search and to enquire into the Nature of Bodys within our power, of which ⟨as yet possibly⟩ we know very Little as we ought; that is Determinately Experimentally and Scientifically. ⟨for instance⟩ Even the Air Though it may seem to have been exhausted even to a vacuity; may yet be found a plenum of other qualitys, Vertues ⟨uses⟩, or powers than those we have ⟨hitherto⟩[415] touched upon or

---

[411] Deleted: the.

[412] Corrected: conceive.

[413] Cf. Bacon (2004), 450–1.

[414] Deleted: (…).

[415] Corrected: yet.

‹thought of›. the like may be sayd of the water. And much more of the earth and Earthly bodys. And if we Comprise Mineralls, Metalls, Stones, etc. It seems boundlesse. What have we yet concerning the Nature and Vertue of Plants but what is meerly Conjecturall, Unlesse it be ‹the naming and description of their outward form and› the application of Some parts of them to ‹some› Mechanical uses, or such uses as Nature it Self has dictated for the food of Animalls. But as to their vertues or qualitys for[416] Medicine or ‹for many› other physicall ‹and mechanicall› uses we are yet in the Dark and the utmost that is known concerning them seems to be very Superficiall and very uncertaine, if compared with what is yet unknown which I Conceive Lyes within the Limitts of possibility to be detected and made Evident.

Againe if we inquire for the Quaesita or the Desideranda about the Parts of Animalls. And above all the rest[417] concerning the true forme[418] of the parts of Men the qualifications of humors and the use of All the more solid as well as the more fluid ‹parts› we shall find that Even this part of Science which seems to have been the most Cultivated and the most examined of any, because of the future reward and benefit that is Expected from such a knowledge, ‹is yet imperfect and that› yet there are many parts of this thick ‹and Spatious wood or›[419] forrest; that have Never been passed or Seen, which the many Late Discoverys that have been made will make somewhat the more probable. But the ‹stil[l] Remaining› Praevalency of Diseases Notwithstanding all the Doctors[*] Recipes will make ‹it yet more›[420] Evident. The discovery of all which I suppose to ly[e] within the Limits of possibility

'f. 173r' Againe for the forme Reason and Nature of all Sensible qualitys of Bodys, how much are we still to Seeke? where have we a full and satisfactory account of Light and Colours and of their causes, effects, powers and operations? Or of Sounds with all the varietys of Harmonious or inharmonious whether musicall or Not musicall with their powers and causes and the phaenomena they produce? The like may be ‹inquired›[421] of tast[e]s ‹2›[422] and Odours ‹1›[423] and of all tangible qualitys such as heat and Cold drynesse and moysture Gravity and Levity[424] Fluidity and Solidity, Expansion ‹contraction. Elasticity and mouldablenesse›.

But Lastly for the Insensible qualitys ‹and proprietys of bodys› that is such qualifications and powers of bodys as doe not Immediatly affect the sense, but are to be Discovered by other meanes and contrivances of art, these are as yet innumerable and of such a Nature as 'tis not in the power of any yet Living to Limit or Determine,

---

[416] Deleted: Physick.

[417] Deleted: of.

[418] Deleted: of.

[419] Corrected: and Large.

[420] Corrected: (…).

[421] Corrected: sayd.

[422] Inserted above "tasts".

[423] Inserted above "Odours".

[424] Deleted: and the like.

but there is Left Room enough to take up and imploy the Industry[425] art and Invention of as many as Can be found[426] well willers ‹to›[427] Act[428] the[429] Search ‹after them›. Such inquirers would meet with hints enough to suggest new methods and ‹new› mediums of Examination and those would increase the Limits of Possibilitys beyond those of which we ‹now though erroneously› conceive we have the utmost prospect. For the Limits of Naturall knowledge are as infinite and boundlesse as the quantitys ‹or numbers› of Nature['']s productions which admitts[430] of a maximum,[431] noe[432] more than ‹they doe› of a minimum[433]; and After all there[434] will be a plus ultra ‹left› for future ages to exercise their Industry ‹in[435] attaining to and passing through and beyond any pillars that shall be erected by the greatest Hercules in Natural philosophy›.[436] And there will Still be a praemium to excite hope and Expectation of the usefullnesse and[437] Benefit of that which remains undiscoverd beyond thule.

Now for the making of all these Inquirys as they ought there will be necessary the assistance of the two handmaids (as the Lord Verulam is pleasd to call them) of Physick namely Logick and mathematicks, without which, Experiments, tryalls and Examinations or Inductions from them, will be very Lame and Insignificant. And therefore though he be pleased to make Mathematicall Learning to be but an appendix ‹or auxiliary› to physicks metaphysicks mechanicks and magicks, which he sayes he was in a manner compelld to doe for the wantonnesse and arrogancy of mathematicians who could be content that this Science might even command and Overrule physicks, and did boast its certainty beyond ‹them›[438] and Soe took upon ‹it›[439] a kind of Power and dominion. Yet ‹That science› is[440] not however Lesse to be Regarded because of the Praejudice he ‹seems to have› had[441] against it, nor from his decree ‹against it, is to be slighted or› degraded from its Due Rank and quality ‹as one of the first and most necessary of all humane inventions for exercising the

---

[425] Deleted: and.
[426] Deleted: to be.
[427] Corrected: and.
[428] Corrected: Active in.
[429] Corrected: this.
[430] Deleted: not.
[431] Corrected: minimum.
[432] Deleted: of a.
[433] Corrected: maximum.
[434] Deleted: (…).
[435] Deleted: (…).
[436] Corrected: (...) or inquiry concerning it.
[437] Corrected: of.
[438] Correted: it.
[439] Corrected: them.
[440] Corrected: tis.
[441] Deleted: to the.

faculty of Reasoning›. For[442] whoever will Duely Consider of it, will find physicks to be but some branches springing from The Stock of Mathematicall knowledge ‹in the same manner› As Opticks, musicks, Astronomy, Geography ‹mechanicks› and all the other parts of Mixed mathematicks ‹are now acknowledged to be, they being all indeed but sproutes shooting forth and growing from the stock›[443] and Root of Geometry. ‹whence›[444] it will follow that all Such Experimenters or Inquirers as are not first fitted where those parts of knowledge will be of Little Significancy, And their productions ‹will prove but› like ‹the seed springing up from stony ground which will soon wither and dye›.[445] Of which Plato was well aware when he forbid all not Qualifyd with Geometry to enter into his Schole.[446] And ‹therefore also› Aristotle esteemed of it as Highly usefull and necessary to produce a practicall and Mechanicall or Reall knowledge, for till our physicall knowledge comes to be of that nature it is for the most part imperfect and Unintelligible and ‹either only›[447] verball or at best Empiricall. And though we can give the[448] Derivation of the Name, or can ‹fore›tell some certaine effects yet we have '**f. 173v**' noe Notion or Conception of the true cause and Reasons of those effects. and by what power and in what manner those effects come to be produced. And thence it comes to passe that such knowledge is generally barren and not productive of any further light or knowledge. Whereas if the causes be examined into and Detected those will foreshew what other effects are to be expected from the concurrence of the like agents upon the like patients. Thus one that tells me the Names of all the Various kinds of figured petrifications, and whence those names are derived and why they have been given to this or that kind of figured body As the Learned Aldrovandus has done in his Museum metallicum and afterwards tells me that they are formed by a plastick faculty that Immites in Sports the formes of the animall and vegetable kingdomes.[449] tells me noe more than Hesiod does in his Theogonia that every particular effect had a particular daimon or deity that tooke care of Producing it.[450] whereas if I can be informed certainly that these productions are nothing els but the Reall Shells, bones, teeth, or parts of Animalls or Vegetables praeserved by meanes of a petrifying water or Juice and that this petrifying water is produced by the effects of fire[451] by the calcination of certaine Stones or other substances that will produce such a kind of fixed and fixing Solution, and that wherever those substances are found there are also found other Symtomes of fiery eruptions, as vetriolate Salts and various Sorts

---

[442] Corrected: And.

[443] Corrected: (...).

[444] Corrected: And therefore.

[445] Corrected: castles built in the air or upon a Sandy and Sandy and uncertaine foundation.

[446] Biancani (1615), 45.

[447] Corrected: (...).

[448] Deleted: pedigree and.

[449] Aldrovandi (1648), 440.

[450] Hesiod (2006), vol. I, 8–59.

[451] Deleted: or the

of pyritae Or vast masses of Chalk which is[452] nothing els but a[453] Lime by Length of time and the Sokes of Raine slacked and concreated into a kind of Stone. this will give me a great stock of further inquirys and Discover the causes and Reasons of a great many other phaenomena, and direct me for the finding out the proper Indications of such metamorphosed Substances, besides many other hints to judge of the praeceding Mutations and Catastrophys that have hap[pe]ned in the world, and besides the explanation of severall phaenomena in the Celestiall bodys. Againe he that tel[l]s me that the fire is a particular Element like the Air but Lighter and more Rarefyed and placed above the air and under the Concave of the Solid orb of the moon, and that the Reason of its ascent upward is[454] its appetite to be in its proper place and that by a decuple condensation it becomes air as the air by a Decuple condensation becomes water but that these elements soe soon as they are sett at Liberty Return to their naturall constitution and to their proper places, he I says that tells me all this and more of the like kind as the Peripateticks doe, does seeme to tell me something but when I examine into severalls particulars of it and finde that there is noe ground for many of these assertions, as[455] when I find that the air though ten times ten times rararefyed is noe more fire[456] nor productive of it then it was before but rather the Contrary contributing much to the extinguish of fire, and againe the same air ten times ten times condensed is not any whit nearer of the nature of water but of the quite contrary contributing much more then before to the increasing[457] augmenting and prolonging of fire this tells me that this knowledge is not only barren but it is unsound and Rotten and tends only to error[458] darknesse and corrupting the truth. Whereas if by examination of severall phaenomena I perceive that what we call fire is nothing els but the Dissolution of certaine bodys by the air which works upon those bodys much after the same manner as Aqua Fortis [459] or other corrosive menstruums doe upon the metalls or other bodys they dessolve This explaines the reasons of[460] most of the phaenomena of fire and gives me many Hints or Heads of Inquiry for other Similar effects, the Like I might Instance in about the use of Lungs of Animalls or the use of the air for maintaining the Vital fire.[461] But I should[462] [not] Exceed the Limits of Examples which are sufficient when they Explaine evidently enough the Doctrine praemised.

---

[452] Deleted: noe other.

[453] Deleted: Stock.

[454] Deleted: by.

[455] Deleted: the.

[456] Deleted: than.

[457] Deleted: and.

[458] Deleted: and.

[459] Nitric acid (HNO_3).

[460] Deleted: almo.

[461] Deleted: of life.

[462] Deleted: wear.

'**f. 174r**' I shall therefore proceed to shew that[463] ‹many›[464] things may Lye within the[465] limits of possibility in the matters of Art and in mechanicks ‹as well as in physicks›. And here it is that the Industry and Labour of Men should be diligently Employed, for here is scope enough for usefull discoverys.

This doctrine, the benefits and advantages that the praesent age injoyes ‹from mechanicall Inventions› beyond any of the former will sufficiently Confirm; ‹as› witnesse the[466] Infinite number of Books produced by printing, the new ‹countrys›[467] discovered by the Magneticall compasse, the prodigious effects produced by Powder, the Great Discoverys in Astronomy by telescopes and of the texture of bodys by Microscopes, the Rectification of Clocks by Pendulums. and the like these I say doe sufficiently testify of what great benefit mechanicall inventions may be to[468] the praesent and future times wherein they are Discovered, and[469] are a suffi-cient argument to shew that there may be many others yet behind as Usefull which may not have ever been ‹yet› known to the world ‹nay nor soe much as thought of.› How many things are there in the possibility of mechanicks that the world is know-ingly desirous of performing; which yet none hath been able ‹hitherto›[470] to Effect as the finding of the Longitude of places on the Earth, both upon the sea and upon the Land ‹had been much desired›, and yet noe one has ‹hitherto›[471] produced it, how much is the Decuple improvement of telescopes to be desired. and has been Laboured for,[472] by divers Ingenious mechanists and ‹yet none has›[473] attained it. ‹Thus many things in shipping and navigation have been heretofore and they still remain Desiderata›[474] especially in cases[475] where its use is of greatest necessity as where the wind [476] drives a ship Directly upon a Dangerous Coast or shore ‹or when a ship is very Leaky in a storm and many such other cases›. Againe many have attempted to find ways of accelerating the motion of Ships, and they would doubt-lesse be of Great adveantage for[477] Discoverys as well as for trade, yet the advan-tages that have been hitherto produced have not very much exceeded the common

---

[463] Corrected: what.

[464] Corrected: (…).

[465] Deleted: possibility.

[466] Deleted: Great benefit of Printing, the Magneticall needle, The use of Powder the use of telesc.

[467] Corrected: worlds.

[468] Deleted: (…).

[469] Deleted: I.

[470] Corrected: yet.

[471] Corrected: yet.

[472] Deleted: the.

[473] Corrected: not yet.

[474] Corrected: the way of Sayling against the wind has been also much told of, and has been in some measure attained by gaining upon the wind by Sayling near it, but it may be certainly further improved.

[475] Deleted: of.

[476] Deleted: be.

[477] Deleted: trade.

swiftnesse of other vessells. There are multitudes of other desiderata in the bussinesse of navigation, which have not yet proved data or Inventa, As there are also in Severall other mechanick Arts, which yet I conceive to Lye within the compasse of possibility. ‹now if it›[478] be asked why[479] are they not effected,[480] I can only answer that[481] I[482] conceive[483] that those that have attempted[484] to discover them have either failed in the successe of Inventing, or in the expectation of Profiting themselves by them But neither of those ought to Discourage others from using their Indeavours[485] nor to make them despair of obteining a more prosperous Issue Since every one has the successe that is peculiar to his own fate.

Read before the Royall Society July the 24th 1689.[486] Present Mr Henshaw. Mr Lodwick, Dr Mullius, Dr Mills, Dr[487] Mr Saint George Ash. Mr Perry. Mr Waller.[488] Mr Hall[e]y, Mr Weaks, Mr Hunt.

Royal Society Classified Papers vol. XX, ff. 178r–179v

'f. 178r' In my Late Lecture[489] I have treated concerning two of the Branches of Naturall Knowledge which I Reckond among the Number of Deficient, ‹i.e.› which need to be further prosecuted and cultivated in order to their augmentation and Increase. The first was the procuring and Collecting of Information and Descriptions of All such Inventions, Arts, manifactures, or operations as are Extant and practised in Severall parts of the world, of which notwithstanding we are for the most part ignorant here at home. And ‹the› Second was the procuring the Description of the Naturall, Geographicall, and Artificiall History of Regions countrys Island Seas Citys, towns, Mountains, Deserts, plains, woods and the like places, parts, and Regions of the Earth, whether Inhabited and cu[l]tivated, or Not Habitable or unfrequented, which may yet afford great Information for the Explication of various phaenomena highly necessary to the augmenting and perfecting of Naturall philosophy. For that such instances doe often afford the Limitations and boundarys of

---

[478] Corrected: But ‹there› it may possibly.

[479] Deleted: inventions.

[480] Deleted: to which.

[481] Deleted: the only Reason.

[482] Deleted: can.

[483] Deleted: (…).

[484] Deleted: it have not gone.

[485] Deleted: (…).

[486] Cf. Journal Book of the Royal Society, vol. VIII, 271.

[487] Void space.

[488] Void space.

[489] Royal Society Classified Papers, vol. XX, ff. 176r–177v; Journal Book of the Royal Society, vol. VIII, 283.

Naturall operations and Effects. and were to be sett as terminj bound stones or as the Lord Verulam stiles them *Experimenta Crucis*, to inform the traveller in Naturall inquirys that here is the utmost limit, and that this way or that way there is a *ne plus ultra*.[490] and therefore he must turne an other way. But these parts of knowledge as yet Deficient must be obteined by the help of Such information as can be procured from such as may have opportunity to See and acquaint themselves with the truth of the praesent state of those things that are inquired: After for that the *autopsia* of any single person will make very small dispatch in soe Multifarious and busy a subject of inquiry. And I doe not at all doubt but that if fitting Means were made use of there might in a short ‹time› be reaped a plentifull harvest of Such Substantiall[491] Proper and pertinent materialls as would (when the chass[492] and tares were separated) be fitt to be put up into the Storehouse or Repository of usefull ‹memorialls or› provisions for the Magazine of Naturall Philosophy. But though the Harvest be great yet the Labourers are but few. Many are willing to Reap the Benefit, but few will Lend a helping hand, to alleviate the trouble. However on this occasion I shall produce one thing not impertinent to this particular to shew what might be done if there were an industrious prosecution of all opportun[i]ty[s] for collecting the Artificiall and Naturall productions of severall forrein parts.

But to Leave those to a more propitious Aspect. I shall proceed to Mention Some other Deficients for the producing of Naturall Knowledge which doe seem to Lye more within our ‹owne› power to Supply then the former and yet they are not Supplyed. and the[493] ‹first I shall name now› is ‹the collecting or producing› the Inventions and Discoverys that are made or might Or[494] ought to be made by Such as are either members of this Honourable Society, or Such other Inquisitive and Learned persons of this[495] City or Nation who have bent their thoughts Bodys and proprietys towards the advancement of this Noble and Usefull Designe of Improving Naturall Knowledge: for that such productions might be most properly appropriated to this Nation this Society, or this ‹or that› person, and is not to ‹be› accounted of forrein growth or manufacture as both the other will at best be. *Nam quae non fecimus ipsi, vix ea nostra voco*,[496] the Labour, Industry and art spent in obteining the production of forrein parts is very commendable and Deserves great Incouragment, but to the producing of[497] the Same or as good or possibly much better here at home Deserves much more and greater. This ‹I humbly conceive› is principally and before all to be done, though the other also be not to be Left undone ‹for that›, this

---

[490] Cf. Bacon (2004), 318–21.

[491] Deleted: and.

[492] Corrected: (…).

[493] Corrected: that.

[494] Corrected: and.

[495] Deleted: (…).

[496] Ovid (1916), vol. I, 238–9.

[497] Deleted: as goo.

produces a Stable Commodity that will passe in exchange[498] for, and will fetch in all ‹other›[499] Forrein productions[500] whether naturall or artificiall. And the perfecting of one such of our own production does I conceive more advance the Interest[501] Stock and Power or Riches of the Society then to the other way produced. Though I cannot but aknowled[ge] also that it may be more Difficult to Obtein then (it may be) to the other Way, and soe possibly some may say then that they are of aequal Value. and soe the Industry or Labour Imploy'd in either kind is aequally to be incouraged. Others possibly there are who (will doe neither the one nor the other themselves) will praeferr a forrein before a Native production, And 'tis a humor too frequent to be found even amongst Ingenious men and it was soe of old. A prophession not esteemed soe in his own Country or in his own time. But in[502] a forrein ‹country› and ‹in a time› after he is dead. **'f. 178v'** However I ‹humbly› conceive it will be much for the Benefit of this Society that as many inventions of their own as can be procured might be here produced and perfected, and not only Soe but Recorded and asserted to the true authours of Such inventions, Since that alone will be one Inducement to others to produce other ‹like› and ‹may be› better, as the contrary practice will be a great Discouragment, and hinderance to soe usefull and benefitiall an effect[503] for that few will produce any thing [of] this value if thereby they give others opportunity to arrogate the same to themselves. And the true author Looseth that which is the Least of what is due to him the Reputation and praise of the Successe of his Studies. There remain enough yet undiscovered usefull things which[504] be hoped to be procured from the Inquisitive men of this Learned age. if there be not encouragement wanting to prompt them to such inquirys. For Art as well as Nature has[505] power of producing ‹many› wonderfull as well as usefull effects, such as before they are Discovered are thought impossible, yet when they are soe it is much more our wonder that they had Layn soe Long undiscovered. What was easier to be thought of then some way of printing to make many coppys of the Same[506] Letters words or Sentences, at Least after the manner that we are told the Chinese practice. Since the use of Scales and Stamping of many has been soe long known and practised in the world. Yet we do not find any[507] hint of it till about 250 years since. And then 'tis sayd it came from China. Roger Bacon has in his Letter to the then pope and in his treatise De mirabilis potestate artis et Naturae[508]

---

[498] Deleted: for all forr.

[499] Correcterd: other.

[500] Deleted: and.

[501] Deleted: and Powers.

[502] Deleted: smaller.

[503] Deleted: but for the Honor and Advantage of Soe excellent an Instiution.

[504] Deleted: may.

[505] Deleted: (…).

[506] Deleted: writing.

[507] Deleted: (…).

[508] Bacon (1659); cf. *Bibliotheca Hookiana*, 53 n.75.

reckond up many wonderfull things which he at that time thought possible to be done yet we find noe mention of this, though he has in another Discourse mentiond the Invention to be practised by the Chinesse in an embassy to Russia which one would have thought had been intimate enough to a person soe inventive and knowing in Meckanicks as he seems to have been.

However since it hath proved Lucrifeorous as well as Lucifeorus the use and practice thereof has been improved to the Great benefit of mankind. Whereas others that are Lucifeours only doe seem to stand as alloy if not to be neglected. as the use of telescopes and Microscopes which are certainly instruments of Great benefit for the Discovery of the operations of Nature and might be yet more more improved[509] and used for those ends. Again the[510] contrivance of a watch to observe and keep an account of the varietys of the weather was once thought[511] very desirable yet the consequences of such an account being not praesently, to be perceived I feare will make it have the same fate. And yet most will grant that if from the help of such accurate certaine and Constant accounts of the Variety of Seasons and the Dependence or Consecutions of them a theory could be made to foreknow the[512] States and mutations thereof some considerable time before they arrive, it would be of Great[513] benefit to most men.

The Designe was first suggested by Sir Christopher Wren, as feasable to make a watch to mark the quarters of the wind[514] and the Degrees of Heath and cold in the air. But at the Desire of the Society I have made one by other contrivances to keep an account ‹not only› of the mutations of those two but also of Divers others to witt of the Strength or Swiftnesse of the wind as well as of the quarter ‹of it›, of the Drynesse and moysture of the air, of the quantity of the rain that falls and the times of its falling, of the pressure of the air and its varietys, of the heat and cold of the air ‹adjusted to the standard of freezing› and if the place were convenient of the Sunshine and cloudinesse of the Sky. and could have added some other if the watch had had a place adapted for their observation. But these I conceive are abundantly more than ever were yet observed, and 'tis impossible almost that they should be for a constancy both Day and Night by any other meanes but this of a watch. I deny not but that the particular contrivances of[515] Remarking them and also of applyng them to a watch may be varied and possibly for the better, that is they may be made to Distinguish both the degrees of such Variations and the parts of the time wherein they happen more minutely, but those possibly may be too curious or more then is necessary, and whatever is soe ought to be omitted however if that should be found necessary the same contrivance will performe the observations as often and as

---

[509] Deledet: for.

[510] Deleted: use.

[511] Deleted: a.

[512] Deleted: the.

[513] Deleted: (…).

[514] Deleted: thoug.

[515] Deleted: applying.

Distinct as shall be requisite but 'tis not good to charge it with too many operations unless they ‹be› of absolute necessity. As it might be made to Sound all the varietys that happen and when they happen as a clock doth strike the houres or quarters of time. But that would be nothing to the praesent Design of it and ought to be omitted as impertinent to the keeping a Register, the same might be made to mark the Day of the month or the month of the year etc. but that is Soe Easily Supplyed by the keeper thereof that an addition of that kind must be looked on as impertinent. Again it might be made to keep it self[516] **'f. 179r'** wound up. but that is Likewise soe easy to be done by him that shall Supply it with papers that that also is to be omitted it needing to be wound up but once in a week. It has in it some advantages[517] and New contrivances, which were never made ‹use› of before and were looked upon as Disadvantages, one of which is the hanging of the Pendulum by a Stiff and strong spring, which is soe far from hindering or clogging or Disturbing its motion that it mightely contributes to the contrary effects. Insoemuch that by that contrivance only a pendulum will continue to Move of it Self without the help of Clock work for 24 houres, though the first Vibrations too be but of a Small arch. The clock it Self will shew Divers others. Upon the whole I humbly conceive that It were very desirable that it were once again adjusted and that it might be applyed to its proper use. and this not only for producing that Desirable effect of Registring the constitution of the weather and Seasons, but for that I understand it has been much admired and valued by strangers that have either seen it here or heard of it abroad.

There are two other instruments which I have heretofore mentiond to be worth the Imploying for the making of very usefull Discoverys in Nature which are as yet undetermined but may be by tryalls made with them be adjusted In the next place I have formerly Suggested an hypothesis of the ovall figure of the Earth and thence of the Decrease of the power of the gravity towards the æquinotiall I suggested also a way how the Same Might be Examind and adjusted, and though it were at first Lookd upon as a paradox and Not to be regarded yet since some others have been pleasd Since my Demonstrating of it ‹here› to assert it Likewise as ‹probable›[518] and as their own (as they have done Severall other Discoverys I had first made) it has been lookd upon with a more favourable aspect, to determine the matter and put it out of Doubt I would humbly propose that an Instrument may be sent by a person who will carefully make the observations and bring back a true account thereof to be here Recorded as a Discovery first made and adjusted by this Society. the Same person will also make exact observations of the Dipping and variation of the magneticall needle if he be fournisht with one fitt to make the observations and will give an account thereof at his Returne from China which I conceive May be of Great use to Navigation as well as of Single information as to the perfecting the Theory of the magnetisme of the Earth, ‹of› which I conceive noe one has yet ‹Suggested›[519] a

---

[516] Deleted: noe.

[517] Deleted: (…).

[518] Corrected: likely.

[519] Corrected: Imagined.

probable cause: ‹now› for both these enquirys I conceive I can adapt instruments much better then any I have yet met with if this Society shall think fitt to supply the charges of them which will not neither be very much and will[520] be abundantly recompensed by a true Discovery of the matters of fact beside severall other advantageous Discoverys to be made by them, which I shall in due time manifest.

These are Experimenta crucis which will give[521] positive arguments for Determining both the Qualitys and quantity of two of the most powerfull powers of the body of the Earth hitherto undetermined, and for that end I[522] conceive they may be worthy the consideration of this Honourable Society. however I humbly submitt my Sentiments to their Determination.

This Lecture was Read before the Royall Society Dec:18.1689. Praesent Sir Joseph Williamson, Mr Pepys, Mr Bemde, Mr Aubrey, Lodwick, Dr (...), Dr Tyson, Dr Sloane Mr Evelyn, Mr Pillfield and severall others.

Royal Society Classified Papers, vol. XX, ff. 183r–184v

'f. 183r' The uses and advantages of microscopes[523]

Read before the Royall Society Nov. 29.1693

Among the various Methods of inquiring into the Latent and internall structure and composition of Bodys, that by the help of *Microscopes* has not been the least Significant, but when we consider the nature of the informations it has and may yet afford, it may possibly[524] appear to Deserve to be ranked even among the most Considerable, and that upon this account that it doth immediately inform the ‹Sight›[525] of the true and naturall construction of the minuter and otherways wholy insensible constituent parts of bodys, discovering their forms and shapes and how they serve to make up the more grosse and Sensible parts both by their texture and motions, which carryes the Inquirer soe much further into the Latent and internall mechanism of Bodys, whilst they are yet intire and undisturbed in their Naturall State whereas the other ways of Examining of the Nature of Bodys ‹as› by fire or chymistry[526]) the naturall constitution of the bodys to be examined is wholy vitiated and Destroyed, and torne all to peices and Scarce soe much Lost Intire as may Deserve the name of the ruines of it, but ought Rather to be called the Rubbish and Corruption thereof, for that noe one part of the concreat is Left Intire but every part is as it were ground to Dust and attomes by the Action of the fire ‹or menstruum›

---

[520] Corrected: I.

[521] Deleted: a.

[522] Deleted: humbly.

[523] Endorsed in lefthand, in different ink.

[524] Deleted: Deserve.

[525] Corrected: Sense of.

[526] Deleted: ‹or›.

and not only Soe but even those attoms or Dust yet blended and compounded by other heterogeneous Substances insinuated and mingled with them, that are properly parts of the Fire or *Menstruum* that Dissolved them. Now as the method of Dissecting and anatomising a body is a much more probable and ‹has›[527] Experimentally proved a more effectuall way to Discover the construction make and use of the Constituent parts of animated bodys and the uses to which they are subservient, and as living Dissections, or inspections and Experiments and observations made whilst nature is yet acting, are more informing than by Destroying the Life of the animated body and beating it all to and mash ‹in a morter› with water or any other substance and then examining of the *Compositum*: Soe the Discoverys to be made by the *microscope* are upon both those accounts to be praeferrd before those other that I have named, for that ‹in› a multitude of Such kind of observations as are to be made by the *Microscopes* you may not only Discover the texture and fabrick of the part by Dissecting or anatomising them, but you may with pleasure and admiration behold the wonderfull construction, motions, operations and uses of the parts whilst Nature is still ‹and at peace›[528] undisturbed, and working in its Direct and Naturall course, without any such Dissecting and without any Dislocation of Parts or any alteration or Disturbing of the motions, or ‹the› effects thereof. This is a Prospect that is wholy due to the *microscope*, and is hardly to be found in all the visible *phaenomena* to the Naked Eye for that the texture and constitution of most animate bodys ‹as to the parts›[529] thereof that are visible to the ‹naked› eye are of Bulk enough to '**f. 183v**'[530] make them opaque or not of transparency enough to Discover the internall make and motions thereof through them, whereas there are thousands of Objects that by the *microscope* may be found whose Skin, Rind or inclosing teguments[531] are Soe transparent as to admitt a free prospect for the sight to Discover through them the fabrick, figure, texture ‹and›[532] motion[533] thereof: For As most of the parts of animated ‹bodys› are *in minimis* transparent enough to permit a[534] passage for the light free enough to Discover pretty cleerly the differing shapes and lineaments of the parts behind them and the variety of the Refraction and reflection of Differing parts is sufficient to make a Sensible difference in the appearance where too many of them are not Confounded and blended together as in those smaller fabricks of animalls and vegetables, visible only by the assistance of the microscope Soe in most others that are Discoverable by the bare eye, there is soe

---

[527] In different ink.

[528] Corrected: (…).

[529] Corrected: in such that ‹the› Part the.

[530] Deleted: Some bodys big enough to be (...).

[531] Corrected: tegumetall parts.

[532] In different ink.

[533] Deleted: and effects.

[534] Deleted: free.

great a masse of such transparent ‹particles[535],[536] joyned together that those varietys of Refractions and Reflections[537] blended all together doe confound each other and by that time they are made big enough to be visible to the Eye they appear opaque and as it were in a fogg or cloude. Of these kinds of Discoverys I have given Severall Specimina in my *Micrography*, as in the Gnat, Mite, Louse, and some others, but had not there opportunity of instancing or Relating all the Severall observation[s] I had made. my Designe in that being rather to show in what Variety of Subjects[538] Discoverys might be advantagiously made by the help of microscopes ‹and other optical glasses[539]›, by exhibiting some one or other instance thereof, then to[540] persist in prosecuting any one Species of all those varietys. That designe indeed as it would Require much time and Labour, soe if it were well performed would not want[541] its benefit for the explaining the progresses and operations of Nature and would prove as instructive a piece of Naturall history as any yet extant, and possibly for some uses much more than any other that has yet been published ‹as› to exhibit the structure[542] fabrick and Contexture of animated bodys. For that most of the History we have either of Plants or animalls give us only a superficiall and outward Description of their visible shapes and of the more grosse appearances of them, But tell us not the inward Fabrick operations[543] vertues and uses of their parts, And even there where anatomy has been applyed for that Purpose, we are gone noe further then to the forme and marks of the greater constituent parts such as are big enough to be visible to the Eye and tractable to the hand, But all the organs that are Lesse then such a Bulk they remain in their Primitive Obscurity and are only the Objects of Conjecture and Imagination.[544] '**f. 184r**' Now how for such kinds of Discoverys

---

[535] Deleted: (…).

[536] Corrected: parts.

[537] Deleted: are.

[538] Deleted: the.

[539] Corrected in different ink: optic glasses.

[540] Deleted: In.

[541] Deleted: (…).

[542] Deleted: and.

[543] Deleted: and.

[544] Deleted in different ink: Now how wide such conjectures would ‹probably› be from the truth of the thing under consideration may be conceived from this Similitude. If to a wild Indian that had never heard of or seen clock watch or wheel one should Describe or show him a watch inclosed in a Leather case of the bigness and shape of a small turnep and tell him that this did keep a certain account of the time and did divide all the time from noon to noon into 24 aequall parts or houres and at every hour[']s end tell by soe many distinct sounds which hour it then was, he would certainly conclude that there was some very cunning creature included in that box. if then you should open the Lid '**f. 184r**' of the case and shew him the face of the watch and shew him the Diall plate where he might see the Diall Ring and the hand pointing to the Divisions of the Day, he would then understand somewhat more of what the effects of the watch then what by seeing only the out case and hearing the noyse of the Strokes on the Bell he did conceive, but still he would be apt to think some living thing was kept within the yet unopend part, for that he could hitherto see noe more then the Diall face and moving hand and hear the Pulse or beating of the Ballance and at the hours

may be made into the curious fabricks of the plants and animalls 'tis not for me to Determine, but it seems very probable that the Microscopes will show us many essentiall and constituent parts of which we had before noe Imagination, some *specimina* of this kind I have former[l]y shewn, and Mr *Leuwenho‹ek›*[545] has since prosecuted the Inquiry and made Divers considerable additions.[546] Some other[s] also have indeavourd to Doe somewhat that way as *Cherubine* at *Paris* and *Greindelius* in *Germany* but have made but Little progresse either for the Improvement of *microscopes* or for New discoverys by the help of them.[547]

That which Revived this Discourse at Praesent is a new treatise Published in Latine by *Sig.r Bonanj* the same that published a book of the Description of the Shells of Fishes. He calleth it *Micrographia Curiosa sive Rerum Minutissimarum observationes quae ope microscopiae Recongnitae at Expresse Describuntur.* wherein he indeavours to Revive the practise of it in *Italy* shewing it to be very useful for the Convincing of *Atheists* and to bring them to the Acknowledgment of *God* by contemplating the wonderfull works of his providence, But at the same time indeavours through his whole Discourse to prove Spontaneous Generation, which seems to have a contrary tendency.[548] He seems to have perused most of the authors who have written any thing considerable concerning *microscopes* and Micoscopicall Observations, and upon the whole has given the Result of his own sentiments, first concerning the instrument itself Describing which kind he doth most approve of and which he made use of for his own observations, and therein he Describes which way he thinks best for grinding of Glasses to a true figure which he Delivers in a Cyphaer aenigmatically in 3 Directions or Rules which I having Decyphred read thus[549]: the tools are to be made of brasse or tin to grinde the ‹glasses›[550] of their true formes. Which tools or Dishes must be ‹of›[551] a Due form to procure which he says care must be had *ut non valde amplae sint sed lentis faciendae ita commensurentur, ut lens fere triplicem diametrj portio‹nem› occupet.* That is the dish must be 3[ce] the Diameter of the glasse. Next the glasse must be ground to its figure ‹in› a tool or Dish of a Little bigger Sphaere. Then perfected in the Lesser, which is thus per-

---

end the noyse and strokes of the clock part. 'tis easy to conclude[e] that his conjectures concerning the fabrick of the watch would be differing enough from what they really were, if the watch should further be opend to him to see the wheels move and the Ballance beat and all the make of them as they are to be seen when the watch is opend he would still be at a Lesse what was that made them move, and would think that the Spirit of the watch was either in the wheels or in the Barrell of the Spring. And soe as he could further and further Discover the conceald parts he would more truely be informed of its excellent contrivances:.

[545] In different ink.

[546] Deleted: somewhere.

[547] Deleted: I.

[548] Buonanni (1691b), 2–5; Id. (1691a), 19–23.

[549] Endorsed in lefthand in different ink: Bonanis' Cypher about microscopes and grinding glassed decyphered.

[550] In different ink.

[551] In different ink.

formed. *Utrasque manu simul concurrente, ita ut radente Vitro scutellam percurrat Leva antem Scutellam vitro adaptet, alterâ obsecundante alterj.* Both hands must be imploy'd the ‹right›[552] to hold the glasse and move it in the Dish and the Other hand ‹to›[553] adapt the Dish to the glass and Soe both to move them true one in the other. When by this means it be ground to its true figure **'f. 184v'** then it Remains only to be polished and of all the ways for Doing this he praeferrs that by paper Stuck fast whit glue in the tool or dish in which it was[554] Last ground and by Spreading upon that paper the fire powder of tripolj,[555] and thereon to work the glasse till it has acquired a Due polish which you may the better perform, *Sj instrumento tornalj utaris quod beneficio Rotae maioris velociter circumversatur,* that is if the tool or dish be fixt to a mandrill and that be made to run Swift Round by a larger either foot or hand wheel.[556] both which ways I have Seen commonly made use of hand for[557] above 35 years since, and also a better way which is by a reciprocating motion: both which ways our workmen here very well know and commonly practice yet I am apt to think the Last polish from the Bare tool without the Paper to be the more exact, at least by Severall tryalls I have Soe found it for the object glasses. But for the eye glasses I judge[558] the polish by tripolj or paper as he praescribes to be ‹sufficiently exact.›[559]

The microscope he prescribes has too much apparatus and clutter and yet is wanting of many accommodations for examining[560] or of it were handling and turning the object into all postures and for all lights, and therefore I shall not Spend time in the Description of it. nor shall I Repeat here the names of the multitude of all those Authors which he has mentioned ‹who have spoken of› the ways of making microscopes or the Descriptions of some kinds of them: but in short to note that each of them is referable[561] either to the Simple or the Compounded form, the Simple is by one glasse only but the Compound by 2, 3, 4, 5 or more glasses, but still the more the worse, however he thinks every kind some ways or other usefull: As[562] concerning that of *Griendelius*[563] published in 1687[564] which that author would have the world to beleive[565] exceeded all that had been ever made in England,

---

[552] Corrected: left.

[553] In different ink.

[554] Deleted: gro.

[555] Word of Latin origin indicating the rotten stone.

[556] Buonanni (1691b), 33.

[557] Deleted: Much.

[558] Deleted: it sufficiently exact.

[559] In different ink.

[560] Deleted: and.

[561] Deleted: or.

[562] Corrected: (...).

[563] Griendel (1687); Cf. Journal Book of the Royal Society, vol. VIII, 151.

[564] Corrected: 1689.

[565] Deleted: that it.

France, Italy or Holland, this author upon examining it is quite of a contrary opinion and thinks it much inferior to them which I ‹also› was before sufficiently Satisfyd of when I Read the Said *Griendelius* his Description of it.[566] But this Author seems most to approve that which I have described in my micrography. *Omnes fere expertum esse sine fuco medacij affirmabo: nec ulla plene satisfactum Singula enim vitio aliquo laborabant nullamque abunde commodam judicavj ad observationes, praesetim si oculo attente respiciente, manu delineare vellet quidquid observabatur. Inter omnes utiliorem modum exstimavj quo suas observationes fecit Cl Hookius etc.*[567] however he has made some additions to it for the fixing it more Steadely to observe the object whilest it is in drawing which I conceive may be too troublesome and yet not sufficient for all purposes.

The instances he has mentiond to have observed are ‹many›[568] taken out of other authors upon[569] which he has added his own remarks, but he excuseth his inability for Delineating them soe well as he could see them however he has copyed Severall of those I have Describd in my micrography and some also out of others those of mine are the gnat worm in both formes, the foot of the fly and the wing as also the eyes the sling of the bee, the louse and the flea, the Stinging points of nettles and the Scales of fishes, and he has both to these and all his other added his own adnimadversion among which there are severall very curious. But it would be too Long to mention them at this time, and much more to adde my objections to some of them, which I may have a more proper occasion to doe in an other discourse: I shall at present only shew the figure of the stings or thornes of the prickly pare or indian figge which I mentioned before ‹3 weeks since› when the plant it Self was here produced.

The Brown tufts on that prickly Pare I found to consist of a great number of very Small and Sharp pointed thorns or needles being abundantly Smaller then the finest Needells I ever saw. These thornes being Soe very small Soe sharp and yet Soe Stiffe do easily pierce the Skin of whoever toucheth them, and which makes them the more troublesome they being all over barbed with thornes like a bramble almost or a bee's Sting they Stick fast in the flesh and cannot be easily gott out where they are once entred.

British Library MS Sloane 1039 ff. 112r–113v

'f. 112r' The Designe of the Royal Society being the improvement of Naturall knowledg, they Pursue that Designe by all meanes they conceive to conduce thereunto, And knowing that much of it Lyes Dispersed here and there amongst Learned

---

[566] Griendel, *op. cit.*, pp.3–4.

[567] Buonanni (1691b), 26.

[568] Corrected: much.

[569] Deleted: ‹something›.

and Experiencd men, when it is oftimes Little Regarded because not enquired after and soe generally Lost by the Death or forgetfullnesse of the possessors; they conceive many Excellent and usefull observations may this way be collected into a generall Repository where Inquisitive men may be sure to find them Safely and Carefully praeserved both for the Honour of those that communicate them and to the Generall Good of mankind which is their Principale and ultimate aime. And though a various action be a sufficient reward to it self, and that it is oftimes a greater pleasure to communicate them ‹to› conceale an invention, yet they Resolve to Gratify all ‹that communicate›[570] with suitable Returnes of such Experiments Observations and Inventions of their own, or[571] advertisements from other of their correspondents as shall in some kind make them amends. And that you may understand what parts of ‹naturall› knowledge they are most inquisitive for at this praesent, and soe hereafter from time to time, they Designe to print a paper of advertisements once every week or fortnight at furthest, wherein will be conteind[572] the heads or substances of the Inquirys they are most solecitous about together with the progress they have made and the Informations they have Receivd from other hands, together with a short account of such other philosophicall matters as accidentally occur, such as ‹the› character of such new books or Discourses as are published here or els where soe soon as they can procure them[573] A brief account of what is new and considerable in their Letters from all parts of the world, as which ‹the› Learned and Inquisitive are doeing or have done in[574] Physicks Mathematicks, mechanicks, opticks, Astronomy, medicine, Chymistry, anatomy, both abroad and at home.

First It is earnestly Desired that all observations that have been accurately made of the variation of the ‹magneticall› needle in any part of the world might be communicated together with all the circumstances remarkable in the making thereof, as the coelestiall observations for knowing the true meridian, or by what other meanes it was found, the time of making it, by whom and in what manner, with what kind of needle, whether a shipbord or ‹upon›[575] Land, Ice, etc. But from a considerable Collection of such observations A theory might be made of that Admirable Effect of the Body of the Earth upon A needle toucht by a Loadstone

'f. 113r' That if it will (as tis probable it may) be usefull for the Direction of Seamen or others for finding the Longitude of Places, the observations soe collected[576] together with the theory thereof may be Published for the Generall good of Navigation. which they Ingage to Doe soe soon as they have a sufficient Number of

---

[570] Corrected: such.

[571] Deleted: Inqui.

[572] Deleted: such.

[573] Deleted: (…).

[574] Deleted: Natur.

[575] Corrected: at.

[576] Deleted: (…).

such observations wherein mention shall be made of every person soe making and communicating his observations.

now that such observations may be made the more effectuall for the purposes aforesaid it is thought fit to[577] communicate those Instructions following for the better Directions of such as may not otherwise be soe well versed in the performance therof

British Library MS Sloane 1039, f. 114r

'f. 114r'

The Invention of Archimedes[578] to Discover the cheat of the goldsmith in the making of Hiero's Golden Crown was esteemd soe ‹great›[579] by that Eminent[580] Geometrician that in a great transport he cryd ευρεκα ‹έ›[581] ευρεκα ‹έ›,[582] and It has not been otherwise esteemed by all Learned Persons since his time though the manner how he Did it hath not been soe well and fully deliverd by Historians as it were to be wisht. Ghetaldus[583] upon the same principle hath improved the way by taking the weight of the bodys first in the air and then in the water in order to find the co[m]parative weight of the bodys soe weighd first to water and then to one another multitudes that have writ of this Subject since have followd his way and have (as he hath done) given us the comparative ‹specifique› weight of severall bodys, and all agree that this Invention of Archimedes was sufficient to performe what was Intended the reason of which I suppose was that ‹the geometricians who understood the theory and calculating part› wanted[584] Experiments and tryalls to see whether Nature did really follow those methods that Artists had supposed from their theorys. On the other side the Experimenters (Among which may be Reckonned the Lord Verilam) made the tryalls but out of some kind of awrsenesse to Geometricall and arithmeticall Speculations made not the Calculations and Soe were deficient on the other side: between both the matter is taken from Granted, but whether soe or not is not yet proved. The Experiments which I have hitherto made Doe seem to call it in question, and shew that the compound or mixed mettall made by melting two together hath a compound ‹specifique› gravity not truly proportiond to the Specifique Gravitys of the two compounding bodys but sometimes exceeding sometimes defective: for By the first Experiment that was made with the mixture of Copper and tin

---

[577] Deleted: publish.

[578] Vitruvius (1931–36), vol. I, 202–7.

[579] Corrected: much.

[580] Corrected: Great.

[581] Inserted above the letter ε.

[582] Inserted above the letter ε.

[583] Ghetaldi (1630), 299.

[584] Deleted: of.

it was found that the[585] specifique gravity of the compound made of aequall parts of tin and copper was not only heavier than it ought to have been supposing those mettals only joynd together but even heavier than either of the two even than the heaviest. Whence It followes necessarily that there is a penetration of the Dimensions of each other in that compositum On the other side by the Later Experiment made by the Mixture of Lead and tin ‹the specifique gravity of› the Compositum Is considerably lighter than it ought to be according to the bare composition of those two bodys. Whence it necessarily follow[s] that there is an aversion in the joyning of those two bodys and a kind of recesse and the parts acquire[586] a greater rarefaction of texture than they had before their union.

Read in the Society Jan. 29, 1679[587]

---

[585] Deleted: compoun.

[586] Corrected: require.

[587] 29 January 1679/80 according to Julian calendar, Cf. Birch (1756–57), vol. IV, 6.

Bibliography

Aldrovandi, Ulisse. 1648. *Museum metallicum*. Bononia.

Bacon, Roger. 1659. *Of the miracles of art, nature, and magik*. London.

Bacon, Francis. 2004. *The Instauratio magna. Part II: Novum organum*, ed. Graham Rees with Maria Wakely. Oxford: Oxford University Press.

Biancani, Giuseppe. 1615. *De mathematicarum natura dissertation*. Bononia.

Birch, Thomas. 1756–57. *The History of the Royal Society of London*, 4 vols. London.

Buonanni, Filippo. 1691a. *Observationes virca viventia*. Rome.

———. 1691b. *Micrographia curiosa*. Rome.

Descartes, René. 1964–74. *Oeuvres*, 12 vols., eds. Charles Adam and Paul Tannery. Paris: Vrin.

Galilei, Galileo. 1890–1909. *Opere*, 20 vols, ed. Antonio Favaro. Florence: Barbera.

Ghetaldi, Marino. 1630. *De resolutione et compositione mathematica*. Rome.

Griendel, Johann Franz. 1687. *Micrographia nova*. Nuremberg.

Harriot, Thomas. 1631. *Artis analyticae praxis*. London: Walter Warner.

Hérigone, Pierre. 1634. *Cursus mathematicus, 4 vols*. Paris.

Hesiod. 2006. *Works*, 2 vols., ed. Glenn Most. Cambridge, MA: Harvard University Press.

Hooke, Robert. 1678. *Lectures and collections*. London.

———. 1705. *Posthumous works*, ed. Richard Waller. London.

———. 1726. *Philosophical experiments and observations*, ed. William Derham. London.

Hunter, Michael. 2007. *Editing early modern texts: An introduction to principles and practice*. London: Palgrave.

Lawson, Ian. 2015. *Robert Hooke's microscope: The epistemology of a scientific instrument*. PhD diss., University of Sidney.

Oldroyd, David. 1987. Some writings of Robert Hooke on procedures for the prosecution of scientific inquiry, including his 'Lectures of Things Requisite to a Ntral History'. *Notes and Records of the Royal Society of London* 41: 145–167.

Oughtred, William. 1631. *Clavis mathematicae*. Oxford.

———. 1652a. *Elementi decimi Euclidis declaratio*. Oxford.

———. 1652b. *Theorematum in libris Archimedis de sphaera et cylindro declaratio*. Oxford.

Ovid. 1916. In *Metamorphoses*, ed. Frank J. Miller, vol. 2. London: Heinemann.

Pugliese, Patri. 1982. *The scientific achievement of Robert Hooke*. PhD diss., Harvard University. Ann Arbor: University Microfilm International.

Rahn, Johann Heinrich. 1668. *An introduction to algebra … much altered and augmented by D. P.* London.

Recorde, Robert. 1557. *The whetstone of witte*. London.

© Springer Nature Switzerland AG 2020   261

F. G. Sacco, *Real, Mechanical, Experimental*, International Archives of the
History of Ideas   Archives internationales d'histoire des idées 231,
https://doi.org/10.1007/978-3-030-44451-8

Rigaud, Stephen Jordan (ed.). 1841. *Correspondence of scientific men of the seventeenth century*, 2 vols. Oxford.

Rudolff, Christoff. 1533. *Die Coss*, ed. Michael Stifel. Königsberg.

Stifel, Michael. 1544. *Arithmetica integra*. Nuremberg.

Viète, François. 1591. *In artem analyticam isagoge*. Tour.

———. 1646. *Opera mathematica*. Leiden.

Vitruvius. 1931–36. *On architecture*, 2 vols., ed. Frank Granger. London: Heinemann.

Wallis, John. 1685. *A treatise of algebra both historical and practical*. London.

Index

© Springer Nature Switzerland AG 2020
F. G. Sacco, *Real, Mechanical, Experimental*, International Archives
of the History of Ideas Archives internationales d'histoire des idées 231,
https://doi.org/10.1007/978-3-030-44451-8

Printed in the United States
by Baker & Taylor Publisher Services